8th International Conference on Transportation and Traffic Engineering (ICTTE 2019)

MATEC Web of Conferences Volume 308 (2020)

Auckland, New Zealand
19 - 21 December 2019

Editors:

Alan Nicholson
Luc Wismans

ISBN: 978-1-7138-0683-7

Printed from e-media with permission by:

Curran Associates, Inc.
57 Morehouse Lane
Red Hook, NY 12571

CURRAN ASSOCIATES INC.
proceedings
.com

Some format issues inherent in the e-media version may also appear in this print version.

No changes have been made to the content of these proceedings. There may be changes to pagination and minor adjustments for aesthetics.

Printed with permission by Curran Associates, Inc. (2021)

For additional information, please contact EDP Sciences – Web of Conferences at the address below.

EDP Sciences – Web of Conferences
17, Avenue du Hoggar
Parc d'Activité de Courtabœuf
BP 112
F-91944 Les Ulis Cedex A
France

Phone: +33 (0) 1 69 18 75 75
Fax: +33 (0) 1 69 28 84 91

contact-edps@webofconferences.org

Additional copies of this publication are available from:

Curran Associates, Inc.
57 Morehouse Lane
Red Hook, NY 12571 USA
Phone: 845-758-0400
Fax: 845-758-2633
Email: curran@proceedings.com
Web: www.proceedings.com

TABLE OF CONTENTS

Preface

The 8th International Conference on Transportation and Traffic Engineering (ICTTE) was held in Auckland (New Zealand) from 19-21 December 2019.

Around the world, there is rapidly increasing concern about the impacts of transport on global warming and climate change, as well as the social costs of accidents. ICTTE 2019 provided a great opportunity for discussions of methods, technologies, systems and practices in the different areas of transportation and traffic engineering from different places around the world, for mitigating the adverse effects of transport, while assisting economic development. ICTTE 2019 enabled delegates to exchange ideas with other researchers, and develop connections upon which international collaborations, aimed at solving transportation and traffic problems, may be based.

The Technical Programme Committee of ICTTE assembled an excellent programme, which included presentations from four Keynote Speakers: Prof. Alan Nicholson (University of Canterbury, New Zealand), Prof. Monteiro Figueira (University Lusofona, Portugal), Prof. Vinayak Dixit (University of New South Wales, Australia) and Prof. Khair Jadaan (University of Jordan, Jordan). The programme included 31 papers, which were accepted for presentation and publication in the ICTTE conference proceedings, after a rigorous review of the 62 papers submitted for presentation and publication.

On behalf of the Organizing Committee, thanks are extended to the keynote speakers, the authors of papers selected for presentation and publication. Thanks are also extended to the paper reviewers and the members of the organizing committee, for their contribution and commitment to putting together a programme of paper presentations that conference participants should find beneficial in their professional development.

Conference Chair

Professor Alan Nicholson

edp sciences

E3S Web of Conferences

Statement of Peer review

In submitting conference proceedings to E3S Web of Conferences, I certify to the Publisher that I adhere to the **Policy on Publishing Integrity** of the journal in order to safeguard good scientific practice in publishing.

1. All articles have been subjected to peer review administered by the proceedings editors.
2. Reviews have been conducted by expert referees, who have been requested to provide unbiased and constructive comments aimed, whenever possible, at improving the work.
3. Proceedings editors have taken all reasonable steps to ensure the quality of the materials they publish and their decision to accept or reject a paper for publication has been based only on the merits of the work and the relevance to the journal.

Title, date and place of the conference

2019 8th International Conference on Transportation and Traffic Engineering (ICTTE 2019)

December 19-21, 2019

Auckland, New Zealand

Proceedings editor(s):

Alan Nicholson

Luc Wismans

Date and editor's signature

Mode shift at bottleneck increasing transit dispatch in peak time

ZHONG Lan-fen [1,2] , WANG Wen-zhong[2,a]

[1]*College of Computer Science, Inner Mongolia University, Hohhot 010021, China*
[2]*Department of Information Science and Technology, Baotou Teachers' College Inner Mongolia University of Science &Technology, Baotou 014030, China*

Abstract: For attracting more auto commuters to shift to transit mode and mitigating congestion at a bottleneck in morning &rush hour, additional dispatches of transit are operated during peak time. Classical bottleneck model combining with Logit-based discrete choice formula is extended to investigate commuters' mode choice behaviors between private car and public transit. The existence of bi-mode user equilibrium when tolling auto commuter is proofed, and waiting time and time delay costs are formulated in two modes when additional buses are dispatched. Numerical experiments are conducted to examine mode split patterns and aggregate travel cost when additional dispatching service interval varying. Our results show the system aggregate trip cost would reduce prominently when extra buses are added into runs in appropriate time. Especially when waiting time equals 0.65 hour for auto commuters, the system aggregate trip cost would reduce by 70% in theory.

1. Introduction

As the development of economy and society, the Traffic Authority (TA) around the world are challenged by using limited road space in green way to provide transportation service in city. For reducing congestion and pollution, various traffic management strategies have been proposed by researchers, such as congestion toll, parking policy, auto restraint, and transit service. Effective transit service is one of empirical ways to reduce congestion and protect environment. The challenge for transit system is to attract people, besides transit captive users, out of cars, especially in congested bottleneck, since that Traffic jam always appears at bottleneck in morning and evening rush hour. One of the adopted strategies is to encourage park-and-ride travel, namely leading auto commuters to park around bottleneck, then they shift to a public transit to go through congested corridor of road and accomplish the rest of trip. More commuters are to switch over to public transport when increasing the frequency of bus dispatch. However, increasing either dispatch frequency or expanding capacity is not the most cost-efficient instrument, especially when commuting is strongly concentrated in peak period, for increasing capacity will be largely idle outside peak period. In the most of the literatures of bi-mode study, transit dispatch frequency is fixed in all service time for management convenience, but realistic commuting demand is varying with time in a day. So, in this paper, what brings forth new idea is we consider to take some additional bus dispatch into routine dispatched transit service in morning rush hour at bottleneck to attract auto commuter to shift to take bus, thereby mitigating congestion at a bottleneck in morning rush hour.

Vickrey proposes one of the most insightful and tractable bottleneck model to analyze morning commuting way at bottleneck [1]. The work on the bottleneck congestion in morning commuting with two modes (car and transit) has also been developed. Tabuchi studies pricing and mode split in a bi-mode competitive system containing transit and high way [2]. Huang extends Tabuchi's study of mode choice by introducing in-vehicle crowding congestion and analyze how to optimize bottleneck capacity utilization by adjusting car congestion toll, transit fare, mode split, and dispatch frequency [3-5]. Kraus studies the optimality of fare, serving number of train unit and capacity of individual train unit [6-7]. Li and Zheng proposes a time-dependent pricing scheme to motivate mode shift and increase the efficiency of pricing to attain equitable savings for all users[8-9]. All these research work reveal the rule of commuting behavior and the performance of tolling

[a] Corresponding author: wwenzhong@126.com

regimes at bottleneck. But, the effect of additional transit dispatch on decreasing travel cost and the degree of congestion at bottleneck has rarely been considered in these literatures. Our outstanding work is focused on it.

Besides, mode choice between transit and auto is a complex decision process, which is influenced by many factors. To account for these factors in practice, mode split functions have been developed. One of the most early and widely used mode split functions is Logit-based discrete choice formula. Recently, Lu proposes a nested Logit-based stochastic user equilibrium model to characterize the mode choice of commuters and finds that public subway and private parking lot can approximately generate the maximum expected net benefit of transport system while motivating transit travel[10]. More recently, Kitthamkesorn develops a Logit-based model to alternate the weibit-based model for the combined modal split and traffic assignment problem, which is effective to quantitatively evaluate and examine the behavioral modeling of travelers' mode shift between private motorized mode and go-green modes [11]. From lane perspective, Logit formula is applied to investigate two types mode split problems: physically separated rail and together rail. When the rail of train or subway is viewed as parallel lane with auto road, Lu and Kitthamkesorn's results are suitable to analyze competitive bus and auto traffic system with dedicated bus lane. About dedicated bus lane strategies, please refers to existing literatures [12]. Here, it is assumed that buses take its own lane and what we concern is how dispatch additional buses to find a system cost optimum solution.

Gonzales and Daganzo provided a system optimum solution with two modes, and presented pricing method to achieve the system optimum in the condition that commuters arrive and pass a bottleneck in order of wished departure [13]. The study has made progress in competitive transportation system modeling in peak time; nevertheless, its mode split, pricing and system optimum are based on the uses equilibrium which is derived from the principle that transit user and auto user have the same cost in equilibrium, which is impractical for all cases, since mode choice results from a complex decision process, not following user equilibrium principle. we will further the bi-mode equilibrium study with Logit-based formula, meanwhile system user optimum cost will be discussed when additional transit service starts to be provided at a certain time point.

The paper is organized as follows. Section 2 formulates the equilibrium commuting patterns of two modes. Section 3 analyzes the travel costs of both private auto and public transit at equilibrium. Section 4 gives numerical results for intuitively demonstrating the mode split and commuting patterns with respect to some different input parameters. Section 5 concludes the paper.

2. User equilibrium

In the bottleneck model, N commuters have to go through a bottleneck from O (home place) to D (work place) in morning rush hour. The bottleneck's capacity is fixed at s . If arrivals at the bottleneck ever occur at a rate exceeding s , then a queue develops. Here all commuters are assumed to be identical in perceiving time value, and congestion would happen at bottleneck only.

When only auto mode is used, there exists a highly congested duration around (t_1, t_2) , as shown in Figure 1. When an alternative public transit service becomes available, from simple geometrical considerations, the transit service operation aims to shift the portion of commuters represented by the area of triangle BCD (shown in Figure.1.) from auto mode to transit mode. It is assumed in this section that the TA charges a fixed fare and operates buses at a fixed headway. There are two kinds of user by transit. One is captive user, and the other is choice user. For simplifying problem, we assume captive transit users can be served by routine transit dispatches. Here only considers the choice users who can alternate two modes. The transit service mentioned following refers to the additional dispatched buses besides routine transit buses. So we make three assumptions. (*i*) Commuters are choice users, who can change their travel mode arbitrarily. (*ii*) Routine transit is available in whole service time, and additional transit service is only available during highly congested period of rush hour. (*iii*)When transit is operating, it is fully segregated on its own lane, so that transit service is not subject to traffic congestion. All choice travelers adjust their departure time and travel mode to reduce their travel cost. An equilibrium is obtained when no individual has an incentive to change his departure time and travel mode. Actually, two types of equilibrium arise at the bottleneck: one is experienced by auto commuters, and the other is mode split pattern which means the ratio of auto commuters to bus commuters is invariable when equilibrium state comes up.

In Figure 1, t^* denotes the work starting time. Without restriction, the free flow travel time form O to D is assumed to be zero. Let $\tilde{t}$ be arrival time at bottleneck for punctuality for work. So commuters arriving at bottleneck before $\tilde{t}$ are early, and commuters arriving at bottleneck after $\tilde{t}$ are late. Each unit time of earliness is associated with a penalty of β and each unit time of lateness is

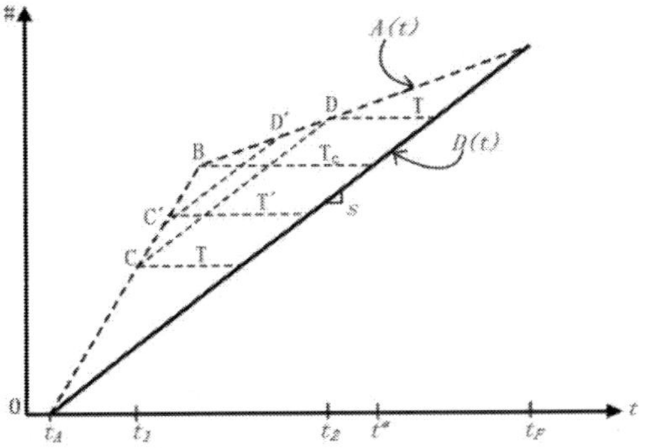

Figure 1.User equilibrium for a fixed capacity bottleneck in single auto mod

$\gamma . \alpha$ is waiting time cost rate. The slope of the arrival curve in equilibrium, shown in Figure 1, satisfies:

$$A'(t)=\begin{cases} s/(1-\beta/\alpha), & t\in[t_A,\tilde{t}), \\ s/(1+\gamma/\alpha), & t\in(\tilde{t},t_F]. \end{cases} \quad (1)$$

The queuing time increases at rate $\beta/(\alpha s)$ with each additional commuter in the time interval $[t_A,\tilde{t})$ and the queuing time declines at the rate $\gamma/(\alpha s)$ with each departing commuter in $(\tilde{t},t_F]$. Let $W(t)$ denote the waiting time when a commuter who arrives at bottleneck at time t would experience. $W'(\mathrm{t})$ representing waiting time rate, meets:

$$W'(\mathrm{t})=\begin{cases} 1/(\alpha/\beta-1), & t\in[t_A,\tilde{t}), \\ 1/(\alpha/\gamma+1), & t\in(\tilde{t},t_F] \end{cases} \quad (2)$$

$D(t)$ denotes cumulative arrivals at the bottleneck at time t and $D'(t)=s$. When only auto mode is used, in equilibrium, a critical commuter with departure time $\tilde{t}$ arrives at work on time but experiences the maximum waiting time $T_c=\beta\gamma N/(\beta+\gamma)\alpha s$ as a queuing delay, and $\tilde{t}=t^*-T_c$. The waiting time for commuters who arrival at t_1 or t_2 (shown in Figure 1) is the same, and it is assumed as T.

From preceding assumptions, if the transit service is provided during $[t_1,\ t_2]$, there must have a certain number of commuters who will take it, and the waiting time $W(t)$ for any auto commuter, would be reduced. When bi-mode transport system gets its equilibrium, $W(t)$ is supposed to be equal to T. However, it follows that other commuters will choose to pass the bottleneck in this time interval due to the decrease of the waiting time, which in turn results in $W(t)$ increased. Therefore, a toll varying wite time t should be implemented during $[t_1,\ t_2]$. Toll, denoted by $\tau(\mathrm{t})$, satisfies: satisfies:

$$\tau(\mathrm{t})=\begin{cases} \beta(t-t_1), & t\grave{o}[t_1,\tilde{t}), \\ \gamma(t_2-t), & t\grave{o}(\tilde{t},t_2] \end{cases} \quad (3)$$

Proof. In equilibrium, the auto commuters departing to join in the bottleneck experience the same waiting time T, which means $A'(t)=0,\ t\in[t_1,t_2]$. The amount of toll is corresponding to the amount of cost which is lessened on account of available additional transit service Knowing from formulate (1), the number of queuing commuters increases at $s/(1-\beta/\alpha)$ per unit time, and the queuing time increases at rate $\beta/\alpha s$ with each additional commuter in the time interval $[t_1,\tilde{t})$, so in per unit time the queuing time for auto commuter increases at $s/(1+\gamma/\alpha)$. Toll should balance the sum of decreased queuing time cost and increased early schedule delay cost. Therefore, $\tau(\mathrm{t})=(\alpha\beta/(\alpha-\beta)-\beta^2/(\alpha-\beta))(t-t_1)=\beta(t-t_1)$. Likewise, when $t\grave{o}(\tilde{t},t_2]$, we can get $\tau(\mathrm{t})=\gamma(t_2-t)$.

At equilibrium the mode split at aggregate demand level is governed by a Logit formula specified below

$$N_t=N/(1+\exp(\theta C_t-\theta C_a)). \quad (4)$$

Where θ is a positive parameter relating to the standard deviation of random variables and its value can be estimated from survey data; and C_a and C_t are trip cost for each auto and transit commuter respectively. N_t and N_a are the distributing number of commuters for mode transit and auto respectively. And $N=N_t+N_a$.

When more bus runs are dispatched in peak time, some auto commuters are supposed to be attracted to take bus. An equilibrium is easy to find when the mode split is governed by Logit formula (4).Here, it is assumed that the population choosing to take bus, represented by N_t, is geometrically shown by the triangle area BCD in Figure 1, and it follows that the maximum queuing time for auto commuter related to N_a is T.

For demonstrating commuters no intensive to change their mode, we consider the trip cost when transit service is provided on a later time, t_1' (corresponding to C'), as shown in Figure 1. According to no-toll equilibrium, presented by Arront [14-17], the quantity of T does not change the trip cost of auto user who arrives at the bottleneck when the transit operating begins. So it is intuitive to know $C(t_1')=C(t_1)$. But when transit service and the implementation of time-dependent toll rule actually starts at time t_1, the trip cost $C(t_1')$ becomes larger. That is $C(t_1')=C(t_1)+\tau(\mathrm{t}_1')$, $\tau(\mathrm{t}_1')>0$. i.e. $C(t_1')>C(t_1)$, so commuters have no intensive to change their mode split situation, and here comes equilibrium state.

3. Travel costs

For determining equilibrium mode split pattern, it suffices to figure out auto traveler individual cost and transit user individual cost, respectively.

3.1 Auto mode

Individual auto commuter's total travel cost(ATTC) depends on waiting time, schedule delay, toll and vehicle expenses (e.g., fuel consumption and vehicle depreciation). To simplify, we assume a linear travel cost function:

$$C_a(t) = \alpha T_w(t) + d(t) + \tau(t) + \eta_1. \quad (5)$$

Where $C_a(t)$ is ATTC who arrives at the bottleneck at time t , $T_w(t)$ is the waiting time in the queue, $d(t)$ is the schedule delay cost, $\tau(t)$ is toll, and η_1 represents vehicle expenses. In the equilibrium, any auto commuter is unable to find an arrival time at the bottleneck) which can reduce ATTC. In other words, $dC_a(t)/dt = 0$. Hence, all they have the same ATTC, C_a .The first and last auto commuter of arrival at bottleneck in the time interval $[t_1, t_2]$,face the same waiting time, T , and have the same travel cost C_a , i.e.,

$$C_a(t_1) = C_a = \alpha T + \beta(t^* - t_1 - T) + \eta_1. \quad (6)$$

$$C_a(t_2) = C_a = \alpha T + \gamma(t_2 - t^* + T) + \eta_1. \quad (7)$$

During $[t_1, t_2]$, the numbers of cars going through the bottleneck are N_a , and we then have

$$N_a = s(t_2 - t_1). \quad (8)$$

Combining equation (6) and (7), we can determine the following times unknown:

$$t_1 = t^* - T - \gamma N_a/(\gamma + \beta)s \quad (9)$$

$$t_2 = t^* - T + \beta N_a/(\gamma + \beta)s. \quad (10)$$

Substituting (10) and (11) into (7) and (8), respectively, C_a can be expressed as:

$$C_a = \alpha T + \beta\gamma N_a/(\gamma + \beta)s + \eta_1. \quad (11)$$

Furthermore, when t_1 and t_2 are determined, we can get t_A and t_F using the below expressions.

$$t_A = t_1 - (\alpha - \beta)T/\alpha s^2. \quad (12)$$

$$t_F = t_2 + (\alpha + \gamma)T/\alpha s^2. \quad (13)$$

3.2 Transit mode

Now, we consider the travel cost happens on commuters who select dispatched buses as their commuting mode during $[t_1, t_2]$. The cost should depend on waiting time at original bus stop, schedule delay, bus fare and discomfort generated by body congestion in bus carriage. Then, individual transit total travel cost (TTTC) at time t is

$$C_t(t) = \alpha T_w(t) + d(t) + f(\mu) + \eta_2. \quad (14)$$

$T_w(t)$ is waiting time at origin bus stop, $d(t)$ is schedule delay cost, $f(\mu)$ is body congestion in bus carriage with μ capacity limit, and η_2 is bus fare. From transit commuter costumer view, the average cost is crucial for the determination of mode choose. So here we use C_t represents the average trip cost for a transit user.

$$C_t = \alpha T_w + \delta + f(\mu) + \eta_2. \quad (15)$$

T_w , δ is waiting time at the original stop and schedule delay cost respectively in average. Each transit run has a strict capacity of μ , the number of runs is denoted by R , $R = N_2/\mu$, and the number of physically distinct buses used to make the runs is denoted by K . The time length for a bus to make an around trip is d minutes. Following the optimal schedule of running buses in clusters [7], we expend the clusters from before t^* to the whole peak time. From what has been discussed above, for any given T , in equilibrium, C_a is the function of the number of commuters N_a ; C_t is determined by N_t in term of given $\{\sigma, \text{K}, \partial, d\}$. Then C_a and C_t can be written as $C_a = C_a(N_a)$ $C_t = C_t(N_t)$ .Using the logit formula described in Section 2, N_1 and N_2 , the split pattern, can be determined under the condition of given T . Note that the problem of finding a mode split is in fact to solve a fixed point problem because the equilibrium travel costs C_a and C_t depend on N_a and N_t , respectively. Then passenger flows references, t_1 , t_2 , t_A , and t_F , can be determined by the equations (8)– (12).

4. Numerical results

The input parameters of our numerical example are: $N = 5000$ (persons), $(\alpha, \beta, \gamma) =$ $(6.2, 3.7, 15.2)$ $(\$/h)$, $t^* = 8{:}00Am$, $s = 2500(cars/h)$, $\partial = 5(mins)$, $\eta_1 = 14(\$/trip)$, $\eta_2 = 2$ $(\$/person)$, $\theta = 0.05$. We assume $d = 45(minutes)$, $K = 8(units)$. $f(\mu) = -5\ln\left(1 - \sqrt[4]{\mu/80}\right)$ .When T equals 0.5 hours, we got (N_1, N_2) $= (1351, 1071)$ .Figure 2 depicts the number of commuters using auto mode and mode split patterns when T varies from 0 to T_c , respectively. Here, $T = 0h$ means transit service is available in all morning rush hour, and the rate of auto users to transit users is 3 to 2. As waiting time T increases, the two-mode period decreases and the rate gradually approximates to 1 to 1. The number of transit users is decreasing because of shorter and shorter service period $t_2 - t_1$. Finally, when T is approaching 1, both N_1 and N_2 are approaching zero. Figure 2 shows aggregate trip cost of commuters is different when transit service is operated in different waiting time *T*.

The aggregate trip cost of commuters without transit service is 3.1×10^4 , while it decreases when transit is available. It can be seen that a local minimum for the aggregate trip cost is 0.9×10^4 at bottleneck if transit is provided when waiting time T equals 0.65 h, shown in Figure 3. The aggregate trip cost drops by 2.2×10^4 theoretically, about 70% cost saving.

Besides, compared with the data of the numerical experiment in Huang [5] literature, the aggregate trip cost in our experiment is roughly less than the Huang's .

Our dispatching headway is 5 minutes, but Huang's is 2 min; and our model's peak duration becomes shorter duo to additional dispatches. It means more buses are needed in Huang's model than our model dose, and the more running buses inevitably cause the more heavily environmental pollution and the more transit agency cost.

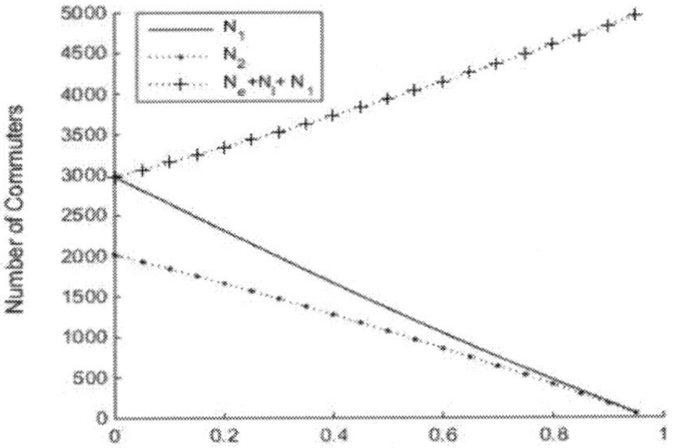

Figure 2.The mode split patterns when *T* varying

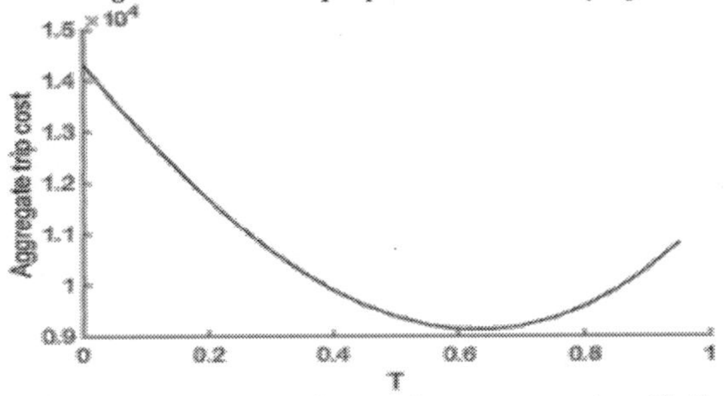

Figure 3.The aggregate trip cost of commuters varies with *T*

5. Conclusion

In this paper we study the mode choice and commuting behaviors at a bottleneck with a bi-mode transportation system based on Vickery's bottleneck model and Kraus's study when an additional transit service is provided in rush hour. The bottleneck model are developed to depict peoples commuting behaviors by auto and transit modes, respectively. We develop Kraus's qualitative results to quantitative experimental conclusions. The mode choices is governed by a logit-based model. We analytically and numerically investigate the properties of equilibrium state and the consequences of trip cost by providing additional transit service at different time duration.

Through this study, we give new insights into peoples' mode choice and aggregate trip cost at the bottleneck with time-constrained transit service. The main results gained can be summarized below: (i) Everyone experiences a lower cost when additional transit service is provided than if it is not. Theoretically more than half of aggregate commuter trip cost is saved. (ii) Commuters aggregate trip cost gets its optimum value when additional transit service begins when waiting time is a half of the maximum waiting time T_C , i.e. $T = T_c/2$. Looking at our findings, two stand out. One is that dispatching additional buses during the peak time of rush hour is an effective way to reduce congestion at bottleneck, which can save commuters aggregate trip cost a lot. The second is that with numerical experiment help we find out the optimum service duration of additional dispatches, which has practical value in reality.

The above results rely on our analyses carried out in a simple bi-mode, exclusive bus lane transportation system. Hence, the generality of the results is restricted to some extent. For simplicity, we assume that all commuters are homogeneous choice users. The modeling of multiple modes can be further improved by considering additional runs among pre-existing runs serving captive bus users. For example, only one part of bus passenger capacity remains for choice users, and the other part is default for captive users.

Acknowledgments

This project has received funding from National Natural Science foundation of China (grant agreement 71622005) http://www.nsfc.gov.cn.

References

1. W S Vickrey. AER, J.**59** (2) (1969)
2. T. Tabuchi. J URBAN ECON, J.**34** (3) (1993)
3. H J Huang. TRANSPORT RES E, J. **36**(4) (2000)
4. H J Huang. EUR J OPER RES, J. **140**(3) (2002).
5. H J Huang, et al. TRANSPORT RES E, J.**43**(5) (2007)
6. M Kraus, Y Yoshida. J URBAN ECON, J. **51**(1) (2002)
7. Marvin Kraus. J URBAN ECON, J. **4**(3) (2003)
8. Li S. Prompt-Traffic & Transportation, J. **24**(5) (2012)
9. N Zheng, et al. TRANSPORT RES, J. **62**(2) (2016)
10. X S Lu et al. Transport Policy. **44** (2015)
11. S Kitth, A Chen. TRANSPORT RES C, J. （2007）
12. Bagloee et al.Public Transport, 1-29(2017)
13. E J Gonzales, C F Daganzo. TRANSPORT RES B J. **46**(10) (2012)

14. R Arnott, A D Palma. J URBAN ECON, J.**27** (1) (1990)
15. R Arnott, et al. REG SCI URBAN ECON, J.**22** (1) (1992)
16. R Arnott. Boston College Working Papers in Economics. **28**(2)(1993).
17. R Arnott, A Yan. RURDS, J. **12**(3) (2000)

Improved Memetic Algorithm for Multi-depot Multi-objective Capacitated Arc Routing Problem

Jie Wan[1], Xinghan Chen[1,a] and Ruichang Li[1]

Hebei University of Technology, School of Economics and Management, 300401 Tianjin, China

Abstract. The capacitated arc routing problem (CARP) is a challenging vehicle routing problem with numerous real-world applications. In this paper, an extended version of CARP, the multi-depot multi-objective capacitated arc routing problem (MDMOCARP) is proposed to tackle practical requirements. Firstly, the critical edge decision mechanism and the critical edge random allocation mechanism are proposed to optimize edges between depots. Secondly, a novel adaptive probability of local search with fitness is proposed to improve the Decomposition-Based Memetic Algorithm for Multi-Objective CARP (D-MAENS). Compared with the D-MAENS algorithm, experimental results on MD-CARP instances show that the improved memetic algorithm (IMA) has performed significantly better than D-MAENS on convergence and diversity in the metric IGD and the metric HV.

1 Introduction

Capacitated Arc Routing Problem (CARP) [1] has wide real-world applications such as snow removal [2], urban waste collection [3-4], pipeline repair [5] etc. Today, these practical issues that are closely related to the logistics transportation have become very important to the government administration. For example, in order to prevent roads from freezing and reduce the traffic accidents, the UK spends millions of pounds on salting in the relevant streets each year.

With the development of society, the typical CARP model cannot meet the actual needs of complexity. Therefore, the application of the CARP model has been widely concerned by researchers. To fulfil the needs of application, some extended versions of CARP were considered, such as the multi-objective CARP [6], the multi-depot CARP [7], the periodic CARP [8] and so on.

Recently, a number of researchers have done a lot of work on multi-depot and multi-objective. Multi-depot CARP (MDCARP) is the CARP with more than one depot or one center. Amberg et al. [9] described a minimal Euler graph, and transformed the problem into the capacitated minimum spanning tree to solve MDCARP early. However, the method only applied to the undirected graph and found the local optimal solution. Kansou et al. [10] insert the ant colony into the sequence of tasks and uses heuristic information to insert the appropriate location, thus proposed a new ant colony algorithm (ACs) to solve MDCARP. Xing et al. [11] improved augment-merge, path-scanning, and the Ulusoy-split, which seeks good quality feasible initial solutions. Krushinsky et al. [7] proposed a dual-index MILP model based on asymmetric multi-depot problem. Oliveira et al. [12] proposed a problem decomposition approach. The task or customer is allocated to the depot based on the distance so that each sub-problem becomes a single depot problem. But the repair operator needs to remove duplicate tasks which lead to the new cost.

On the other hand, in the multi-objective CARP (MOCARP), Lacomme et al. [13] studied the two-objective capacitated arc routing problem model to meet the needs of Troyes in waste collection, and solved the problem by using NSGAII algorithm. Mei et al. [14] proposed the decomposition based memetic algorithm for multi-objective capacitated arc routing problem (D-MAENS), which combines the MAENS algorithm with a multi-objective evolutionary algorithm framework based on problem decomposition. Shang et al. [15] improved the update mechanism and intended to an improved decomposition-based memetic algorithm for MOCARP. Shang et al. [16] used the routing distance grouping strategy to reduce the scale of the problem in large-scale MOCARP.

However, little attention has been paid to the multi-depot multi-objective capacitated arc routing problem (MDMOCARP). In part, these observations motivated our study. Therefore, this paper contributes to constructing a MDMOCARP model. An improved memetic algorithm (IMA) with the critical edge decision

[a] Corresponding author: chenxh2017001@163.com

mechanism and the critical edge random allocation mechanism is proposed. In order to control local search frequency, we develop an adaptive probability of local search. In order to better solve the MDMOCARP model and increase its application range.

This paper is organized as follows. The next section gives the mathematical description of the problem and the improved memetic algorithm. The third section presents the results of our experimental studies, which compare the D-MAENS algorithms using the MDCARP instances. The conclusion is provided in section 4.

2 Model establishment and algorithm improvement

2.1 The MDMOCARP model

In this section, the mathematical model is described for the multi-depot multi-objective CARP. The main purpose of the model is to minimize the total cost of all the tours and to minimize the largest consumption of m vehicles. Because in the actual CARP, if the area of the vehicle requiring service is expanded, it is necessary to use multiple-depot for service. At the same time, only minimizing the total cost cannot meet the actual demand. In addition, minimizing the maximum loop cost is needed to ensure the optimization effect of arc path. For convenience, the following necessary assumptions are given:

1. Each route must start and end the same depot.
2. The vehicle only serves once for the task side.
3. The total demand of served tasks cannot exceed the capacity Q of the vehicle.
4. All edges in the connected graph are allowed to traversed by vehicles of different depots, but only serviced by the vehicle of the belonging depot.

Table 1. The definition of variables.

Variables	Definition
v_i	The vertices of the graph
$tc(v_i,v_j)$	The cost of traversing the edge
$sc(v_i,v_j)$	The cost of serving the edge
$d(v_i,v_j)$	The service demand of the edge (v_i,v_j)
$x^d_{h(v_i,v_j)}$	The 0-1 variable, $x^d_{h(v_i,v_j)}=1$ is the depot d and the vehicle h serves from v_i to v_j; otherwise $x^d_{h(v_i,v_j)}=0$
$y^d_{h(v_i,v_j)}$	The 0-1 variable, $y^d_{h(v_i,v_j)}=1$ is the depot d and the vehicle h traverses from v_i to v_j; otherwise $y^d_{h(v_i,v_j)}=0$
d	The depot ID number
h	The vehicle ID number
n	The depot total number
m	The vehicle total number
$\lvert L_h^d \rvert$	The length of the task sequence of the vehicle h in the depot d
$Tc(\lvert L_h^d \rvert)$	The total cost of the h vehicle in the depot d

As Table 1 shows, the key variables are defined. The graph is presented by $G=(V,E)$, as shown in Figure 1. $V=\{v_1,v_2,\cdots v_{(w+n)}\}$ is the set of vertices in a graph. $V_1=\{v_{w+1},v_{w+2},\cdots v_{(w+n)}\}\subseteq V$ indicates that there are n depots in the graph. An edge $(v_i,v_j)\in E$ is directed edge from v_i to v_j.

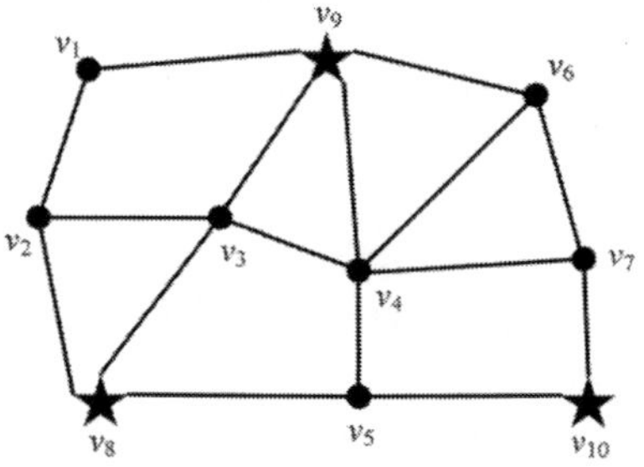

Figure 1. The schematic diagram of MDMOCARP model.

Based on that, the mathematical model can be given as follows.

$$\min f_1(x)=\sum_{d=1}^{n}\sum_{h=1}^{m}Tc\left(L_h^d\right) \quad (1)$$

$$\min f_2(x)=\max_{1\le d\le n}\left(\max_{1\le h\le m}\left(Tc\left(L_h^d\right)\right)\right) \quad (2)$$

s.t.

$$Tc\left(L_h^d\right)=\sum_{(v_i,v_j)\in E}\left(sc(v_i,v_j)x^d_{h(v_i,v_j)}+tc(v_i,v_j)y^d_{h(v_i,v_j)}\right) \quad (3)$$

$$x^d_{h(v_i,v_j)}+y^d_{h(v_i,v_j)}=x^d_{h(v_k,v_i)}+y^d_{h(v_k,v_i)}$$
$$1\le h\le m, 1\le d\le n \quad (4)$$

$$\sum_{d=1}^{n}\sum_{h=1}^{m}x^d_{h(v_i,v_j)}=1,\ \forall(v_i,v_j)\in E \quad (5)$$

$$\sum_{d=1}^{n}\sum_{h=1}^{m}y^d_{h(v_i,v_j)}\ge 1,\ \forall(v_i,v_j)\in E \quad (6)$$

$$\sum_{(v_i,v_j)\in E}d(v_i,v_j)x^d_{h(v_i,v_j)}\le Q,\ \forall(v_i,v_j)\in E \quad (7)$$

$$x^d_{h(v_i,v_j)}\le y^d_{h(v_i,v_j)},\ 1\le h\le m, 1\le d\le n, \forall(v_i,v_j)\in E \quad (8)$$

Here, the target f_1 is to minimize the total cost and the target f_2 is to minimize the largest consumption of m vehicles. The constraint (3) is the total cost of the vehicle h of the depot d. The constraint (4) ensures that vehicle h start at the depot and return to it when the service is completed. The constraint (5) means that the task side is served only once. The constraint (6) is that all sides in the graph allow the vehicle to pass through indefinitely. The constraint (7) requires that the total demand cannot exceed the capacity Q of the service vehicle. When the vehicle is servicing a task edge, it must pass through the task edge by constraint (8).

2.2 Improved memetic algorithm

For MDMOCARP, an improved memetic algorithm (IMA) is proposed based on D-MAENS [14]. In the section, we first described the critical edge decision mechanism considering two indicators of distance from the task side to depots and demand on the task side. Then we design the critical edge's random allocation mechanism. Finally, we present the adaptive probability of local search.

2.2.1 Critical edge decision mechanism

In the multi-depot problem, we generally need to convert multiple depots into a single depot. Different conversion strategies affect the final optimization results. In order to gain the global optimization, this paper uses the critical edge optimize. The distance between the task edge and the depots and the demand of the task edge are the criterion for determining the critical edge.

$$\theta < \frac{\Delta task(v_{w1},(v_i,v_j))}{\Delta task(v_{w2},(v_i,v_j))} + \rho * (d(v_i,v_j)/capacity) \quad (9)$$

Where, θ is the threshold. $\Delta task(v_{w1},(v_i,v_j))$ and $\Delta task(v_{w2},(v_i,v_j))$ are the distance from the edge to depot w_1 and w_2 . $d(v_i,v_j)/capacity$ denotes the proportion of the demand of the edge and the capacity of vehicle. ρ refers to the weight parameter.

2.2.2 Critical edge random allocation mechanism

In order to efficiently handle MDMOCARP, we propose the critical edge random allocation mechanism. The critical edge random allocation mechanism is described as follows:

Algorithm 1: Critical edge random allocation mechanism
Input: Critical edge set $T_{(v_i,v_j)} \in E$ The present best solution: S Output: The best solution: S^* 1. Randomly select n edges from the set $T_{(v_i,v_j)}$ 2. Assign n edges randomly to a depot 3. Use the IMA algorithm to solve the problems of each depot, and generate a global new solution 4. Replace if the offspring solution is better than the previous solution 5. Count the number of new solutions without replacement with $numb$ 6. If $numb$ reaches MAX_numb , it will exit or it will be converted to step 1

2.2.3 Adaptive probability of local search

In the process of evolution, local search occupied a lot of computer resources, and it is not efficient that the same local search probability is applied to the search individuals. we design an adaptive probability of local search with the fitness of the individual for multi-objective optimization. The main ideas are local search is only allowed when needed and adaptive probability which increases with the fitness of the individual.

$$\varphi = \frac{fit_{ave} - fit_{best}}{fit_{best}} * 100\% \quad (10)$$

$$S = \frac{1}{1 + k * \sqrt{\frac{\sum_{i=1}^{popsize} (fit_i - fit_{ave})^2}{popsize}}} \quad (11)$$

$$pls = \sin(\frac{\Pi}{2} * \frac{fit_{best}}{fit_i} * S) \quad (12)$$

$$P = \begin{cases} pls, \varphi > \gamma \\ 0, \varphi \le \gamma \end{cases} \quad (13)$$

Here, fit_{ave} and fit_{best} represent the average value of fitness of the individual and the best value of fitness. the deviation rate denoted by φ in the formula (10). fit_i is the fitness of the i . k means adjustment factor, $k = 0.1$. $popsize$ is the population size. P represents the adaptive probability of local search. γ is set as 0.01.

3 Experiment analysis

3.1. MDCARP instances and parameter settings

The MDCARP instances [10] called md-Golden are obtained by adding depots in the basic CARP instances. It includes the 23 CARP instances. The size of instances is from 7 to 27 vertices and 11-55 required edges. It has two depots, at the first node and at the final node.

The IMA was implemented in Visual C language, and all experiments were run on a personal computer with the 3.0GHZ processor and 8GB memory. The parameters were: The population size is 30. The maximum number of iterations was 1000. The maximum number MAX_numb of depot cycle was 100. The parameter settings of the probability of local search for D-MAENS was 0.2.

3.2 IMA algorithm threshold value θ analysis

The IMA algorithm uses threshold θ to determine critical edge, so we need analyze it. Seven kinds of thresholds θ were analyzed, which were 0.5, 0.6, 0.7, 0.8, 0.9, 1.0, 1.1, respectively with the instance g12. IMA algorithm was run 10 times independently, and gained the average of the total cost. The results are presented in Figure 2.

In the Figure 2, it can be found that when the thresholds are in the middle of 0.6 and 0.8, the total cost can get better optimization results. Therefore, the threshold of 0.7 is more reasonable in the IMA.

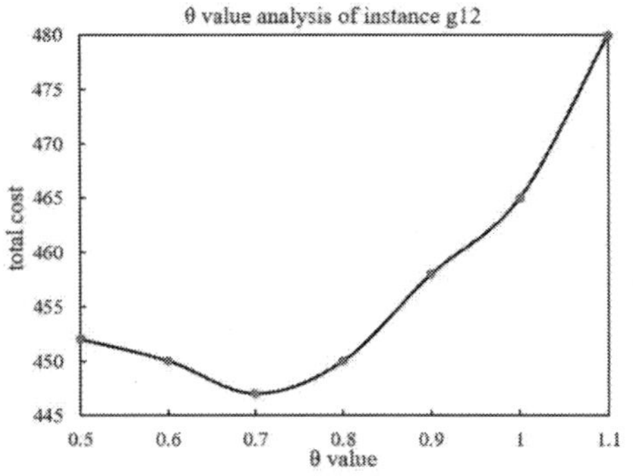

Figure 2. The total cost for different thresholds value θ .

3.3 Comparison between D-MAENS and IMA

The two performances metrics evaluate the performance of the two algorithms. one is the inverted generational distance from reference set (IGD) [17], which indicates the convergence and diversity of a solution set in multi-objective problem. Another one is hypervolume (HV) [18], which mainly describes the convergence and diversity of a solution set in a sense. Each algorithm was run 30 times on each test instance and the average of the indicators was calculated for comparison. If an indicator of the algorithm is better than other algorithms, it will be bolded.

Table 2. Results of D-MAENS and IMA.

Instances	V	E	d	D-MAENS		IMA	
				IGD	HV	IGD	HV
g1	12	22	1, 12	8.64	658	**8.28**	**735**
g2	12	26	1, 12	10.75	**633**	**9.71**	577
g3	12	22	1, 12	9.86	**900**	**8.79**	719
g4	11	19	1, 11	**35.06**	**434**	35.31	414
g5	13	26	1, 13	15.42	851	**13.09**	**1056**
g6	12	22	1, 12	17.46	**1121**	**15.25**	962
g7	12	22	1, 12	18.43	1862	**18.31**	**1977**
g8	27	46	1, 27	23.36	291	**23.11**	**380**
g9	27	51	1, 27	**5.76**	356	5.91	**384**
g10	12	25	1, 12	16.17	**1620**	**16.08**	1502
g11	22	45	1, 22	**13.55**	**5129**	14.60	1312
g12	13	23	1, 13	7.27	900	**7.15**	**970**
g13	10	28	1, 10	**80.38**	7624	93.98	**9500**
g14	7	21	1, 7	**6.34**	274	6.56	**365**
g15	7	21	1, 7	1.37	**412**	**1.22**	150
g16	8	28	1, 8	7.07	221	**6.55**	**253**
g17	8	28	1, 8	**1.15**	76	1.68	**78**
g18	9	36	1, 9	14.64	**569**	**13.11**	315
g19	8	11	1, 8	**0.02**	55	0.023	55
g20	11	22	1, 11	13.03	96	**1.56**	**343**
g21	11	33	1, 11	14.97	**495**	**9.23**	352
g22	11	44	1, 11	13.29	514	**12.34**	**683**
g23	11	55	1, 11	**10.13**	363	10.64	**432**

Table 2 shows the attributes of the instances and the experimental result for the IMA performance. It can be seen that the IMA performs better than D-MAENS on 15 out of the 23 instances in the metric IGD. The average HV result is obtained by IMA and D-MAENS. It can be found that IMA obtains a better solution on 13 out of the 23 instances.

Table 3. The best solution of D-MAENS and IMA.

Instances	V	E	d	D-MAENS		IMA	
				f_1	f_2	f_1	f_2
g1	12	22	1, 12	320	76	**300**	**63**
g2	12	26	1, 12	352	62	**321**	**59**
g3	12	22	1, 12	307	63	**263**	**59**
g4	11	19	1, 11	310	63	**266**	**59**
g5	13	26	1, 13	379	63	361	64
g6	12	22	1, 12	323	68	**301**	**60**
g7	12	22	1, 12	345	72	**326**	**72**
g8	27	46	1, 27	350	38	334	39
g9	27	51	1, 27	305	37	**286**	**35**
g10	12	25	1, 12	284	66	**282**	**63**
g11	22	45	1, 22	397	84	**387**	**84**
g12	13	23	1, 13	480	77	446	81
g13	10	28	1, 10	536	151	**515**	**120**
g14	7	21	1, 7	107	21	**92**	**21**
g15	7	21	1, 7	58	15	58	15
g16	8	28	1, 8	125	29	**125**	**20**
g17	8	28	1, 8	91	13	91	13
g18	9	36	1, 9	168	26	**140**	**24**
g19	8	11	1, 8	55	21	55	21
g20	11	22	1, 11	**121**	**30**	121	34
g21	11	33	1, 11	158	30	**128**	**28**
g22	11	44	1, 11	201	29	184	31
g23	11	55	1, 11	233	26	223	28

In the Table 3, the nondominated solutions of the multi-objective functions are indicated. It can be seen that the IMA gains the better solutions for 14 out of 23 instances, while the D-MAENS only gains one better solutions. Figure 3 shows the nondominated solutions of the two algorithms after 30 independent runs. Due to limited space, the test instances of g1, g10, g14 and g21 only are displayed. It can be seen clearly that the IMA solution sets are better close to the Pareto-Optimal Front than D-MAENS.

Therefore, it can conclude that the performance of IMA is much better than the D-MAENS in the convergence and diversity of a solution set for MDMOCARP. The IMA considers the critical edge assignment problem in arc path optimization. The overall optimization of the arc path can effectively reduce the total cost and the largest consumption of vehicles.

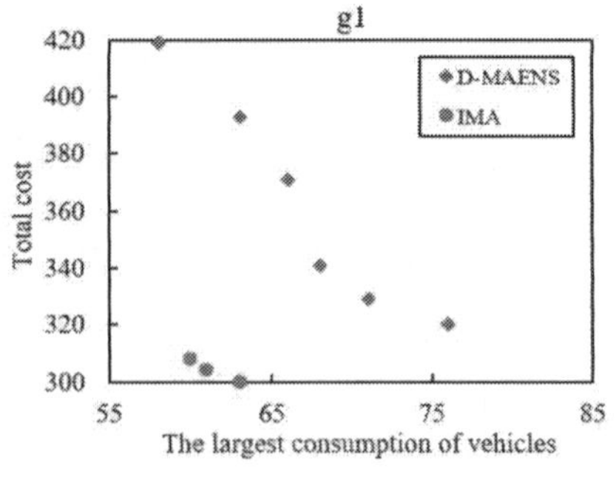

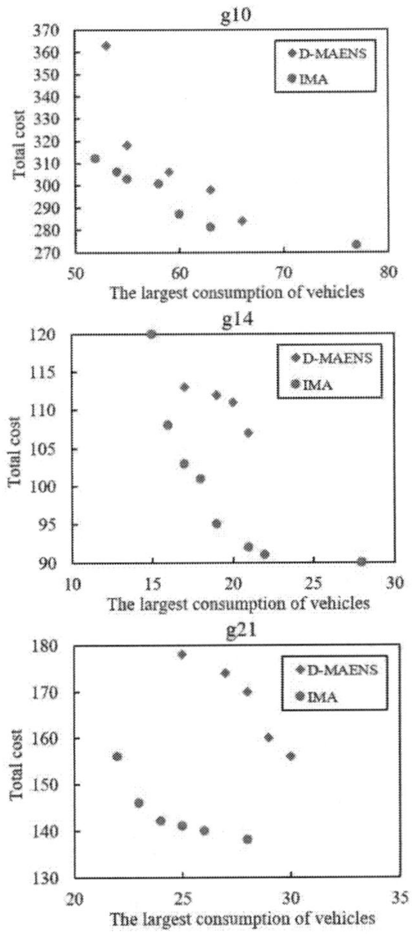

Figure 3. The nondominated solutions of the two algorithms.

4 Conclusion

In this paper, we study the multi-depot multi-objective Capacitated Arc Routing Problem, which is a more practical version of the CARP, and propose the multi-depot multi-objective CARP model. this model has two objective functions in the multi-depot, minimizing the total cost and minimize the largest consumption of vehicle. The IMA is proposed to solve the problem, which has the critical edge decision mechanism, the critical edge random allocation mechanism and a novel adaptive probability of local search with fitness of the individual.

Simulation results show that the IMA is competitive with respect to improving the quality of the non-dominant solution. Appropriate selection and distribution of critical edges can better optimize the routing, and reduce costs greatly.

In the future, some practical problems remain unsolved. Firstly, we will study the arc routing problem with the fuel consumption or the carbon dioxide emissions. Secondly, more efficient local searches probability needs to be studied to reduce computational costs and time. Finally, we only analyse the threshold, so we need to analyse more parameters in the future so that the algorithm achieves better results.

Acknowledgment

This paper is supported by No.18BGL012 from the China Social Science Foundation Project, No.2019031201028 from the China Hebei Social Science Development Research Project and No.2017GJJG021 from the China Hebei Higher Education Teaching Reform Research and Practice Project.

References

1. B. Golden, R. Wong. CARP, Net., **11**, 305-315 (1981)
2. M. Tagmouti, M. Gendreau, J. Y. Potvin. DCARP., TRC., **19**, 20-28(2009)
3. S. Nesmachnow, D. G. Rossit, J. Toutouh. Compari., ENDM., **69**, 93-100(2018)
4. T. P. B. Vecchi, D. F. Surco, A. A. Constantino, et al. A seq., PSEP., **102**, 238-250(2016)
5. R. D. Gabriel, V. Daniele, T. Fernando, et al. Visual., COR., **103**, 13-34(2019)
6. R. Shang, B. Du, H. Ma. Immune., ASOC., **49**, 748-758(2016)
7. D. Krushinsky, T. V. Woensel. An app., EJOR., **244**, 100-109(2015)
8. E. B. Tirkolaee, I. Mahdavi, M. M. S. Esfahani. A robust., WasMan., **76**, 138-146(2018)
9. A. Amberg, W. Domschke, S. Voß. Multi., EJOR., **124**, 360-376(2000)
10. A. Kansou, A. Yassine. A two., ICCIE. IEEE, (2009).
11. L. Xing, P. Rohlfshagen, Y. Chen, et al. An evo., TEVC., **14**, 356-374(2010)
12. F. B. D. Oliveira, R. Enayatifar, H. J. Sadaei, et al. A coo., ESWA., **43**, 117-130(2016).
13. P. Lacomme, C. Prins, W. R. Cherif. Com., ANOR., **131**, 159-185(2004)
14. Y. Mei, K. Tang, X. Yao. D-MAENS., TEVC., **15**, 151-165(2011)
15. R. Shang, J. Wang, L. Jiao, et al. An impro., ASOC., **19**, 343-361(2014)
16. R. Shang, K. Dai, L. Jiao, et al. Impro., TCYB., **46**, 1000-1013(2016)
17. Q. Zhang, H. Li. MOEA/D., TEVC., **11**, 712-731(2008)
18. T. Liu, Z. Jiang, N. Geng. A gen., FSM., **26**, 540-564(2014)

Research on Passenger Flow Early Warning of Urban Rail Transit Station Based on System Dynamics

Hui Chen[1], Bo Wang[2, a], Wei He[1] and Jianhu Zheng[1]

[1] *Minjiang University, 350108Fuzhou, Fujian Province, China*
[2] *Beijing Transportation Information Center,100044Beijing, China*

Abstract. Large-scale passenger flows occur frequently during the peak hours of urban rail transit stations and on holidays. Thus, the timely and accurate early warning of impending large-scale passenger flows can positively impact the operational safety of the entire station. By further deepening the definition of passenger flow warnings in stations, a new model of urban rail transit station passenger flow based on system dynamics is constructed. The method of determining the key area of passenger flows in the early warning stage based on streamlines is proposed; the key indicators and thresholds affecting early warnings are studied. Finally, taking a typical station as an example, a station model is built using Anylogic software. The parameter sensitivity analysis is used to determine the impact of each key indicator on the passenger flow in the key area of the station early warning, and the reference threshold of each indicator is determined.

1 Introduction

Due to the special geographical locations and building structures of urban rail transit stations, passengers' personal safety and station operation safety can be compromised if very large passenger flows occur without warning. In such situations, station management personnel may be unable to issue warnings and take passenger flow control measures in time.

This important topic recently emerged in the research community and garnered significant attention. D. Li et al. [1] proposed the concept and framework of passenger flow early warning, and studied the level, rating basis, and indicator system of passenger flow early warning. L. Ding, B. Zhao et al. [2,3] studied early warning classification methods for passenger flow in the stations. X. Xu et al. [4] studied the early warning methods of passenger flow risk in the business district from four aspects: regional density, section flow, bidirectional flow ratio, and average flow rate. Y. Zheng et al. [5] established a fuzzy synthetic evaluation model based on passenger flow density and evacuation time. M. Li et al. [6] proposed a correlated early warning analysis process based on system dynamics (SD) equations. A pedestrian evacuation model considering residence time, crowd density, and exit distance was proposed by L. Fang et al. [7]. X. Ding et al. [8] promoted a passenger flow status identification method based on image and WiFi detection.

It should be noted that the early warning research on passenger flow in urban rail transit stations by domestic and international scholars is mostly focused on early warning levels, early warning indicators, safety assessments, etc. However, there are few studies investigating the thresholds of the early warning indicators.[1] The early warning of passenger flow at stations is a complex system, which is closely related to the station layout, facility equipment capacity, organization of train operation, and passenger flow. Therefore, a novel station passenger flow model and simulation platform based on SD theory is proposed in this paper. Meanwhile, the station early warning methods and key indicator thresholds are studied to provide a rigorous foundation for the research.

2 Definition of Passenger Flow Early Warning in Urban Rail Transit Stations

According to the structure of urban rail transit stations, the actual operation of stations, the characteristics of passenger flows in the station, and the literature [1], the passenger flow early warning of urban rail transit station is defined as follows. The station management personnel shall monitor the key indicators in key areas, and predict potential large passenger flows by comparing the indicator values to the indicator thresholds. Meanwhile, there should be sufficient time to take relevant measures to control the station passenger flow, and the safety of the station passengers is ensured during this period.

3 Early Warning Analysis of the Station Passenger Flow Based on an SD Model

3.1 SD model construction for station passenger flow

[a] Corresponding author: bowang@bjtu.edu.cn

By analyzing the causal relationship between the three elements of the station facilities, passenger flow, and operational organization, a station passenger flow model based on SD theory is established. The system flow diagram is shown in Figure 1 and the relevant parameter variables in the model are shown in Table 1.

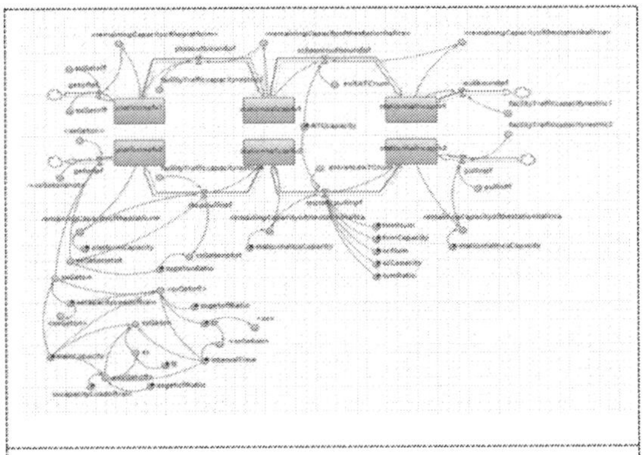

Figure 1. System flow chart of the station passenger flow model

Table 1. Variable descriptions

Variable/Parameter Code	Variable/Parameter Name	Variable/Parameter Definition
Q_1	platformpfv1	The real-time outbound passenger flow at the platform.
Q_2	Platformpfv2	The real-time inbound passenger flow at the platform.
Q_3	Stationhallcapfv1	The real-time outbound passenger flow in the paid area of the station hall.
Q_4	Stationhallcapfv2	The real-time inbound passenger flow in the paid area of the station hall.
Q_5	Stationhallncapfv1	The real-time outbound passenger flow in the non-paid area of the station hall.
Q_6	Stationhallncapfv2	The real-time inbound passenger flow in the non-paid area of the station hall.
F_{1in}	getoffpf	The real-time outbound passenger flow for getting off the trains.
$F_{1out}\|F_{3in}$	ptosoutboundpf	The real-time outbound passenger flow from the platforms to the station hall.
$F_{3out}\|F_{5in}$	sctosncoutboundpf	The real-time outbound passenger flow from paid area to the non-paid area in the station hall.
F_{5out}	outboundpf	The real-time outbound passenger flow.
F_{6in}	pullinpf	The real-time inbound passenger flow.
$F_{6out}\|F_{4in}$	snctoscpullinpf	The real-time inbound passenger flow from the paid area to the non-paid area in the station hall.
$F_{4out}\|F_{2in}$	stoppullinpf	The real-time inbound passenger flow from the station hall to the platforms.
F_{2out}	getonpf	The real-time inbound passenger flow for getting on the trains.
F_{in}	pullinPf	Number of inbound passengers.
$RQ_1\|RQ_2$	remainingCapacityoftheplatform	The real-time remaining capacity of the platform.
$RQ_3\|RQ_4$	remainingCapacityofthestationhallca	The real-time remaining capacity of the paid area of the station hall.
$RQ_5\|RQ_6$	remainingCapacityofthestationhallnca	The real-time remaining capacity of the non-paid area of the station hall.
RF_{5out}	facilityTrafficcapacityrestric1	The restricted capacity of the equipment at the connection of the non-paid area of the station hall and the outside of the stations in the outbound streamline.
RF_{6in}	facilityTrafficcapacityrestric3	The restricted capacity of the equipment at the connection of the non-paid area of the station hall and the outside of the station in the inbound streamline.
$RF_{1out3in}$	facilityTrafficcapacityrestric2	The restricted capacity of the equipment at the connection of the platform and the paid area of the station hall in the outbound streamline.
$RF_{4out2in}$	facilityTrafficcapacityrestric4	The restricted capacity of the equipment at the connection of the platforms and the paid area of the station hall in the inbound streamline.
CQ_{12}	platformCapacity	The platform capacity.
CQ_{34}	stationhallcaCapacity	The capacity of paid area in the station hall.
CQ_{56}	stationhallncaCapacity	The capacity of the non-paid area in the station hall.
θ	intervalTime	The departure interval time.
FS_{1in}	sxGetoff	The passenger flow getting off the upward direction trains at arrival.
FX_{1in}	xxGetoff	The passenger flow getting off the downward direction trains at arrival.
SF_{2out}	sxGeton	The available capacity for the upward direction trains at arrival.
XF_{2out}	xxGeton	The available capacity for the downward direction trains at arrival.

SPF_{2out}	sxGetonptot	The number of passengers for the upward direction trains at the platform.
XPF_{2out}	xxGetonptot	The number of passengers for the downward direction trains at the platform.
N_1/ N_2	exitATCnum/ entranceATCnum	Number of exit/ entrance ticket gates.
N_3/ N_4	tvmNum/ scNum	Number of ticket vending machines (TVM) / number of security checks.
C_{N12}	ATCcapacity	The ticket gate capacity.
C_{N3}/ C_{N4}	tvmCapacity/ scCapacity	The capacity of ticket vending machines / the capacity of security checks.
ρ_{N3}	tvmRatio	The ratio of passengers who need to purchase tickets at the TVM in the inbound passengers.
ρ_{sof}/ρ_{xof}	sxgetoffRatio / xxgetoffRatio	The ratio of alighting passengers of upward/downward direction trains.
η_s/η_x	sxcapacityLoadoftrain/xxcapacityLoadoftrain	The load factors of upward/downward direction trains.
ρ_{son}	sxgetonRatio	The proportion of passengers who need to get on the upward direction trains at the station.
C_{train}	trainCapacity	The train capacity.
β	tt	The arrival time interval between upward direction and downward direction trains.

(1) The calculation equation of the state variable

$$Q_i = \int \left[F_{iin}(t) - F_{iout}(t) \right] dt, i = 1, 2 \cdots \cdots 6 \quad (1)$$

(2) The calculation equations of the rate variables

- The passenger flow equations at the connected equipment are as follows:

$$F_{iin} | F_{jout} = \min \left\{ RF_{iinjout}, RQ_i, \left(\frac{Q_j}{\Delta t} \right) \right\}, i = 3, 1; j = 1, 4 \quad (2)$$

$$F_{5in} | F_{3out} = \min \left\{ N_1, C_{N12}, RQ_5, \left(\frac{Q_3}{\Delta t} \right) \right\} \quad (3)$$

$$F_{4in} | F_{6out} = \min \left\{ \begin{array}{l} \min \left\{ \left(\frac{Q_6}{\Delta t} \right) \cdot \rho_{N3}, N_3 \cdot C_{N3} \right\} + \\ \left(\frac{Q_6}{\Delta t} \right) \cdot (1 - \rho_{N3}), N_4 \cdot C_{N4}, N_2 \cdot C_{N12}, RQ_4 \end{array} \right\} \quad (4)$$

- The passenger flow equations for the inbound and outbound flows in the stations are as follows:

$$F_{5out} = \min \left\{ \left(\frac{Q_5}{\Delta t} \right), RF_{5out} \right\} \quad (5)$$

$$F_{6in} = \min \left\{ RQ_6, RF_{6in}, F_{in} \right\} \quad (6)$$

- The passenger flow equations for the inbound and outbound flows in trains are as follows:

$$F_{1in} = \min \left\{ F_{S1in} + F_{X1in}, RQ_1 \right\} \quad (7)$$

$$F_{2out} = \min \left\{ SF_{2out}, SPF_{2out} \right\} + \min \left\{ XF_{2out}, XPF_{2out} \right\} \quad (8)$$

(3) The calculation equations for the auxiliary variables

$$RQ_i = RQ_{i+1} = CQ_{i(i+1)} - Q_i - Q_{i+1}, i = 1, 3, 5 \quad (9)$$

3.2 Determine streamline based judging criteria for key areas

The streamlines of passenger flows in urban rail transit stations primarily consist of the outbound streamline and the inbound streamline. For the adjacent equipment passing these two streamlines, congestion can readily form if the passing capacity of the downstream equipment is smaller than that of the upstream equipment. Substantially, these streamlines are key areas for passenger flow early warning in the station. According to the characteristics of the SD model for station passenger flow, combined with the literature [1], this paper only judges the station platform area, the paid area in the station hall, and the non-paid area in the station hall.

3.3 Analysis of key elements and determination of key indicators' thresholds

The analysis of the passenger flow composition, organization of train operation, carrying capacity of the key area, and capacity of passing equipment and facilities in each early warning area is carried out to determine the key factors affecting the area. In particular, the passenger flow density is the most direct factor to represent the risk of passenger flow [9]. The literature [1,3,7,8] uses pedestrian density as an important monitoring indicator for station early warnings or safety assessments. Meanwhile, the station service level for pedestrians is also closely related to passenger flow density [10, 11]. Therefore, the passenger flow density is used as a key indicator for monitoring the station's early warning area in this paper.

The early warning of the station passenger flow depends on the current regional passenger flows as well as the changing trends of passenger flows into/out of the area. It is necessary to comprehensively consider the relationship between the two to determine the key indicators' thresholds. The specific methods for the outbound streamline and the inbound streamline are as follows:

(1) Passenger flow early warning for the outbound streamline due to the lower passing capacity of downstream equipment compared to that of upstream equipment.

If one of the following three conditions is met:

$$F_{S1in} + F_{X1in} > RF_{1out3in} \cdot \theta \quad (10)$$

$$RF_{1out3in} > N_1 \cdot C_{N12} \quad (11)$$

$$N_1 \cdot C_{N12} > RF_{5out} \quad (12)$$

Then, the thresholds of the key indicators must be determined by analyzing the following:

- Study the impact of different β values on the real-time passenger flow of the corresponding early warning area and determine the thresholds of the key indicators.
- Study the impact of different θ values on SF_{2out}, XF_{2out}, and the real-time passenger flow of the corresponding early warning area; determine the thresholds of the key indicators.
- Study the impact of different F_{S1in} and F_{X1in} values on SF_{2out}, XF_{2out}, and the real-time passenger flow of the corresponding early warning area; determine the thresholds of the key indicators.

(2) Passenger flow early warning for the inbound streamline due to the lower passing capacity of downstream equipment compared to that of upstream equipment.

If one of the following three conditions is met:

$$SF_{2out}+XF_{2out} < RF_{4out2in}\cdot\theta \tag{13}$$

$$RF_{4out2in} < \min\left\{N_3\cdot C_{N3}, N_4\cdot C_{N4}, N_2\cdot C_{N12}\right\} \tag{14}$$

$$\min\left\{N_3\cdot C_{N3}, N_4\cdot C_{N4}, N_2\cdot C_{N12}\right\} < RF_{6in} \tag{15}$$

Then, the impact of different F_{6in} values on the real-time passenger flow of the corresponding early warning area is studied and the thresholds of the key indicators are determined.

It can be seen from the above research that the determination of the threshold values of the station early warning indicators involves many factors, which are not fixed values. According to the characteristics of different stations, the early warning schemes of each station under different passenger flow conditions should be prepared well.

4 Case Study

A metro station in Fuzhou, China is selected as a prototype in this paper, and a station simulation model is constructed using Anylogic software. Subsequently, the early warning scheme of the station is studied. In this paper, the minimum unit time for system simulation is minutes, and the total simulation time is 180 minutes. The service level standards for each level of each passenger flow carrying area [10, 11] are shown in Table 2. When the service level F is reached, the passenger flow is blocked [10]. Therefore, the passenger flow carrying capacity of the studied area during the simulation process is defined as the passenger flow carrying capacity at the time when the service level F is reached. The platform area is 696 m^2, the paid area in the station hall is 493 m^2, and the non-paid area in the station hall is 635 m^2. According to the layout and the capacity of the station facilities, the basic parameters of the model are set as shown in Table 3.

Table 2 Service Level Standards for Each Passenger Flow Carrying Area

Service Level / Area	Level A (m^2/person)	Level B (m^2/person)	Level C (m^2/person)	Level D (m^2/person)	Level E (m^2/person)	Level F (m^2/person)
Platform	>1.2	(0.9, 1.2]	(0.6, 0.9]	(0.3, 0.6]	(0.2, 0.3]	≤0.2
Paid Area in the station	>3.3	(2.3, 3.3]	(1.4, 2.3]	(0.9, 1.4]	(0.5, 0.9]	≤0.5
Non-paid area in the station	>3.3	(2.3, 3.3]	(1.4, 2.3]	(0.9, 1.4]	(0.5, 0.9]	≤0.5

Table 3 Basic Parameter Settings of the Station Model

CQ_{12}/Person	CQ_{34}/ Person	CQ_{56}/ Person	C_{N4}/(Person / minute)	C_{N3}/(Person / minute)
3480	986	1270	20	1.3
N_1/per machine	N_2/per machine	N_3/per machine	N_4/per machine	C_{train}/ Person
12	8	12	3	1460
RF_{6in}/(Person / minute)	RF_{5out}/(Person / minute)	$RF_{1out3in}$/(Person / minute)	$RF_{4out2in}$ /(Person / minute)	C_{N12}/(Person / minute)
572	858	234	156	30

4.1 Criteria for determining the key area of the station early warning

According to the method described in Section 3.2, the equipment capacities of the outbound and inbound streamlines of the station are determined separately:

(1) The outbound streamline: Since $RF_{5out} > C_{N12}\cdot N_1 > RF_{1out3in}$, it is easy to form congestion at the platform if and only if $F_{S1in}+F_{X1in} > RF_{1out3in}\cdot\theta$

(2) The inbound streamline: Since $\min\left\{N_3\cdot C_{N3}, N_4\cdot C_{N4}, N_2\cdot C_{N12}\right\} < RF_{6in}$, it is easy to form congestion at the non-paid area in the station hall. In addition, it is easy to form congestion at the platform when $SF_{2out}+XF_{2out} < RF_{4out2in}\cdot\theta$.

Therefore, it can be determined that the platform area and the non-paid area of the station hall are the early warning key areas in the station.

4.2 Threshold study for the key indicators in the station early warning

According to the method described in Section 3.3, the key indicators and their threshold values of the early warning at the platform and the non-paid area in the station hall are studied.

https://doi.org/10.1051/matecconf/202030801003

4.2.1 Threshold analysis of key early warning indicators in the platform area.

(1) The influence of the arrival time interval between the upward direction and the downward direction trains (β value) on the real-time passenger flow of the platform and alighting passenger flow.

According to the traffic statistics and passenger flow statistics of the passenger flow peak hours of Fuzhou Metro Line 1, the model parameters are set to the average values. The details are shown in Table 4. In the model, $\beta \in [0,3]$, and the parameter sensitivity analysis is performed with a one unit step size. The relationship between the β value and the real-time passenger flow of the non-paid area in the station hall is shown in Figure 2. The relationship between the β value and the real-time alighting passenger flow is shown in Figure 3. The relationship between the β value and the real-time passenger flow at the platform is shown in Figure 4.

Table 4 Basic parameters of train operations and passenger flow

ρ_{N3}	θ/ minute	ρ_{son}	η_x	ρ_{sof}	η_s	ρ_{xof}	C_{train}
0.3	4	0.6	0.9	0.5	0.8	0.4	1460

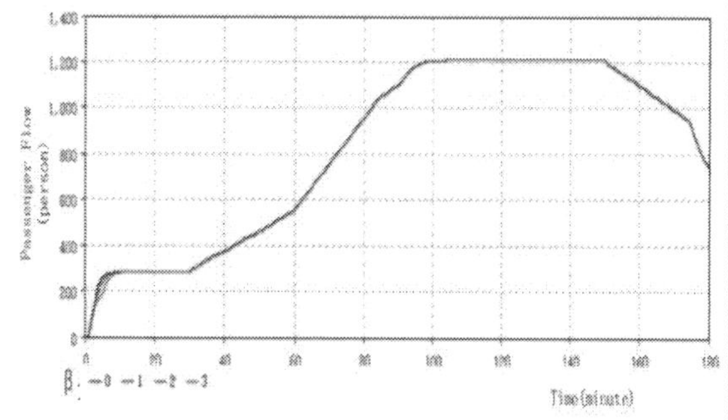

Figure 2. The relationship between the β value and the real-time passenger flow of the non-paid area in the station hall

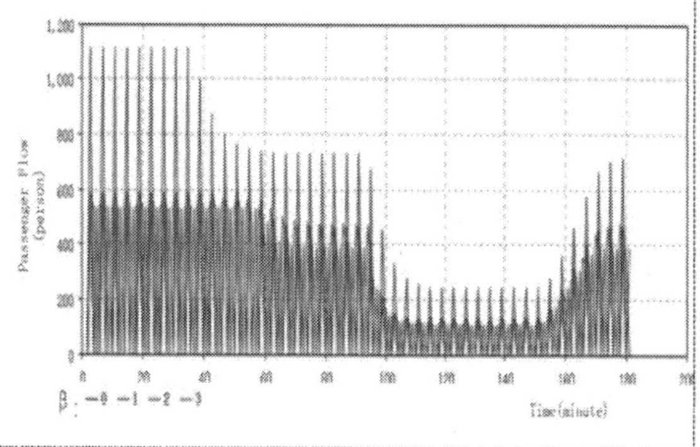

Figure 3. The relationship between the β value and the real-time alighting passenger flow

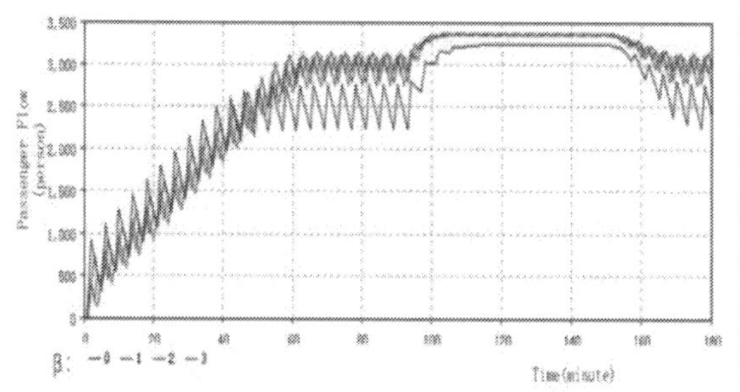

Figure 4. The relationship between the β value and the real-time passenger flow at the platform

It can be seen from Figure 2 that the β value does not affect the passenger flow of the non-paid area in the station hall. Since the inbound passenger flow is too large in the station, the passenger flow in the non-paid area of the station becomes close to its designed capacity at 98 minutes, causing the passenger flow in the station to block.The platform area also approaches the designed capacity in nearly 98 minutes, as shown in Figure 4. Therefore, when analyzing the influence of different β values on the real-time passenger flow at the platform, the influence of the passenger flow blockage caused by the passenger flow in the non-paid area of the station should be excluded.

It can be seen from Figures 3 and 4 that when $\beta = 0$, this parameter can significantly affect the real-time passenger flow at the platform and the alighting passenger flow. Meanwhile, the station area is at the service level D level and above for 165 minutes. From the 39th minute onwards, due to the limitation at the platform capacity, the passengers getting off the train are blocked; some could not even get off the train. This state continues until the end of the simulation. For β=1, 2, and 3, the situation is alleviated. Therefore, β=0 is assumed in the following study of thresholds of early warning indicators.

(2) Influence of the departure interval time (θ) on alighting passenger flow and passenger flow at the platform.

According to a statistical investigation on the departure interval time of China's subway lines, most of the results are in the range of 2-5 minutes. Therefore, $\theta \in [2,5]$, where the parameter sensitivity analysis is performed with a step size of one.

- Influence of departure interval time (θ) on alighting passenger flow

With the station model parameters, the alighting passenger flow at a departure interval θ is calculated to be 1110 persons. If the remaining capacity at the platform is less than the number of alighting passengers, then the congestion prevents some passengers from getting off the train. It can be estimated that alighting passenger flow is affected when the service level at the platform is E or higher. The relationship between θ and the real-time passenger flow at the platform is shown in Figure 5.

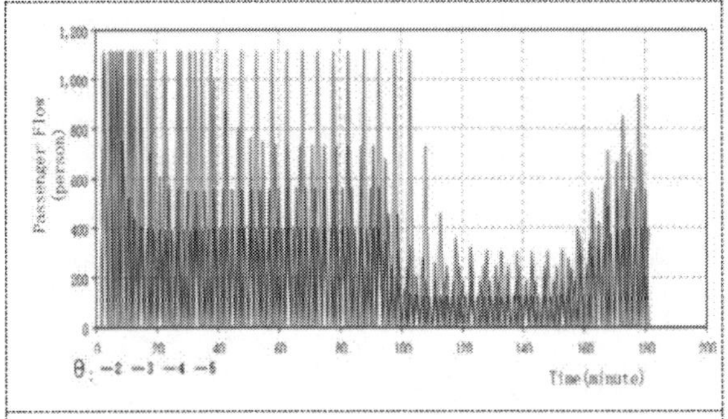

Figure 5. The relationship between θ and the real-time alighting passenger flow

- Influence of the departure interval time (θ) on passenger flow at the platform

The relationship between θ and the real-time passenger flow at the platform is shown in Figure 6.

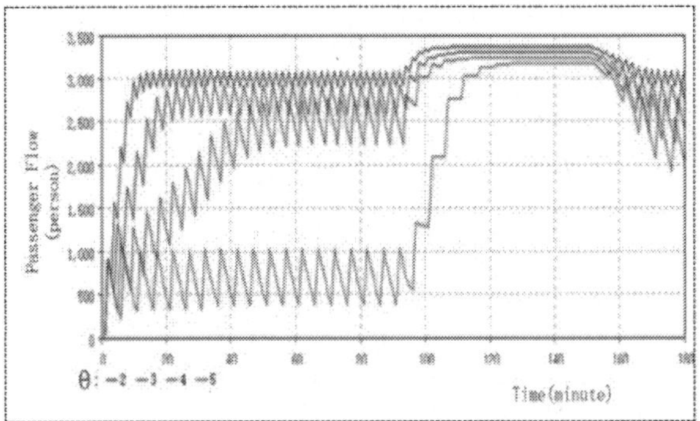

Figure 6. The relationship between θ and the real-time passenger flow at the platform

According to the statistics of the simulation data, the influence of θ on the service level of the platform is shown in Table 5. This table summarizes the average scale of the passenger flow during the peak hours of the station. When θ=5, the triggering times of both the D-level and the E-level are similar to when the passenger flow rates of the non-paid area in the station hall is at or above the capacity limit. Excluding this factor, it can be considered that the service level of the station passenger flow is maintained at or below the D-level at the departure interval time.

Table 5 Correspondence between θ and triggering time / duration time of each station service level

θ	Triggering time / duration time of C- or higher level (minute)	Triggering time / duration time of D- or higher level (minute)	Triggering time / duration time of E- or higher level (minute)	Time of the transition from D-level to E-level /(θ)
2	2/179	4/177	8/173	2
3	2/177	5/174	14/166	3
4	2/17	10/165	34/125	6
5	2/120	97/84	107/67	/

It can be seen from the above analysis that at a fixed passenger flow rate of boarding and alighting, the shorter the departure interval times result in shorter transition times at the platform service level. Considering that the station staff needs time to take necessary measures to control the passenger flow after receiving an alarm, it is recommended that when θ ≤ 3, the warning threshold of the station platform should be taken as the C-level. When θ=4, the warning threshold can take the D-level. When θ ≥ 5, the warning threshold can take the E-level.

- The influence of different alighting passenger flow values F_{S1in} and F_{X1in} on the real-time passenger flow at the platform

In this paper, through a comparative experiment of two setting parameters ρ_{sof} and ρ_{xof}, we study the influence of different passenger flow scales on the real-time passenger flow at the platform. We choose the departure interval θ=3 minutes, which is common in the peak hours for the subways in major Chinese cities. The specific parameter settings and the corresponding passenger flow scale in a departure interval time are shown in Table 6. The effect of different passenger flow scales on the real-time passenger flow at the platform is shown in Figure 7.

Table 6 The relationship between different ρ_{sof} and ρ_{xof} values and the alighting passenger flow scales

ρ_{xof}	0.8	0.7	0.6	0.5
ρ_{sof}	0.9	0.8	0.7	0.6
$F_{S1in}+F_{X1in}$(person/departure interval time)	2106	1857	1608	1359
ρ_{xof}	0.4	0.3	0.2	0.1
ρ_{sof}	0.5	0.4	0.3	0.2
$F_{S1in}+F_{X1in}$(person/departure interval time)	1110	861	612	365

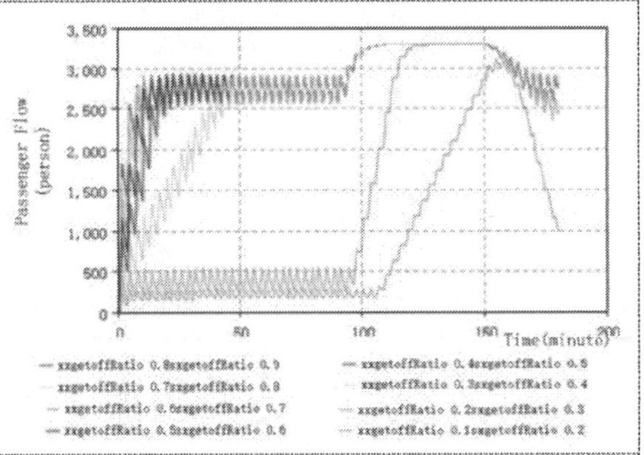

Figure 7. The relationship between F_{6in} and the real-time passenger flow at the platform

It can be seen from the simulation data that sxgetoffRatio=0.4/ xxgetoffRatio=0.3 is the demarcation point. More specifically, when the number of alighting passengers is greater than 861 in a departure interval time, the influence of each passenger flow scale on the passenger flow at the platform is not significant. The transition from the D-level to the E-level occurs in the 2-3 departure interval time. When the number of alighting passengers is less than 861, the impact on the passenger flow at the platform is little and does not cause the passengers to be detained in the station. In this case, we exclude the factor that the passenger flow is blocked in the station when the passenger flow in the non-paid area reaches its capacity limit.

4.2.2 Threshold analysis of key early warning indicators in the non-paid area of the station hall

Based on the method described in Section 3.2, we study the influence of different inbound passenger flow F_{6in} values on the passenger flow of the corresponding early warning area. In this study, $F_{6in} \in [20,90]$, the parameter sensitivity analysis is carried with a step size of 10. The influence of the F_{6in} value on the real-time passenger flow of the non-paid area of the station hall is shown in Figure 8.

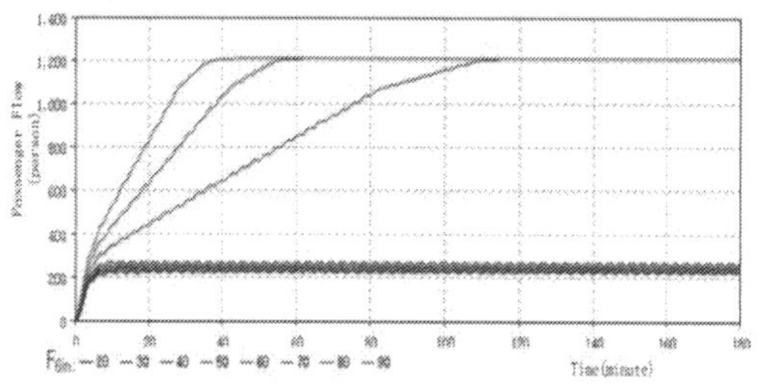

Figure 8. The relationship between different F_{6in} values and the real-time passenger flow of the non-paid area in the station hall

It can be seen from the simulation data that F_{6in}=70 person/min is the dividing line. When F_{6in} <70 person/min, there is no congestion in the non-paid area of the station hall. Therefore, when $F_{6in} \geq 70$ person/min, the warning time should be determined according to the real-time situation of the passenger flow in the non-paid area of the station hall.

5 Conclusion

1) According to the real experience of station operations, we have further enhanced the definition of the station early warning and established the station passenger flow model based on system dynamics.

2) We have proposed a specific method for determining the key areas of the station early warning based on streamlines as well as the threshold values of key indicators.

3) Taking a subway station in a certain city of China as an example, we have verified the feasibility of the station model in a simulation study using Anylogic software.

4) This paper only focuses on ordinary stations in subway lines, and further research is needed for transfer stations.

Acknowledgments

This work is supported by Beijing Natural Science Foundation Key Project(8171003); Fuzhou Science and Technology Plan Project(2018-G-93).

References

[1] D.W. Li, Y.X. Sun , J.L. Huang, Ur. Rap. Rail. Tran.**26**(2)62-66(2013) DOI=10.3969/j.issn.1672-6073.2013.02.016.

[2] L. Ding, Study on passenger flow early warning and coping method of urban rail transit transfer station .Nanjing. Univ. Sci. Technol. 72(2012)

[3] B.F. Zhao, X.L. Zou, X.Y. Qu,.Ur. Mas. Tran. **9** 107-115(2017) DOI=10.16037/j.1007-869x.2017.09.022.

[4] X. Xu, M Y N L, Proc - 2010 IEEE Int. Conf. Emerg. Manag. Manag. Sci. 310-313(2010).

[5] Y.X. Zheng, Y. Qin, J.Y. Guo, L.M. Jia , J.H. Gao, 3rd Int. Conf. Electr. Inf. Technol. Rail. Transp. EITRT 2017 – Transp. **10** 725-734(2017)

[6] M. Li, Y.H. Wang, J.L. Wu, L.M. Jia, China Saf. Sci. J. **24** 22-26(2014).DOI= 10.16265/j.cnki.issn1003-3033.2014.04.012

[7] F.Li, S.Chen, X.Wang, F.Feng , Proced – Soc. Behav. Sci. **138** 314-322(2014).DOI= 10.1016/j.sbspro.2014.07.209

[8] X.Ding, Z.Liu , H.Xu, J. Vis. Commun. Image. Represent. **58** 119-129(2019).DOI= 10.1016/j.jvcir.2018.11.033

[9] T. Li, L.Z. Jin, Y.N. Ma , W.J. Zhu, Saf. Sci. Technol. **8** 75-80 (2012).

[10] J.J. FRUIN, Pedestrian Planning and Design (New York: Metropolitan Association of Urban Designers and Environmental Planners,1971)

[11] D.W. Li, B.M. Han ,Pedestrian Traffic (Bei jing: China Communications Press, 2011). 403- 405.

Impact analysis of new metro line opening on passenger flow: A case study of Shenzhen

Zihao Li[1,2], Yufeng Zhou[1], Zhongwen Li[3], Wei Zhou[3], Maohua Zhong[1,a]

[1] *Institute of Public Safety Research, Department of Engineering Physics, Tsinghua University, Beijing 100084, China*
[2] *Beijing Key Laboratory of City Integrated Emergency Response Science, Beijing 100084, China*
[3] *Shenzhen Metro Co., Ltd, Shenzhen 518026, China*

Abstract. Metro systems, as major components of the transit system, have been significantly developed in recent years and an increasing number of new metro lines are operated in China. This paper not only analyses the short and long-term impact of new lines opening on passenger flow from passenger volume, section imbalance coefficient, direction imbalance coefficient and travel time, with the help of passenger flow data from September 2016 to October 2018 in Shenzhen but also proposes a new method to identify the growth period of the transfer flow in new lines. The result shows that the new metro lines cannot solve the problems of imbalance and congestion in Shenzhen Metro. it increased the travel time in the morning and evening peak, but largely decreased the travel time in the off-peak period. Additionally, there is a significant positive correlation between the length of the growth period and the number of transfer stations of the transfer line.

1 Introduction

With the rapid development of the economy, cities are constantly expanding which increases the pressure on round traffic. Therefore the demand for a multi-model, intelligent, modern, fast and safe transportation system is increasing. The metro system is a mean of public transportation with multiple benefits, such as large capacity, high speed, safety, punctuality, energy-saving, environmental protection, and etc. It plays an important role in the urban transit system and gradually become the main mode of public transportation for people. In China, the total length of the metro system is about 5766.6 kilometers. There are 36 cities owned metro system and 24 of them have formed the metro networks, such as Beijing, Shanghai, Guangzhou, and Shenzhen, etc.

With the opening of new metro lines, great changes have taken place in the network topology, connectivity, and accessibility of the metro system. The volume of passenger flow increased and the passenger flow appeared to exhibit unexpected patterns. However, administrators and researchers lack quantitative understandings of the impact of new metro lines opening on passenger flows especially during peak hours. They are also unfamiliar with the growth patterns of passenger flows in different lines. Characteristics of passenger flow are important for transport operators to provide better service and management, such as timetable improvement, transit planning, and station management. Especially, with the continuous expansion of the metro network, it will be helpful for operation optimization to study the change and growth pattern of passenger flow after the opening of new lines.

In this study, a comprehensive analysis of metro passenger flow and travel time are investigated on the basis of Automatic Fare Collection (AFC) data from Shenzhen, China. The short and long-term impacts of new metro lines on passenger flow are analyzed on the level of the whole metro network by different characteristics of passenger flow and temporal distributions of passenger flow. The outline of this paper is as follows, Section 2 presents a review of previous studies about the implementation of smart card data and the impact of new line opening. Data used in this study are described in Section 3. The short and long-term impact of new metro lines on networks and change patterns of transfer passenger flow investigated in Section 4 and 5. Conclusions are presented in the last section.

2 Literature Review

2.1 Implementation of Smart Card Data

In the past, the methods to collect the data on transit behavior have mainly relied on conventional behavioral surveys such as travel diary surveys[1,2], panel surveys, and travel surveys with tracking devices[3,4]. Nowadays, smart card data provide continuous and long-term travel information such as travel ID, boarding time/location,

[a] Corresponding author: mhzhong@tsinghua.edu.cn

and alighting time/location. Based on that, researchers can analyze travel patterns of the metro system, such as each traveler's trip frequency, travel time and distance, peak hour factor, and so on. Several studies have developed methods that utilized smart card data to improve transit management and planning. Trépanier et al.[5] presented a model to estimate the destination of each individual boarding by using smart card data. Asakura et al.[6] estimated boarding trains of railway passengers by using smart card data and evaluated the effects of changes in train operations. They compared passengers' travel choice behavior before and after the railway company altered the train timetable. Ma et al.[7]developed an efficient data mining method to demonstrate the temporal travel patterns and the pattern regularity for transit riders in Beijing. Gan et al.[8] explored the relationship between daily ridership patterns and characteristics of stations. Moreover, some researchers found that the nature of regularity is considerably more complex, varying within and between individuals on day-to-day basis [9], influenced by socio-economic status and land use activity types[10], trip chaining[11], changes in weather[12], simple ad hoc requirements, such as managing congestion through physical means and the change of topology of metro system.

2.1 Impact of New Metro Line

The opening of the new line will change and influence many aspects from the topology of the network, commuting connectivity of passenger flow to the regional economy and transportation. Some researchers analyzed the impact of the new metro line on the economy. Agostini et al.[13] verify the new metro line has a positive impact on the housing units with closer access to the metro and the price increase depending on the distance from the house to the nearest station. The same result was obtained by Dorantes et al.[14] in Madrid. Mejia-Dorantes et al.[15] uses spatial statistical techniques and real-world data to evaluate the effects of Madrid's metro line 12 expansion on business location patterns, which indicate that the pattern of economic activity location is related to urban accessibility, agglomeration, and economies of scale. Other researchers focused on the impact of the new metro line on transportation. Vuk[16] analyzes the transport impact of the metro based on traffic counts, panel interviews and the forecast model. The result showed that most of the modal shift to the metro derives from bus passengers and 8% to 14% is attributable to car users. Golias[17] evaluates the impacts of the construction of a new metro system as the impacts pertain to traveler behavior and mode choice. He indicates that the increase in travel time and cost for the auto would increase the demand for the metro, but not as much for the bus. The prevalence of big data offers an opportunity to conduct relevant demand analyses. Fu and Gu[18] use the smart card data of the metro system from Nanjing, China to show that the new metro line influences passenger flow, travel time and travel time reliability in the metro network. At the same time, the new line opening has different impacts on different categories of commuters. Wang[19] analyze the mechanism of influence produced by a newly constructed urban rail transit line on urban public traffic network and build a prediction model on passenger flow volume. Liu[20] utilize AFC data to develop proxy variables and propose multivariate regression models to explore station-level ridership growth with network topology change and surrounding land use development.

3 Study Area and Data Collection

Shenzhen is the special economic zone and one of the four first-tier cities in China with a population of about 125 million at the end of 2017. In the past decade, the Shenzhen metro network expanded rapidly from a single line with a length of 14.83 kilometers in 2004 to eight lines with a length of 292.6 kilometers in 2018. The overview of the Shenzhen metro network is shown in Fig 1. Line 7 and Line 9 of the Shenzhen metro system were opened on 28 October 2016 at the same time. After the opening of Line 7 and Line 9, Shenzhen metro is the fifth-largest urban rail transit system in China with 199 stations and 28 transfer stations, and the average daily ridership in 2019 is about 5.85 million.

In order to explore the impact of the new metro lines (i.e., Line 7 and Line 9) on the characteristics of passenger flow and travel time in the whole network, the data used in this study are the record of Automatic Fare Collection System (AFC) data in Shenzhen from September 2016 to October 2018. AFC data contain complete boarding and alighting records (including location and time) of each passenger as shown in Table 1, transfer records of each transfer station and passenger record of each metro section. Data for the two years are utilized because the two years data cover not only the period before and after the opening of Line7 and Line9, but also the development period of passenger flow on the new line.

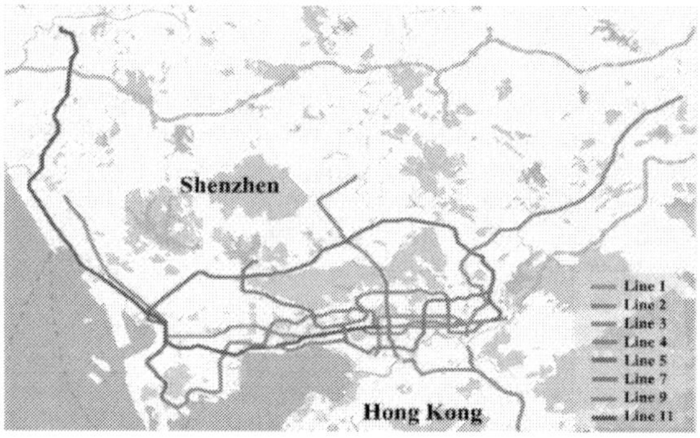

Figure 1. The overview of the Shenzhen metro network in 2019

Table 1. Example of AFC data used in this study

ID	Boarding Time	Boarding Station	Alighting Time	Alighting Station
20000511	2016/11/17 08:26:32	Xiameiling	2016/11/17 08:59:46	Yannan

4 Impact of the New Metro Line on Passenger Flow

In this section, this research compares the boarding/alighting passenger flow, the direction imbalance coefficient, the section imbalance coefficient, transfer coefficient, travel time distribution in the morning/evening peak (the morning peak is at 7 a.m. to 9 a.m. and the evening peak is at 18 p.m. to 20 p.m.) and the average travel time in the time of day one month before and after the opening of new lines to quantitatively analyze the impact of the opening of new lines.

The Direction Imbalance Coefficient (DIC) [21] is the ratio of two times of the maximum section passenger volume on the line l to the sum of the upstream maximum section passenger volume U_{ls}^{max} and the downstream maximum section passenger volume U_{lx}^{max} in the period of t. This value represents the balance of passenger flow in the upstream and downstream direction of a line. The larger the value, the passenger flow more concentrated is in one direction, the more likely it is to have crowed.

$$\varepsilon_l = \frac{2 \times \max\left(U_{ls}^{\max}, U_{lx}^{\max}\right)}{U_{ls}^{\max} + U_{lx}^{\max}} \quad (1)$$

The Section Imbalance Coefficient (SIC) [21] is the ratio of the product of the maximum one-way section passenger volume of the line and the number of line sections to the sum of all section passenger volume in the period of t. The value represents the passenger flow distribution of sections in a line. The large the values, the passenger flow more concentrated is in a section.

$$\delta_l = \frac{U_l^{\max} \times k_l}{\sum_k U_l^k} \quad (2)$$

The Transfer Coefficient (TC) [21] is the ratio of the transfer passenger volume T_l to the sum of the boarding passenger volume E_l, the alighting passenger volume O_l, and the transfer passenger volume T_l. This value represents the proportion of transfer passenger flow in the sum of passenger flow. The large the value, the greater the transport pressure is at the transfer station.

$$\alpha_l = \frac{T_l}{T_l + E_l + O_l} \times 100\% \quad (3)$$

As shown in Fig 2, the distribution of passenger flow with obvious double peaks in a working day and the arrival time of the morning and evening peak have little change. The morning peak is more concentrated than the evening peak passenger flow. The boarding/alighting passenger at the morning peak attracted around 4.2 thousand passengers per 15 minutes more than before, an increase of 6.8% year-on-year. The boarding/alighting passenger at the off-peak period attracted around 3.7 thousand passengers per 15 minutes more than before, an increase of 15.6% year-on-year. The boarding/alighting passenger at the evening peak attracted 8.9 thousand passengers per 15 minutes more than before, an increase of 15.3%.

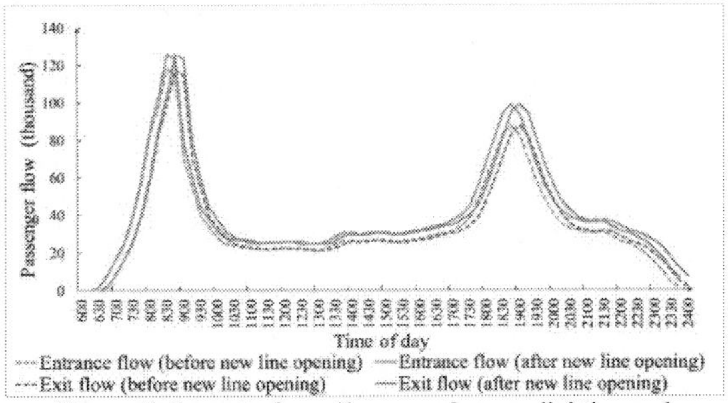

Figure 2. Impact of new lines opening on alighting and boarding passenger flow on weekdays

In Fig 3, the increase of weekend passenger flow distributed in the period after 9 a.m. evenly. Before 9 a.m., the opening of new lines has little impact on passenger flow. However, in the interval from 9 a.m. to 12 a.m., the difference in passenger flow before and after the opening of new lines gradually increased, reaching the maximum at the evening peak. The passenger flow increased by 8.6 thousand passengers per 15 minutes more than before and an increase of 22.9%. In the time of 19 p.m. to 24 p.m., the difference was decreased with the growth of 7.2 thousand passengers per hour more than before.

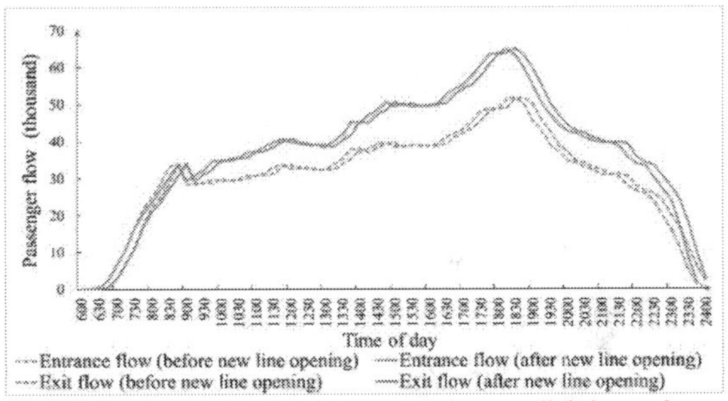

Figure 3. Impact of new lines opening on alighting and boarding passenger flow on weekend

Table 2 shows the changing pattern of the SIC after the opening of new lines. In the upstream direction, the imbalanced coefficient of Line1 and Line4 increases at the morning peak but decreased at the evening peak. That shows that Line1 and Line4 have a strong commuting function and new lines opening intensify the direction imbalance of these two lines. The SIC of Line2 on both directions has been reduced but the max passenger flow has been increased indicating that the opening of new lines can better alleviate the imbalance in Line2. The SIC of Line3, Line5, and Line11 have a large increase in both the upstream and downstream directions and both the morning and evening peak showing that new lines opening increased the transport pressure in those lines. Because those metro lines are mainly responsible for transporting passengers between rural and urban areas, more metro station attracts many travelers for whom it was inconvenient to take the metro in the before.

Since the SIC is a dimensionless number, to analyze the impact of the opening of the new line, the maximum section passenger volume should be compared. For all the metro lines, the growth rate of the morning peak is high with an increased rate of [1.87%, 19.15%]. Therefore, the opening of new lines will not relieve the crowding of the morning peak. At the evening peak, the maximum section passenger volume of Line1, Line2, Line4 has been reduced. At the rest time of the evening peak, the increasing rate is around [0.05%, 12.32%], which is lower than that of the morning peak. Therefore, new lines opening attracted more passengers in the morning peak rather than the evening peak.

Regarding the DIC (see Table 3), no matter in the morning or evening peak, the opening of new lines increased the Average Passenger Volume (APV) of most lines, but it does not affect the maximum passenger flow direction. In the morning peak, for Line1, Line3 and Line4, Line11 the direction imbalance is aggravated. In the morning peak of Line2 and Line5, the imbalance coefficient is declined but the average passenger volume is inclined, indicating that new lines can better improve the utilization rate both upstream and downstream. In the evening peak, as to Line1, Line2, Line3, and Line4, new lines also aggravate the transport pressure of existing lines. However, in the evening peak of Line5, the opening of new lines can better solve the imbalance of direction with the 6.34% increase in passenger volume, which means both the upstream and downstream passenger volumes have grown. So administrators should pay more attention to mass passenger volume incidents. At the same time, new lines can reduce the transport pressure of the Line11 in the evening peak since the number of passenger flow has decreased.

As shown in Fig 4, from the distribution of passenger flow volume and passenger flow rate, it can be seen that the volume of passengers has increased in each time interval after the opening of new lines, especially in the interval of [600s, 2700s]. However, the proportion of passengers in a short time (600s to 1200s) and the proportion of passengers in a long time (greater than 3600s) declined slightly. The proportion of passengers in the interval of [1500s, 3000s] increased. As shown in Fig 5, in the evening peak, the changing pattern is similar to that in the morning peak and the volume of passenger flow also increases in every time period. But the maximum passenger flow rate appears in a shorter time interval. Hence, with the benefit of enhancement of accessibility and connectivity, the opening of new lines reduced the ratio of short-time and long-time passengers.

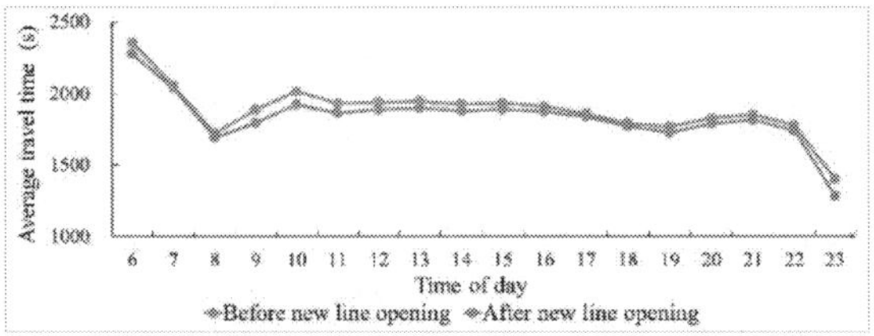

Figure 6. Impact of new lines opening on average travel time in the time of day

As shown in Fig 6, the average travel time of whole networks was shortened by 51 seconds due to the opening of a new line opening, mainly from before and after the morning peak, off-peak period, and after evening peak. However, there is little impact on morning peak and evening peak. This shows that the opening of new lines has not saved travelers' travel time in the morning and evening peak because even if the travel time is shortened, due to the increase in passenger volume, the waiting time and the time of boarding and alighting are increased.

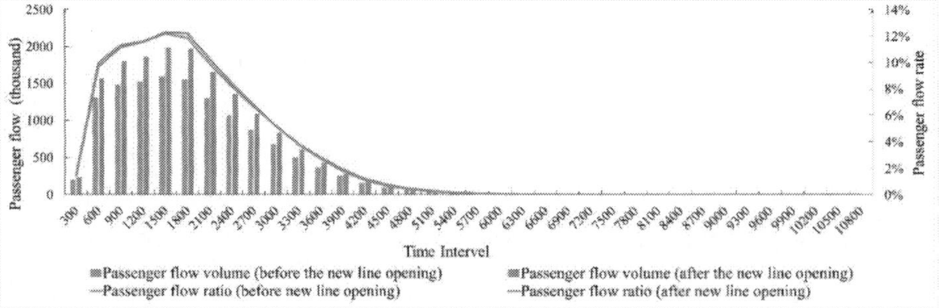

Figure 4. Travel time and travel time ratio distribution in the morning peak

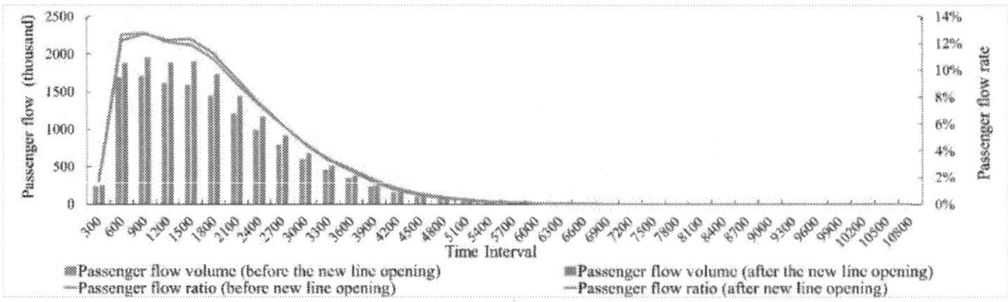

Figure 5. Travel time and travel time ratio distribution in the evening peak

Table 2. The change rate of SIC and max passenger flow of the section after new lines opening

		Upstream Direction				Downstream Direction			
		Morning Peak		Evening Peak		Morning Peak		Evening Peak	
		SIC	Max passenger flow	SIC	Max passenger flow	SIC	Max passenger flow	SIC	Max passenger flow
Line1	Before	1.71	4752	1.95	3495	2.06	3407	1.53	4172
	After	1.86	5665	1.88	3522	2.05	3661	1.61	4695
	Rate	8.39%	18.28%	-3.36%	-0.24%	-0.63%	6.11%	4.63%	11.73%
Line2	Before	1.87	1224	2.22	2123	2.35	2433	1.81	1323
	After	1.86	1466	2.16	2327	2.31	2816	1.65	1332
	Rate	-0.47%	19.15%	-2.48%	8.53%	-1.62%	13.85%	-8.96%	-1.44%
Line3	Before	2.09	4822	1.54	1679	1.53	1236	1.99	4521
	After	2.19	5780	1.54	1716	1.57	1296	2.04	4964
	Rate	4.66%	18.63%	0.05%	1.02%	2.44%	4.44%	2.72%	8.77%
Line4	Before	1.54	5160	1.33	1475	1.39	1232	1.52	4976
	After	1.58	5971	1.26	1399	1.35	1260	1.58	5555
	Rate	2.92%	13.95%	-5.24%	-5.66%	-2.75%	1.87%	3.55%	10.61%
Line5	Before	1.62	2744	1.83	2900	1.92	2921	1.60	2641
	After	1.67	3173	1.90	3283	1.91	3421	1.61	2858
	Rate	2.77%	13.24%	3.53%	12.32%	-0.44%	14.90%	0.61%	6.51%
Line11	Before	1.68	1862	1.59	906	1.66	663	1.63	1943
	After	1.75	2208	1.68	989	1.78	768	1.71	2094
	Rate	4.24%	15.83%	5.63%	8.08%	7.06%	14.20%	4.78%	6.67%

Table 3. The change rate of DIC, APV, and max passenger flow direction of the section after new lines opening

		Morning Peak			Evening Peak		
		Coefficient	APV	Direction	Coefficient	APV	Direction
Line1	Before	1.16	2644	Up	1.09	2671	Down
	After	1.21	2984	Up	1.15	2881	Down
	Rate	4.23%	11.09%		2.59%	7.06%	
Line2	Before	1.30	959	Down	1.21	940	Up
	After	1.30	1149	Down	1.26	1067	Up
	Rate	-0.48%	18.22%		4.08%	11.77%	
Line3	Before	1.54	2257	Up	1.45	2253	Down
	After	1.60	2607	Up	1.48	2417	Down
	Rate	3.36%	14.14%		2.01%	6.34%	
Line4	Before	1.57	3268	Up	1.52	3204	Down
	After	1.62	3695	Up	1.58	3470	Down
	Rate	2.95%	11.62%		4.59%	7.36%	
Line5	Before	1.06	1655	Down	1.07	1645	Up
	After	1.06	1867	Down	1.08	1770	Up
	Rate	-0.62%	10.77%		-1.39%	6.34%	
Line11	Before	1.44	1083	Up	1.35	1404	Down
	After	1.45	1221	Up	1.35	1209	Down
	Rate	0.91%	11.14%		1.73%	-9.23%	

5 Growth Pattern of passenger flow of new lines

The growth period and stabilization period of the new lines are identified by the changing pattern of TC, as shown in Fig 7, 8, 9, 10. Because of the influence of external factors, such as holiday and weather, and the pattern of TC curve, this research performs a logarithmic curve fitting to the changing pattern of TC and obtained the formula and values of R^2 shown in Table 4 and Table 5. The X in the formula represents the number of months after new lines opening. In this study, the change rate of curve slope is less than or equal to -0.0001 taking as the criterion for judging whether it is a stabilization period or not, because under such conditions, the TC is nearly stable. Therefore, the passenger flow stabilization period of different lines can be obtained by calculating the second derivative of the fitting curve.

The calculation result of the growth period is shown in Table 6. The average growth period of Line 7 is generally longer than the average growth period of Line 9. By analyzing the correlation between the length of the growth period, the length of transfer lines and the number of transfer stations in a line, it is found that there is a significant positive correlation between the length of the growth period and the number of transfer stations in the transfer line. For instance, the length of the growth period of Line7 to Line1 is related to the number of transfer stations of Line1. Moreover, there is no significant correlation between the length of the growth period and the length of transfer lines.

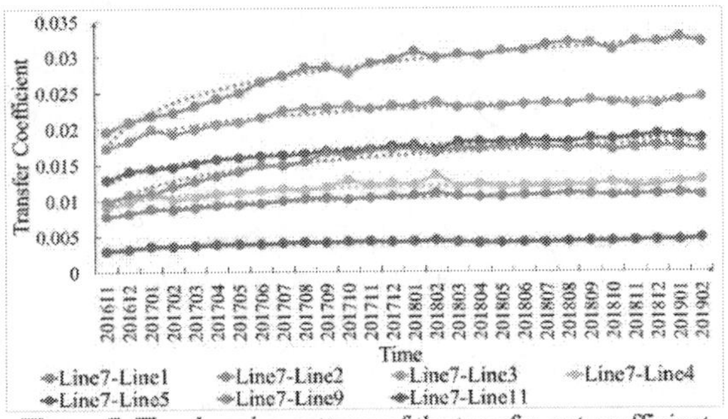

Figure 7. The changing patterns of the transfer-out coefficient on Line 7

https://doi.org/10.1051/matecconf/202030801004

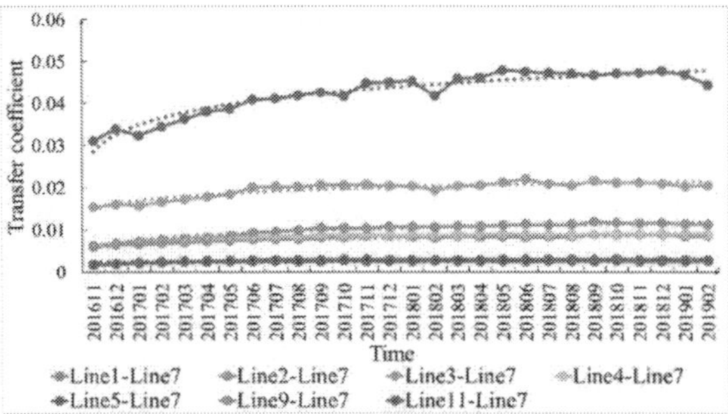

Figure 8. The changing patterns of the transfer-in coefficient on Line 7

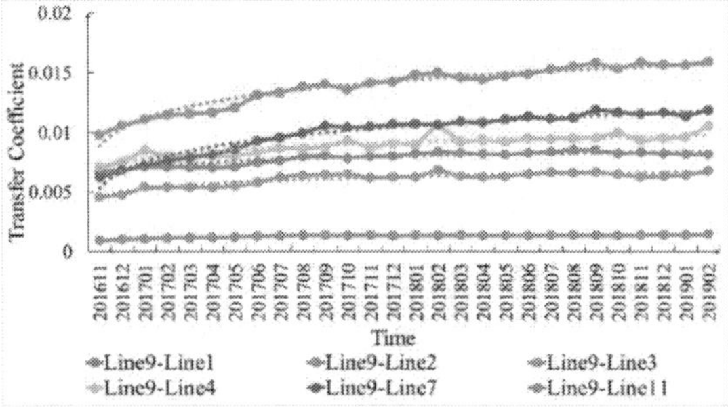

Figure 9. The changing patterns of the transfer-out coefficient on Line 9

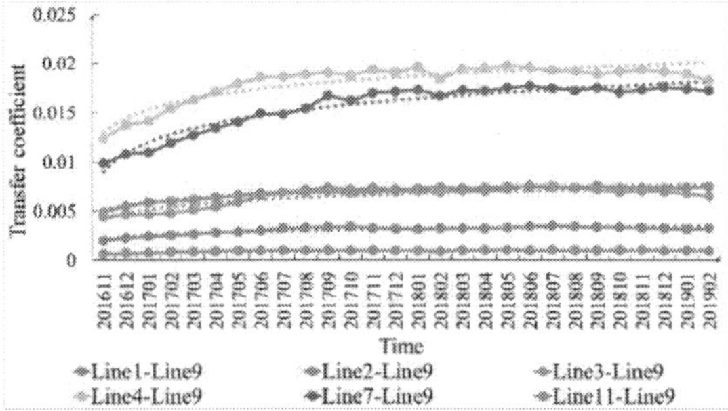

Figure 10. The changing patterns of the transfer-in coefficient on Line 9

As shown in Fig. 11, the passenger flow grows with time in the morning peak, while the total passenger volume in 2018 has a huge increase. With the time pass, the proportion of passenger flow within 600s does not change much. In the interval of [600s, 900s], the proportion of passenger flow slowly increases. In the interval of [900s, 2100s], the proportion of passenger flow grow slowly in 2017 and decline hugely by 12% in 2018, while the proportion of passenger flow in [2100s, 6000s] increase slightly. As shown in Fig. 12, the change of passenger flow and passenger flow rate in the evening peak are similar to that in the morning peak. But in the interval of [900s, 2100s], the rate of passenger flow gradually decreased. The aforementioned change patterns are because, after the opening of the new line, the accessibility and operational efficiency are improved, thus reducing the travel time of passengers and increasing the maximum travel distance.

In the long term shown in Fig. 13, the average travel time of the metro network has increased slightly at the morning peak and the evening peak. The travel time has significantly decreased in the off-peak period by nearly 40 seconds. This phenomenon is because, with the development of the city scale and the opening of the new metro line, passengers are more inclined to travel by metro. In the morning and evening peak, a large volume of passenger flow will cause congestion in the station and increase the travel time. However, in the off-peak period, with the benefit of more route selection and enhancement of accessibility, people can reduce their travel time.

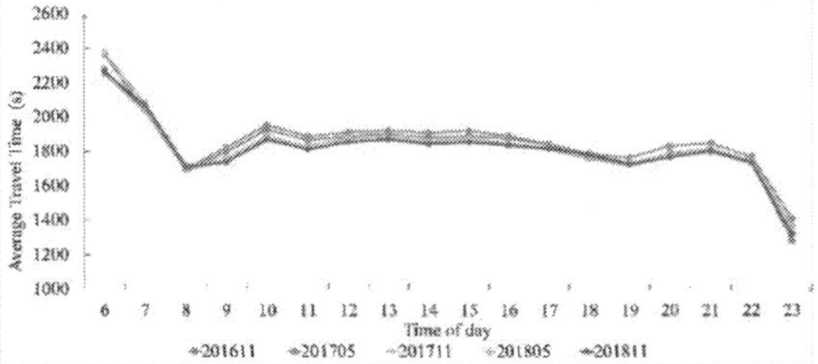

Figure 13. Long term impact of new lines opening on average travel time in the time of day

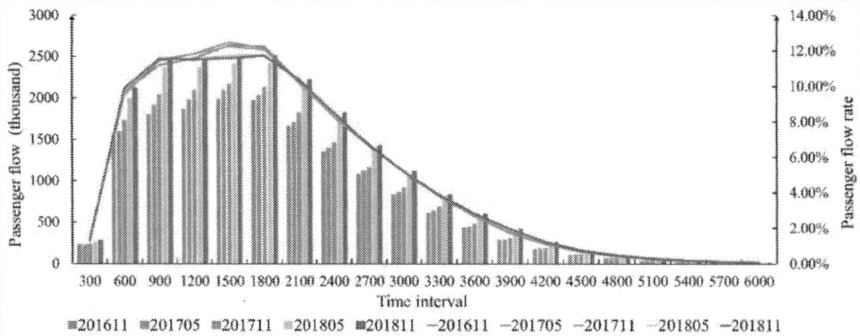

Figure 11. Travel time and travel time ratio distribution in the long term on the morning peak

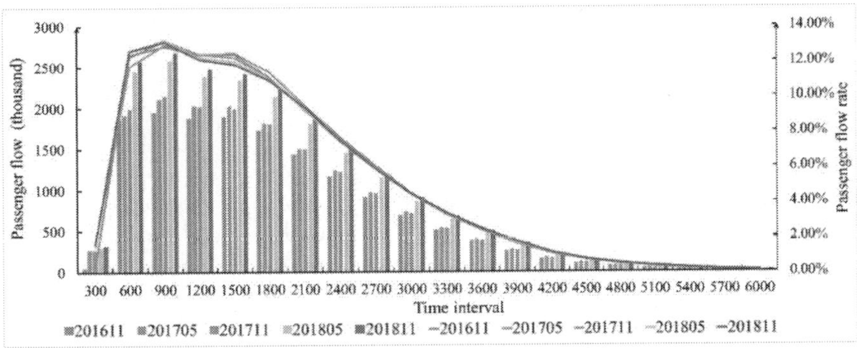

Figure 12. Travel time and travel time ratio distribution in the long term on the evening peak

Table 4. Result of the fitting curve of transfer passenger flow on Line 7

Line	Fitting Curve	R^2	Line	Fitting Curve	R^2
Line7-Line1	0.0011ln(x) + 0.008	0.9427	Line1-Line7	0.0002ln(x) + 0.002	0.6803
Line7-Line2	0.0045ln(x) + 0.017	0.9562	Line2-Line7	0.0009ln(x) + 0.006	0.9301
Line7-Line3	0.0021ln(x) + 0.017	0.9359	Line3-Line7	0.002ln(x) + 0.015	0.8426
Line7-Line4	0.001ln(x) + 0.009	0.764	Line4-Line7	0.0008ln(x) + 0.006	0.9256
Line7-Line5	0.0027ln(x) + 0.009	0.932	Line5-Line7	0.0058ln(x) + 0.029	0.9113
Line7-Line9	0.0019ln(x) + 0.012	0.9635	Line9-Line7	0.0019ln(x) + 0.005	0.9581
Line7-Line11	0.0004ln(x) + 0.003	0.89	Line11-Line7	0.0004ln(x) + 0.002	0.8533

Table 5. Result of the fitting curve of transfer passenger flow on Line 9

Line	Fitting Curve	R^2	Line	Fitting Curve	R^2
Line9-Line1	0.0006ln(x) + 0.006	0.8648	Line1-Line9	0.0004ln(x) + 0.002	0.8437
Line9-Line2	0.0021ln(x) + 0.009	0.9508	Line2-Line9	0.0008ln(x) + 0.005	0.9315
Line9-Line3	0.0006ln(x) + 0.005	0.8497	Line3-Line9	0.001ln(x) + 0.004	0.81
Line9-Line4	0.0009ln(x) + 0.007	0.7701	Line4-Line9	0.0021ln(x) + 0.013	0.8372
Line9-Line7	0.0019ln(x) + 0.005	0.9581	Line7-Line9	0.0027ln(x) + 0.009	0.932
Line9-Line11	0.0001ln(x) + 0.001	0.9233	Line11-Line9	0.0001ln(x) + 0.001	0.7935

Table 6. The growth period of transfer passenger in Line7 and Line9

Line	Growth period (month)	Line	Growth period (month)	Line	Growth period (month)	Line	Growth period (month)
Line7-Line1	3	Line1-Line7	1	Line9-Line1	2	Line1-Line9	2
Line7-Line2	7	Line2-Line7	3	Line9-Line2	5	Line2-Line9	3
Line7-Line3	5	Line3-Line7	5	Line9-Line3	2	Line3-Line9	3
Line7-Line4	3	Line4-Line7	3	Line9-Line4	3	Line4-Line9	5
Line7-Line5	5	Line5-Line7	8	Line9-Line7	4	Line7-Line9	5
Line7-Line9	4	Line9-Line7	4	Line9-Line11	1	Line11-Line9	1
Line7-Line11	2	Line11-Line7	2				

6 Conclusion & Discussion

This study aims to analyze the impact of the new metro line opening on passenger flow from the characteristics of passenger flow and travel time. Using Shenzhen as a case study. The main conclusions are as following:

(1) In the short-term, the opening of new lines increases the boarding/alighting/transfer passenger volume in a short time and the attracting passenger flow in the evening peak is more than that in the morning peak. Moreover, the opening of new lines cannot solve the unbalance and congestion in Shenzhen metro, on the contrary, it increased the transport pressure on lines used for commuting.

(2) In the long term, the opening of new lines changes the ratio of passenger flow. The ratio of short-time and long-time passengers increased because, with the improvement of accessibility and operational efficiency of the metro, the opening of new lines reduce the travel time of passengers and increase the maximum travel distance. . Additionally, the average travel time has increased in the morning and evening peak but there is a significant decreased in the off-peak period.

(3) In the aspect of the growth period of new metro lines, there is a significant positive correlation between the length of the growth period and the number of transfer stations of the transfer line. Meanwhile, there is no significant correlation between the length of the growth period and the length of transfer lines.

Acknowledgments

This paper was supported by the National Key R&D Program of China (2018YFC0809900) and the National Outstanding Youth Science Foundation (51425404).

References

1. Pas, E. I., & Koppelman, F. S. (1987). An examination of the determinants of day-to-day variability in individuals' urban travel behavior. Transportation, **14(1)**, 3-20.
2. Axhausen, K. W., Zimmermann, A., SchöNfelder, S., Guido Rindsfüser, & Haupt, T. (2002). Observing the rhythms of daily life: a six-week travel diary. Transportation, **29(2)**, 95-124.
3. Murakami, E., & Wagner, D. P. A. (1999). Can using global positioning system (GPS) improve trip reporting ?. Transportation Research Part C Emerging Technologies, **7(2-3)**, 149-165.
4. Asakura, Y. , & Hato, E. (2004). Tracking survey for individual travel behavior using mobile communication instruments. Transportation Research Part C Emerging Technologies, **12(3)**, 273-291.
5. Martin Trépanier, Tranchant, N., & Chapleau, R.. (2007). Individual trip destination estimation in a transit smart card automated fare collection system. Intellectual Transportation System. **11(1)**, 1-14
6. Asakura, Y. , Iryo, T., Nakajima, Y. , & Kusakabe, T. (2012). Estimation of behavioral change of railway passengers using smart card data. Public Transport, **4(1)**, 1-16.
7. Xiaolei, M. A., WU, YaoJan, Wang, Y., Feng, C., & Liu, J. (2013). Mining smart card data for transit riders' travel patterns. Transportation Research Part C Emerging Technologies, **36**, 1-12.
8. Gan, Z. , Yang, M., Feng, T., & Timmermans, H. . (2018). Understanding urban mobility patterns from a spatiotemporal perspective: daily ridership profiles of metro stations. the
9. Hanson, S., & Huff, J. (1986). Classification issues in the analysis of complex travel behavior. Transportation, **13(3)**, 271-293.
10. Zhong, C., Batty, M., Manley, E., Wang, J., Wang, Z. , & Chen, F., et al. (2016). Variability in regularity: mining temporal mobility patterns in London, Singapore, and Beijing using smart-card data. PLOS ONE, **11(2)**.
11. Primerano, F., Taylor, M. A. P. , Pitaksringkarn, L., & Tisato, P. (2008). Defining and understanding trip chaining behavior. Transportation, **35(1)**, 55-72.
12. Saneinejad, S., Roorda, M. J., & Kennedy, C. (2012). Modeling the impact of weather conditions on active transportation travel behavior. Transportation Research Part D: Transport and Environment, **17(2)**, 129-137.
13. Agostini, C. A., & Gastón A. Palmucci. (2008). The anticipated capitalization effect of a new metro line on housing prices. Fiscal Studies, 29.
14. Dorantes, L., Paez, A., & Vassallo, J. (2011). Analysis of house prices to assess economic impacts of new public transport infrastructure. Transportation Research Record: Journal of the Transportation Research Board, **2245**, 131-139.
15. Mejia-Dorantes, L., Paez, A., & Vassallo, J. M. . (2012). Transportation infrastructure impacts on firm location: the effect of a new metro line in the suburbs of Madrid. Journal of Transport Geography, **22**, 236-250.
16. Vuk, G.. (2005). Transport impacts of the Copenhagen metro. Journal of Transport Geography, **13(3)**, 223-233.
17. Golias, J. C.. (2002). Analysis of traffic corridor impacts from the introduction of the new Athens Metro system. Journal of Transport Geography, **10(2)**, 0-97.
18. Xiao, F., & Yu, G. (2018). Impact of a new metro line: analysis of metro passenger flow and travel time based on smart card data. Journal of Advanced Transportation, 2018, 1-13.
19. Wang, Z.. (2014). Passenger Flow Prediction Model of the Newly Constructed Urban Rail Transit Line. 2014 International Conference of Logistics Engineering and Management.
20. Liu, S., Yao, E., & Li, B. . (2018). Exploring urban rail transit station-level ridership growth with network expansion. Transportation Research Part D Transport and Environment.**73**, 391-402
21. B. Wang: *Theory and Application for Urban Rail Transit Network Operation* (China Communication Press, China, 2014)

Building and Publishing Fundamental Transit Data for Regional Public Transportation Provided by Municipalities

Chinasa Sueyoshi[1,a], Hideya Takagi[2] , Yoshihiro Yasutake[1] and Kentaro Inenaga[1]

[1] *Faculty of Science and Engineering, Kyushu Sangyo University, 2-3-1 Matuskadai, Higashi-ku, Fukuoka city, Fukuoka, Japan*
[2] *Graduate School of Information Science, Kyushu Sangyo University, 2-3-1 Matuskadai, Higashi-ku, Fukuoka city, Fukuoka, Japan*

Abstract. Public transportation is becoming increasingly important in regions where the residential population is decreasing. In Japan, many regional transportation systems are experiencing financial challenges. It is difficult for fixed-route transportation systems to operate at a surplus when serving only local resident users. We consider one of the problems of these regional public transport systems to be the lack of information on the Internet about regional public transportation routes. For people who are unfamiliar with the region, such as inbound tourists, there is no easy way to conduct a route search. In this paper, we describe our efforts to build a fundamental database of the regional public transportation provided by municipalities and to publish this data for those conducting route searches on the Internet. Specifically, we converted fundamental data regarding regional public transportation to the feed of the General Transit Feed Specification-Japan (GTFS-JP), which is based on the Google GTFS, as formulated by the Ministry of Land in Japan. Then we encouraged municipalities to publish these transit data to enable route searches on Google Maps and other domestic-content providers.

1 Introduction

Domestic regional transportation is positioned as one of the foundations of regional revitalization in Japan. However, contrary to its importance, public transport has not been promoted. According to the Japan Bus Association, the number of passenger buses in Japan has declined, peaking at about 10.1 billion in 1968 and 4.3 billion in 2015 [1]. Although municipalities have made various efforts to address the lack of regional public transportation services that have arisen due to the abolition of private fixed-route transportation, most municipalities manage "community bus" services to improve the welfare of their residents [2]. It's filled the gaps in transport services in less accessible areas, and to promote the outdoor activities of elderly people and the general use of public facilities. Managing regional public transportation efficiently is an ongoing challenge due to the lack of human resources and the current financial constraints on regional public transportation in Japan [3]. Most municipalities have no specialized division of public transportation. As such, it is difficult for fixed-route transportation services to operate at a surplus when serving only local resident users.

One of the problems in regional public transportation is that there is no fundamental transportation data available for making route searches on the Internet. For people who are unfamiliar with a region, such as inbound tourists, the lack of route search capability for regional transportation means that no transportation service is available. Figure 1 shows part of the results of a public-opinion survey on public transport [4] by the Cabinet Office (Government of Japan). The data show that route search services are most often used for public transport checks, route type, high-speed bus, short-to-long-distance railway, etc. For such reports, immediate response by the route-search service regarding regional public transport is desired. We think it is necessary people who do not know the region, it means that transportation does not exist if no regional transportation is found by such a service on the Internet. Solving this problem could revitalize not only regional public transportation services but the region itself.

In this study, we converted fundamental data on regional public transportation to the feed of the (General Transit Feed Specification-Japan (GTFS-JP), which is based on Google's GTFS and was formulated by the Ministry of Land in Japan [5]. The GTFS defines a common format for public transportation schedules and associated geographic information [6]. Figure 2 shows the GTFS/GTFS-JP flow from public facilities to Google and other domestic-content providers. There are municipalities that are managed by paper style such as time schedules so that digitizing the basic data enables easy route search and is expected to be used for the analysis of various operation statuses. This digitalization is a symbol of the promotion in ICT that has tackled regional public transportation in Japan, and it is closely related to the future of transportation such as MaaS

[a] Corresponding author: sueyoshi@is.kyusan-u.ac.jp

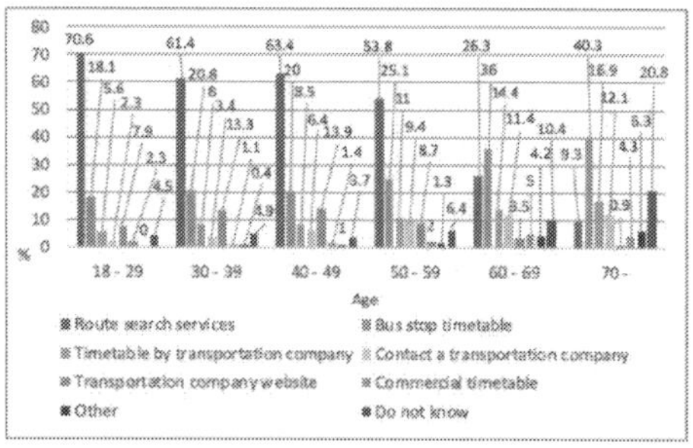

Figure 1. How to check usage method / route in the case of route bus (priority graph created in source data [4])

(Mobility as a Service), which is one of the latest trending important keywords related to transportation.

2 Building and Publishing Fundamental Data

Table 1 shows a list of GTFS configuration files. The difficulty and a large amount of work required to describe these specialized areas of knowledge are obstacles to the creation of a GTFS, which requires a range of ICT support addresses.

Table 1. GTFS configuration files.

File name	Definition (Description)
agency.txt	One or more transportation agencies that provide data
stops.txt	Boarding point
routes.txt	Transportation route
trips.txt	Itinerary for each route
stop_times.txt	Arrival and departure times at each stop for each itinerary
fare_attributes.txt	Route fee information
shapes.txt	Rules for drawing a route line

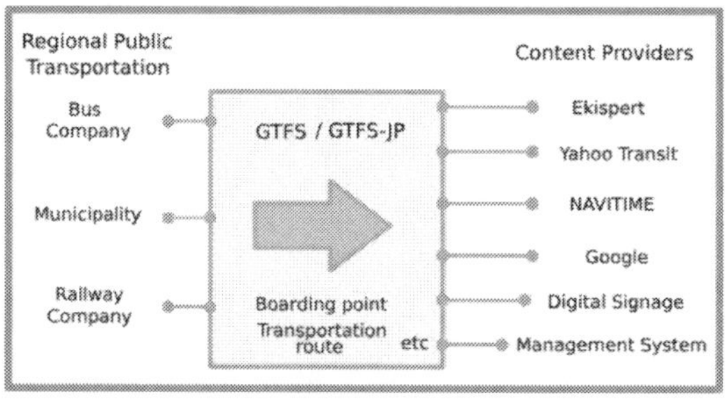

Figure 2. GTFS/GTFS-JP flow from public facilities to Google and other domestic-content providers

The Ministry of Land, Infrastructure, Transport and Tourism, the Kyushu District Transport Bureau Traffic Planning Division, and the Fukuoka Prefecture Transportation Division, Information Policy Division have promoted the building and publishing of GTFS-JP feeds containing transit-site data on the Internet and the providing these feeds to Google and domestic-content providers, such as Ekispert [7], Yahoo Transit [8] and NAVITIME [9]. One example of the use of GTFS for domestic regional public transportation is OpenTrans.it [10] built by Masaki Ito and others at the University of Tokyo for public transportation data feed in Shimada City and Yaizu City, Shizuoka Prefecture. We think that information regional public transportation to content providers will be more promoted than ever if it has been building about on fundamental GTFS/GTFS-JP as shown in Figure 2. As a result, regional public transportation and content providers can be easy to share data for route searches, so that it contributes to improving convenience for regional public transportation users.

Our laboratory team developed a practical solution for regional public transportation services provided by the service management support systems of municipalities in

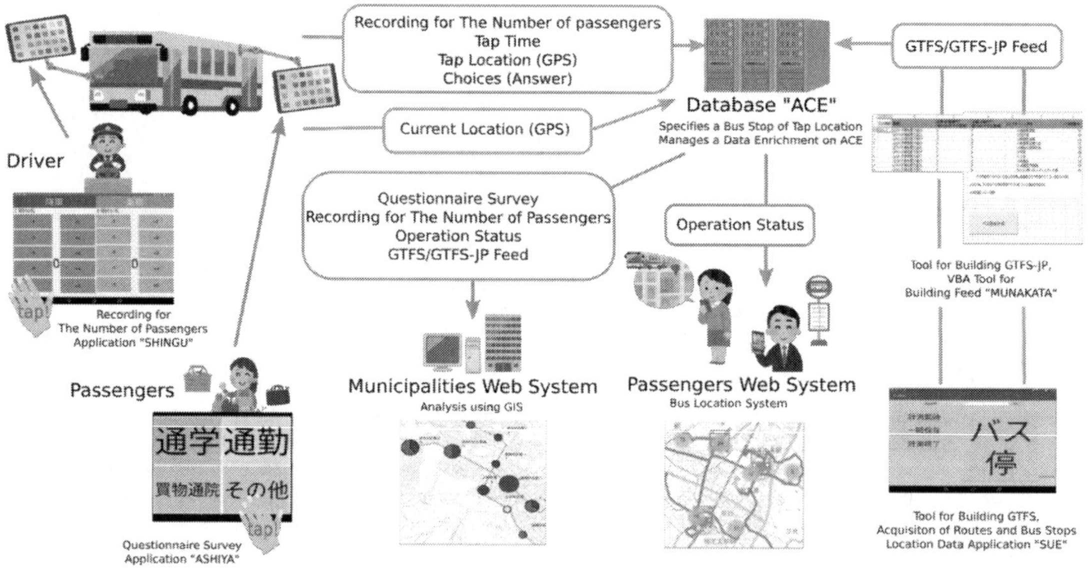

Figure 3. Support system for public transportation management using the applications SHINGU and ASHIYA

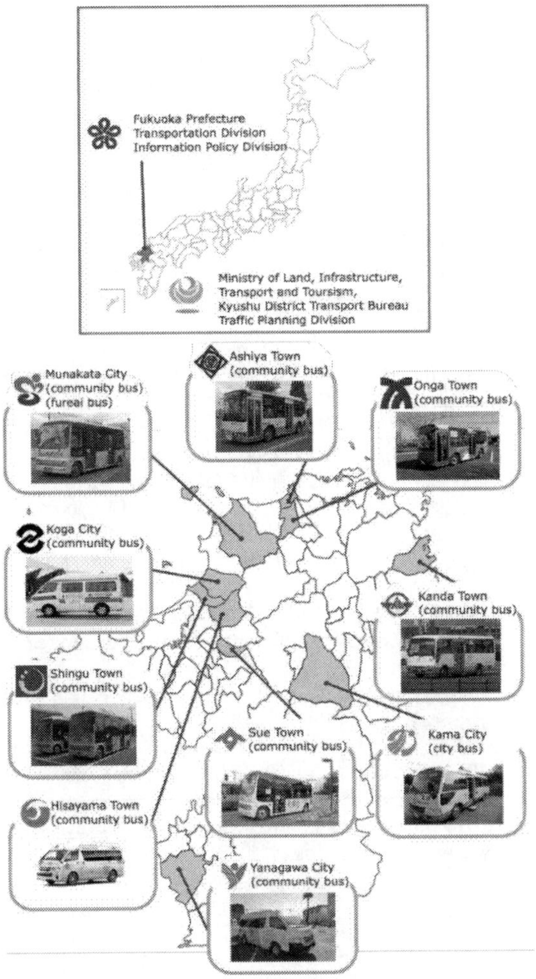

Figure 4. The fundamental data of regional public transportation to the feed of GTFS-JP in cooperation with community bus and local government organ.

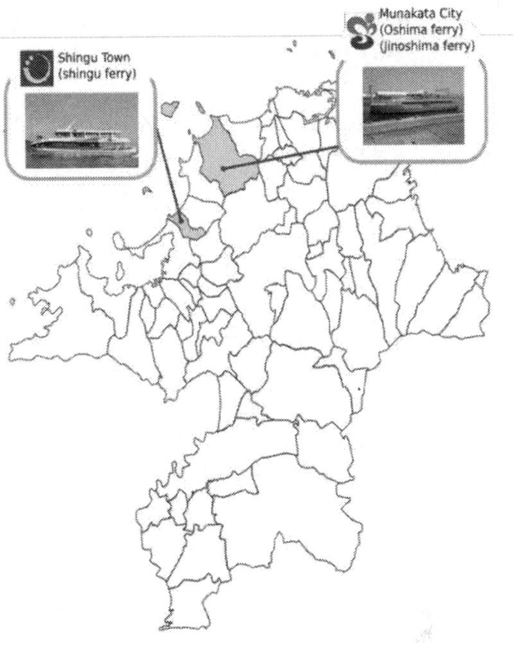

Figure 5. The fundamental data of regional public transportation to the feed of GTFS-JP in cooperation with the ferry.

Japan [11,12]. Figure 3 shows a schematic outline of the system. First, all data is managed by ACE that has GTFS, GTFS-JP, the number of bus passengers and analysis results. We assign two tablets to each bus, one of which uses the application SHINGU to record the number of bus passengers and the second provides a questionnaire survey for passengers using the application ASHIYA. Using SHINGU, the bus driver records the passengers who get-on/get-off the community bus by simply tapping. Passengers use ASHIYA by tapping to answer the questionnaire on each community bus. The obtained data includes the number of passengers, the tap time, the tap location by GPS, the answers provided on the questionnaire by passengers and the current GPS location of the bus. These data are sent to the ACE database. We also developed a tool for building the GTFS, the application SUE used to acquire routes and bus-stop location data, and a VBA tool for building the GTFS-JP and feeding the data to the city of Munakata. The data generated by these tools are stored in the ACE database, which maintains raw tap data and performs data enrichment. In this study, we analyzed the GIS data and some data from the questionnaire surveys, including the number of passengers, the operational states, and the GTFS/GTFS-JP feed. The results enable us to visualize a bus stop with many passengers. Moreover, the bus location system uses the operation status available in ACE. This bus location system is a convenient tool for bus passengers because it contains information such as how many buses have passed through a bus stop.

Tablets equipped with the Android application SHINGU are used by tablets with mobile data communication in our system to record the number of passengers and collect passenger data as a concrete means of utilizing ICT to conduct a long-term and continuous survey of transportation service usage. Previous studies have used SHINGU ver.2 in Android applications and SHINGU mini on smartphones. In this study, we integrated these two systems. SHINGU ver.3 is configured in four modes, which provide transitions for inputting the number of get-off/get-on passengers, transitions for displaying the setup and backup databases, deleting the backup database, and synchronous communication. SHINGU mini is configured in five modes, i.e., inputting the number of get-off passengers or get-on passengers, making transitions on inputting the number of get-off or get-on passengers, making transitions in the display of the setup and backup databases and deleting backup databases. The SHINGU mini separates the input of the number of get-off /get-on passengers.

3 Cooperation with community bus and local government organ

As shown in Figures 4 and 5, we built a fundamental database of regional public transportation for the GTFS-JP feeds in cooperation with multiple municipalities. Figure 4 shows practical examples of bus services and Figure 5 shows those for ferry services. First, we built the GTFS feed for the bus service in Shingu Town in 2016 [13] and that in Munakata City in 2017. Since 2017, we have built GTFS-JP feeds for the bus services in 10 municipalities as well as for the ferry services in the city of Munakata and the town of Shingu. Building and publishing the GTFS-JP feed for the bus and ferry services in Shingu has increased bus usage by about 60.0% by people from outside the area who were unfamiliar with the region, such as inbound tourists [14]. The town of Shingu has a tourist destination, Ainoshima Island, i.e., "cat island."

For further study, it will be necessary to cooperate with local companies to promote regional development and with the Kyushu District Transport Bureau and Fukuoka Prefecture to support the building and publishing of fundamental transit data. Furthermore, to promote future transit development and verify current services, it is necessary for the companies providing transportation services to have access to the data.

4 Conclusion

The conversion of data by local governments, as the main operators of local public transport services, into the required data format for Internet route searches is not practical either in terms of technology or cost. Currently, this service is not supported. Efficient regional public transportation management in Japan is challenging due to the constraints on human resources and finances. In this paper, we describe our efforts to build and publish GTFS-JP feeds for multiple municipalities based on fundamental data to make public transportation services available in route search results by Google and other domestic-content providers.

5 Acknowledgment

The authors would like to thank Shingu Town Kasuya County, Munakata City, Sue Town Kasuya Country, Ashiya Town Onga Country, Koga City, Onga Town Onga Country, Kama City, Kanda Town Miyako Country, Ogori City, Yanagawa City, Fukuoka Prefecture. And the Advanced ICT Human Resource Education University Consortium on the theme of solving regional issues, Fukuoka Prefecture Planning and Regional Promotion Department Transportation Policy Division, Information Policy Division and the Transportation Planning Division, Kyushu Transportation Bureau, Ministry of Land, Infrastructure, Transport and Tourism for their opportunities to us collaborative research and study. This work was supported by JSPS KAKENHI Grant Number JP17K00472.

6 References

1. Nihon Bus Association, "Overview of Japan bus business and Nihon Bus Association 2017," Sep. 2019.[Online].Available:http://www.bus.or.jp/about/pdf/h29_nba_brochure.pdf.
2. Ministry of Land, Infrastructure, Transport and Tourism:Concept and type of community bus. Accesses:Sep.2019.[Online].Available:http://www.mlit.go.jp/singikai/koutusin/rikujou/jidosha/bus/01/images/06.pdf.
3. Ministry of Land, Infrastructure, Transport and Tourism:Current status of regional public transportation in Japan. Accesses : Sep. 2019.[Online].Available:http://wwwtb.mlit.go.jp/kinki/content/000010186.pdf.
4. Cabinet Office, "Public opinion survey on public transportation,"Sep.2019.[Online].Available:https://survey.gov-online.g.,o.jp/h28/h28-kotsu/2-2.html.
5. Ministry of Land, Infrastructure, Transport and Tourism : GTFS-JP. Accesses : Sep. 2019.[Online]. Available:http://www.mlit.go.jp/common/001283244.pdf.
6. Google, GTFS Static Overview. Accesses : Sep. 2019.[Online].Available:https://developers.google.com/transit/gtfs/.
7. Ekispert, Ekispert for WEB. Accesses : Sep. 2019. [Online]. Available : https://roote.ekispert.net/ja/.
8. Yahoo Japan, Yahoo Transit. Accesses : Sep. 2019.[Online]. Available : https://transit.yahoo.co.jp/.
9. NAVITIME JAPAN, NAVITIME. Accesses : Sep. 2019.[Online].Available:https://www.navitime.co.jp/.
10. M. Ito and K. Sezaki, "Current status and prospects of public transport open data in Japan," The 55th Infrastructure Planning and Management Research Presentation and Lectures, 2017, 04-01, Committee of Infrastructure Planning and Management.
11. H. Takagi, C. Sueyoshi and K. Inenaga," Developing and Introducing Vehicle Location System for Regional Public Transportation Service Provided by Municipalities Using Fundamental Transit Data," 14th Japanese Conference On Mobility Management, Kanazawa, Ishikawa, Japan, July 2019.
12. K. Inenaga, H. Takagi and C. Sueyoshi," Practical Use of Management Support System and Supporting to Build Fundamental Transit Data for Regional Public Transportation Service Provided by Municipalities in the Kyushu Region" 14th Japanese Conference On Mobility Management, Kanazawa, Ishikawa, Japan, July 2019.
13. K. Inenaga, J. Nishiyama and K. Harada, "The Development of Tools to Provide Fundamental Data Set of Regional Public Transportation," The 15th ITS Symposium 2017 Proceedings, 2017.
14. K. Inenaga, "A Study on the Effectiveness of Introduction to Open-Data for Regional Public Transportation," Proceedings of the 59th National Convention of JAMS, Japan Association for Management Systems, pp.38--41, 2017.

A High-speed Train Timetable Rescheduling Approach in case of Disturbances

Xinyu Gao[1,a]

[1]*School of Electronic and Information Engineering, Beijing Jiaotong University, China*

Abstract. This paper from a macroscopic viewpoint develops a train timetable rescheduling approach on a single high-speed railway line under disturbances, i.e. inevitable train delays in the duration of the train operation. A mixed-integer linear programming model is formulated to minimize the arrival delay and the departure delay altogether. The commercial optimization software CPLEX is adopted in an effort to seek the optimal solution in an acceptably short time required in the real-time rescheduling process. The proposed approach is further tested on a real-world case study and the numerical results show that compared with the results obtain by the traditional genetic algorithm, using CPLEX to solve the model can yield better solutions and consume the desired computation time, thereby demonstrating its effectiveness and efficiency.

1 Introduction

High-speed railway has developed greatly in recent decades. In China, the large-scale network of railway has been put into use, which poses enormous challenges concerning the safe and efficient operation of high-speed trains when unexpected disturbances occur. In such disrupted cases, the operation of high-speed trains deviates from the planned train diagram, causing train delays which further leads to passenger dissatisfaction as well as economic losses, all of which therefore attach paramount importance to timely timetable rescheduling approaches.The high-speed railway planning process is shown in Figure 1, where the train real time rescheduling, especially the train timetable rescheduling in the operational level has been paid enormous attention by researchers at home and abroad.

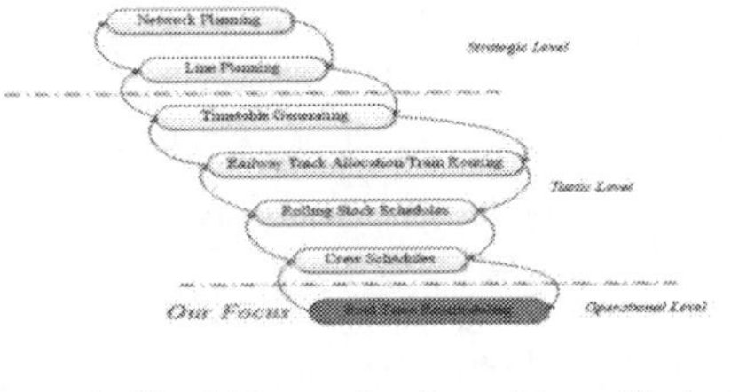

Figure 1. The high-speed railway hierarchical planning process

[a] Corresponding author: 1039789301@qq.com

In the recent study, Higgins et al [1] developed a model for the online train timetable rescheduling and introduced a branch and bound algorithm to minimize the total delay of all trains as well as the train operating cost, which acts as a decision-making tool for train dispatchers to reschedule trains in real time to an optimal outcome. Corman et al [2] proposed an optimization framework by using alternative graphs to model train operation at the microscopic level that considers the role of the block sections and signals. In an attempt to collectively minimize the energy consumption and passenger waiting time, Meng et al [3] formulated and solved a robust single-track train dispatching problem under stochastic segment running times and capacity loss duration. Dong et al [4] integrated artificial systems, computational experiments, and parallel execution, also known as ACP approach, to model complex urban rail transport systems when confronted with emergencies. D'Ariano et al [5] proposed a branch and bound algorithm for solving the rescheduling problem in real time. D'Ariano et al [6] further implemented the alternative graph model in the real-time train rescheduling to support dispatchers to resolve conflicts. Espinosa-Aranda et al [7] introduced a binary integer linear programming to minimize total passenger delays and developed a heuristic method to solve the model. Yang et al [8] developed a novel fuzzy programming model for rescheduling trains based on the fuzzy variable representation of incident duration.

Cacchiani et al [9] presented an overview of the train rescheduling models and algorithms for the real-time adjustment under the disturbances and disruptions. Binder et al [10] integrated three objectives to generate a disposition timetable in the event of disruptions: the passenger satisfaction, the operational costs and the deviation from the original timetable.

In many countries, such as in China, the high-speed train adjustment is mainly dependent upon human operations. In practice, when disturbances occur, dispatchers are expected to act based on some existing contingency plans or adjust the train timetable manually. However, dispatchers tend to find it difficult to opt for the best contingency plan, and adjusting the timetable manually is time-consuming. What is more, when manually adjusting the timetable, the results are highly dependent upon the skills and experience of the dispatcher, which cannot guarantee the optimal quality of the results. Therefore, in this paper, a mix-integer linear programming mathematical optimization model is constructed and the CPLEX solver is used, which can serve as a tool to help dispatchers with the decision-making process.

2 Model formulation

In this section, the detailed mathematical formulation of the rescheduling optimization model is presented in case of disturbances. In the proposed model, the timetable is considered at a macroscopic level, which means that it is represented by the arrival and departure times of each train at each station and detailed information about signaling systems is neglected.

No-early constraints: These following constraints ensure that rescheduled departure and arrival times of all trains cannot be earlier than their scheduled counterparts since it would cause trouble for the travelling of passengers who book the train tickets in advance. Therefore the constraints are presented as:

$$t_{is}^{d} - o_{is}^{d} \geq 0 \;\; \forall i \in I, \forall s \in \{1,2,\ldots,|S|-1\} \tag{1}$$

$$t_{is}^{a} - o_{is}^{a} \geq 0 \;\; \forall i \in I, \forall s \in \{2,3,\ldots,|S|\} \tag{2}$$

Where o_{is}^{d}, o_{is}^{a} are scheduled departure times and arrival times of train i in the intermediate station s, and t_{is}^{d}, t_{is}^{a} are their rescheduled counterparts. I is the set of trains where i, j are indexes of trains, and S is the set of stations where s, h are indexes of stations. $|I|, |S|$ are total numbers of trains and stations respectively.

Train delay constraints: For trains that run properly as scheduled, the arrival and departure delays are formed respectively as follows:

$$d_{is}^{d} \geq t_{is}^{d} - o_{is}^{d} \;\forall i \in I, \forall s \in \{1,2,\ldots,|S|-1\} \tag{3}$$

$$d_{is}^{a} \geq t_{is}^{a} - o_{is}^{a} \;\forall i \in I, \forall s \in \{2,3,\ldots,|S|\} \tag{4}$$

where d_{is}^{d}, d_{is}^{a} suggest the departure delay and the arrival delay of train i at station s. It should be noted that trains departing at the initial station should not be delayed and that trains arriving at the final station are without delays as they are no longer used for transport.

Train running constraints: The time that trains run between stations can be calculated as below, and it is worth noting that train cannot surpass its maximum speed.

$$t_{i,s-1}^{d} \leq t_{is}^{d} \;\forall i \in I, \forall s \in \{1,2,\ldots,|S|-1\} \tag{5}$$

$$t_{i,s-1}^{a} \leq t_{is}^{a} \,\forall i \in I, \forall s \in \{2,3,\ldots|S|\} \tag{6}$$

$$t_{r} \geq t_{i,s+1}^{a} - t_{is}^{d} \,\forall i \in I, \forall s \in \{1,2,\ldots|S|-1\} \tag{7}$$

$$t_{r} \geq l_{s}/V \;\; \forall i \in I, \forall s \in \{1,2,\ldots|S|-1\} \tag{8}$$

where t_r represents the running time between intermediate stations, and l_s indicates the distance between stations.

Dwelling time constraints:If a train is pre-scheduled to stop at a specific station, some time should be given for this train to carry out necessary operations, for example loading and unloading passengers as well as changing crew members, etc. It is obvious that if the arrival and departure times of each train i at each station s, the actual dwelling time p_{is} for train i at station s can be calculated by the difference of the rescheduled arrival and departure times, which is $p_{is} = t_{is}^{d} - t_{is}^{a}$. Therefore, the dwelling time should be no less than a pre-determined minimum time t_{dwe}. The dwelling time constraint at intermediate stations are here as follows:

$$t_{is}^{d} - t_{is}^{a} \geq x_{is} \cdot t_{dwe} \,\forall i \in I, \forall s \in \{1,2,\ldots|S|-1\} \tag{9}$$

Headway constraints: In order to guarantee the safe operations on the railway line, headway constraints need to be formulated to keep a safe distance between the adjacent trains, including the arrival and the departure headway at stations. Specifically, if a train arrives at/departs from a station, a time period should be left to the preparation for the next train's arrival/departure. In other words, this time period is the time distance of two trains for their arrival or departure operations. A binary variable y_{ijs} is here introduced, suggesting if $y_{ijs} = 1$ it means train i departs before train j at station s, and $y_{ijs} = 0$ means the other way round. This binary variable ensures that the overtaking operation can be achieved within stations if necessary. The following constraints therefore indicate that the departure and arrival headway at stations should be assured between adjacent trains that are scheduled to stop. All can be described as follows:

$$t_{js}^{d} - t_{is}^{d} \geq t_{dep} + M \cdot (y_{ijs} - 1) \;\; \forall i,j \in I, \;\forall s \in \{1,2,..|S|-1\} \tag{10}$$

$$t_{js}^{a} - t_{is}^{a} \geq t_{arr} + M \cdot (y_{ijs} - 1) \;\; \forall i,j \in I, \;\forall s \in \{2,3,..|S|\} \tag{11}$$

$$t_{is}^{d} \leq t_{js}^{d} + M \cdot (1 - y_{ijs}) \;\; \forall i,j \in I, \;\forall s \in \{1,2,..|S|-1\} \tag{12}$$

$$t_{is}^{d} \geq t_{js}^{d} - M \cdot y_{ijs} \;\; \forall i,j \in I, \;\forall s \in \{1,2,..|S|-1\} \tag{13}$$

$$y_{ijs} + y_{jis} = 1 \;\; \forall i,j \in I, \; i \neq j, \;\forall s \in \{1,2,..|S|-1\} \tag{14}$$

In the meantime, the departure and arrival headway, which is the headway between the departure time of the former train and the arrival time of the latter train, should also be satisfied as follows:

$$o_{is}^{d} - o_{is}^{a} \leq t_{is}^{d} - t_{is}^{a} \;\; \forall i \in I, \;\forall s \in S \tag{15}$$

Capacity constraints: To ensure that the rescheduled timetable is feasible in real cases, station capacities represented by tracks should better be taken into consideration in the duration of the rescheduling process. It is common knowledge that trains arriving in a station to stop or passing the station should be assigned to a station track. A station track can be occupied by only one train at one time. Note that like in the most literature it is assumed that the origin and destination stations where the train services start and end have the enough capacity to store the starting and ending trains. The capacity constraints are therefore constructed, and it is worth mentioning that the first two constraints represent that if these two constraints are true, they together mean that a train is at the station stopping or passing, which requires an empty track.

$$\left(t-t_{is}^{a}\right)\leq j_{ist}^{a}\cdot M \quad \forall t\in TT,\forall i\in I,\forall s\in S \tag{16}$$

$$\left(t_{is}^{d}-t\right)\leq j_{ist}^{d}\cdot M \quad \forall t\in TT,\forall i\in I,\forall s\in S \tag{17}$$

$$j_{ist}^{p}\geq\left(j_{ist}^{a}+j_{ist}^{d}-2\right)\cdot M+1 \quad \forall t\in TT,\forall i\in I,\forall s\in S \tag{18}$$

$$\sum j_{ist}^{p}\leq Cap_{s} \quad \forall t\in TT,\forall i\in I,\forall s\in S \tag{19}$$

where the binary variables $j_{ist}^{d}, j_{ist}^{a}, j_{ist}^{p}$ are constructed variables for the formulation of the capacity constraints, and Cap_s represents the actual number of available tracks. TT indicates the whole time horizon of the rescheduling process.

Objective function: The objective function is constructed as:

$$\min Z=\sum_{i=1}^{|I|}\sum_{s=1}^{|S|}\left(\alpha_1\cdot d_{is}^{d}+\alpha_2\cdot d_{is}^{a}\right) \tag{20}$$

where in the function the weighted-sum optimization method is used, in which α_1, α_2 are weight coefficients, to aim at minimizing the departure delay and the arrival delays of all trains, so as to restore the normal railway operation of trains as soon as possible and diminish the dissatisfaction of passengers in cases of disturbances.

3 Case study

In order to test and verify the effectiveness and efficiency of the proposed approach, in this section, a real-world case study is carried out based upon the data of Shanghai Hongqiao station to Nanjing South station in the upstream direction. The simulation is coded in Matlab 2016a and performed on an Intel(R) Core(TM) i7-8550U CPU @ 1.80 GHz 2.00 GHz running with a 64-bit windows operating system and 8G RAM. The commercial optimization software CPLEX 12.6 is adopted to solve the proposed the mix-integer linear programming model.

High-speed railway lines generally have double tracks with one track for each direction, and the upstream direction is considered on the assumption that sidings for upstream trains can only be used by upstream trains and vice versa. The number of tracks are set in accordance with half of the reality because only the upstream direction is used. As regards parameters, t_{dwe} is set as 2 min, t_{dep}, t_{arr} are 3 min, t_q, t_t are 0.5 min, V is 300 km/h, M is 1000000, l_s is set according to the data of reality, and α_1, α_2 are set as 5 in the following case study which can be altered to study the influence of the coefficients.

As is illustrated in Figure 2, the real-world planned timetable---from Shanghai Hongqiao station to Nanjing South station within the time horizon from 6:00 a.m. to 11:00 a.m. covering 8 stations and 10 high-speed trains---acts as the benchmark for the rescheduling stage with blue lines.

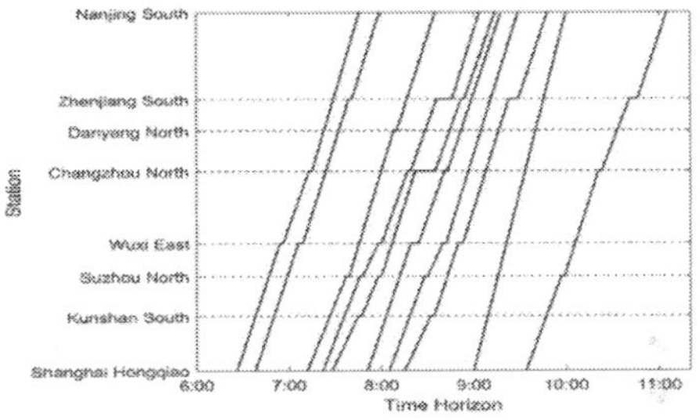

Figure 2. Planned high-speed train diagram.

Two typical disturbance scenario are considered: the fourth train departs from the fourth station with 8 min delays, and the fourth train arrives at the fourth station with 8 min delays.

As for the first departure delay scenario, Figure 3 and Figure 4 show the results obtained by the traditional standard genetic algorithm and CPLEX respectively. It can be seen that both can provide non-conflicting feasible solutions with red lines. However, the total delay calculated by genetic algorithm is 104.75 min, and by CPLEX is 36.2 min, showing that CPLEX can find better solutions by reducing delays by approximately 65%. At the same time, the Gap of CPLEX is 0, meaning that the solution by CPLEX solver is the optimal one. The computation time is 41.6778s and 98.3569s respectively, which can also suggest that using CPLEX solver can get the solutions faster so as to fulfil the real-time performance. The delay reduced and the time shortened therefore demonstrate the effectiveness and efficiency of the proposed approach.

With regard to the second arrival delay scenario, Figure 5 and Figure 6 show the results acquired by the traditional standard genetic algorithm and CPLEX respectively. The total delays calculated by genetic algorithm is 134.4420 min, while by CPLEX is 70.9014 min with the Gap of CPLEX also 0, which indicates that CPLEX can achieve better results by reducing train delays by about 47%. The computation time is 24.3677s and 95.3468s respectively, showing that CPLEX can better achieve the real-time performance as well. All therefore demonstrate the effectiveness and efficiency of the proposed approach.

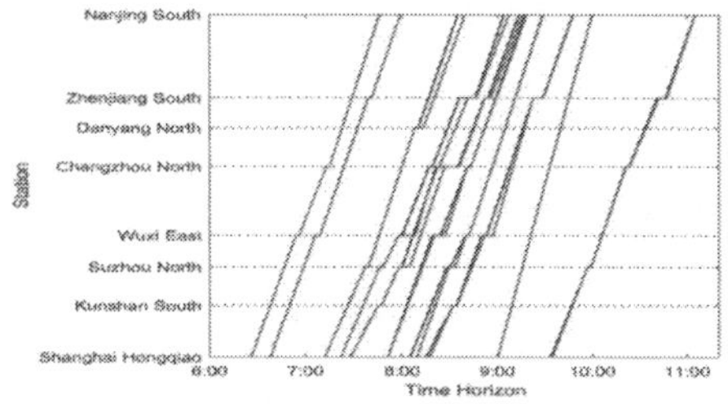

Figure 3. Rescheduled train diagram obtained by the genetic algorithm for departure delay

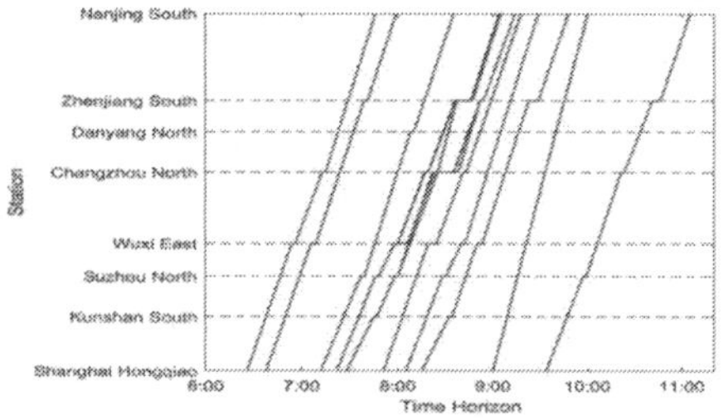

Figure 4. Rescheduled train diagram obtained by CPLEX for departure delay

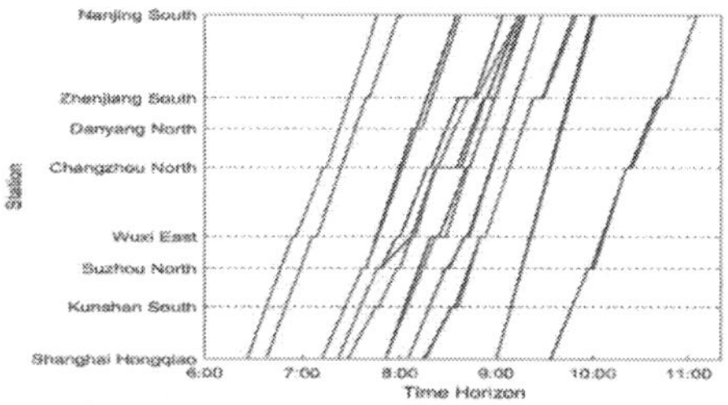

Figure 5. Rescheduled train diagram obtained by the genetic algorithm for arrival delay

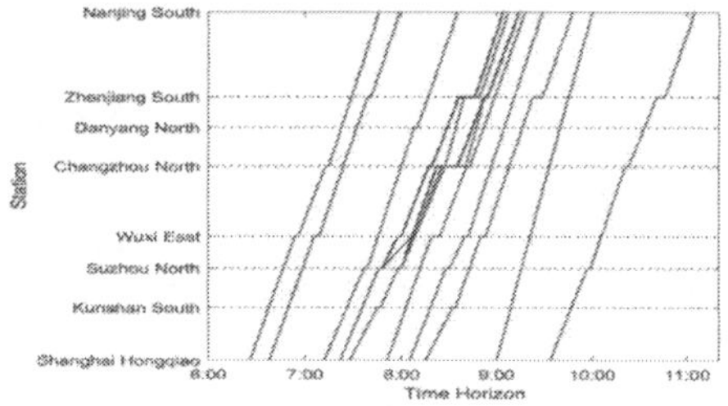

Figure 6. Rescheduled train diagram obtained by CPLEX for arrival delay

4 Conclusion

In this paper, the issue of the high-speed train timetable rescheduling process under disturbances was explored. Specifically, to minimize the departure and arrival delay of all trains, a mixed-integer linear programming model was formulated. The commercial CPLEX solver was used to solve the proposed model. A case study was presented to verify the effectiveness and efficiency of the proposed approach. The simulation results showed that compared with the results attained by the genetic algorithm, CPLEX solver can obtain better solutions in terms of train delays and the computation time. For future research, high-speed train rescheduling models which integrate train timetable, rolling stock and crew staff should be studied, and novel intelligent algorithms should be explored in order to better the rescheduling process.

References

1. A. Higgins, E. Kozan, L. Ferreira, Transportation Research Part B: Methodological, **30**,147, (1996)
2. F. Corman, A. DAriano, I. A. Hansen, and D. Pacciarelli, Journal of Rail Transport Planning & Management, **1**, 14, (2011)
3. L. Meng and X. Zhou, Transportation Research Part B: Methodological, **45**, 1080, (2011)
4. H. Dong, B. Ning, Y. Chen, X. Sun, D. Wen, Y. Hu, and R. Ouyang, IEEE Transactions on Intelligent Transportation Systems, **14**, 627, (2013)
5. A. Dariano, D. Pacciarelli, and M. Pranzo, European Journal of Operational Research, **183**, 643, (2007)
6. A. D' Ariano, F. Corman, D. Pacciarelli, and M. Pranzo, Transportation science, **42**, 405 - 419, (2008)
7. J. L. Espinosa-Aranda and R. Garcia-Ródenas, Journal of Rail Transport Planning & Management, **3**, 1, (2013)
8. L. Yang, X. Zhou, and Z. Gao, Omega, **48**, 75, (2014)
9. V. Cacchiani, D. Huisman, M. Kidd, L. Kroon, P. Toth, L. Veelenturf, and J. Wagenaar, Transportation Research Part B: Methodological, **63**, 15, (2014)
10. S. Binder, Y. Maknoon, and M. Bierlaire, Transportation Research Part C: E-merging Technologies, **78**, 78, (2017)

Research on the relationship between the rail corrugation and the vibration response of the wheel-rail system

Xinyu Jia [1,a], Guoqing Li [2], Xiubo Liu[2] and Peng Chen[1]

[1] *China Academy of Railway Sciences, Beijing, China*
[2] *Infrastructure Inspection Research Institute, China Academy of Railway Sciences Corporation Limited, Beijing, China*

Abstract. Rail corrugation is a very common phenomenon in high-speed railways. Rail corrugation can cause increased vibration of trains and tracks, and may even affect the safety of train operation. Therefore, this paper will analyze the vibration of the wheel-rail system caused by the rail corrugation. The actual transient rolling contact model is established by using explicit and implicit finite element methods. Through the calculation of the wheel-rail contact force and the wheel-rail mode, the vibration relationship between the rail corrugation and the wheel-rail system is obtained. It can provide some references for further analysis of the cause of rail corrugation.

1 Introduction

Rail corrugation is widely found in the railway sector. Whether it is high-speed railway, urban rail transit or heavy haul railway, there is the phenomenon of rail corrugation. Rail corrugation is a kind of short-wave periodic irregularity, which mainly occurs on the top surface of the rail. As shown in Fig. 1, it is rail corrugation that appears on a high-speed railway. Once rail corrugation occurs, the vibration of the vehicle and the track will be intensified, which will make the passenger feel uncomfortable and will also worsen the wheel and the rail, and even affect the safety of driving[1, 2]. Therefore, it is of great significance to study the intrinsic characteristics of corrugation and analyze the vibration law of wheel-rail system caused by corrugation, which will reduce the influence of corrugation and promote the healthy development of high-speed railway.

Fig. 1 High-speed railway rail grinding

Grassie[3] summarized the characteristics, causes and treatment methods of rail corrugation, and classified the rail corrugation in different states and discussed their characteristics separately. Knothe[4] proposed to have a deeper understanding of the formation mechanism of rail corrugation, and must have a deeper understanding of the transient wheel-rail rolling contact.

In the previous research on corrugation, it is often impossible to analyze the real wheel-rail contact model, which is a corresponding simplification of the wheel-rail system. For example, the contact between the wheel and the track is simulated using a Hertz spring, the track is simplified to a beam, and the wheel is simplified to a mass point[5, 6]. These methods can provide a good reference for us to understand the corrugation, but many problems cannot be solved. The finite element method can simulate the relationship between the wheel and rail more realistically. Zhao[7, 8] and so on have verified the finite element simulation of wheel-rail contact.

In this paper, an explicit and implicit combination method is used to simulate three-dimensional transient rolling contact, analyze the vibration response of the wheel rolling over the rail corrugation section. By analyzing the relationship between wheel-rail vibration response and corrugation, it provides a certain reference for understanding the cause of rail corrugation.

2 Finite element model

In this paper, the three-dimensional transient rolling contact finite element model is established by ANSYS/LS-DYNA software. The model can analyze the relationship between high-speed train and ballastless

[a] Corresponding author: jiaxinyu_cars@163.com

track. The model consists of spring mass, primary suspension, wheel, rails, fastenings, slab and mortar layers, as shown in Fig. 2. The wheel profile is selected as the LM_A type, the profile of the rail top is CN60, and the rail inclination is 1:40. In China's high-speed railway, the minimum curve radius of the line is 7000m, the traverse of the vehicle can be controlled within a very small range, and the length of the model analyzed in this paper is also short, therefore, regardless of the lateral displacement of the wheel, only the forward movement along the straight line is analyzed. In order to reduce the number of elements and improve the efficiency of calculation, this paper only uses half of the wheels and tracks for analysis.

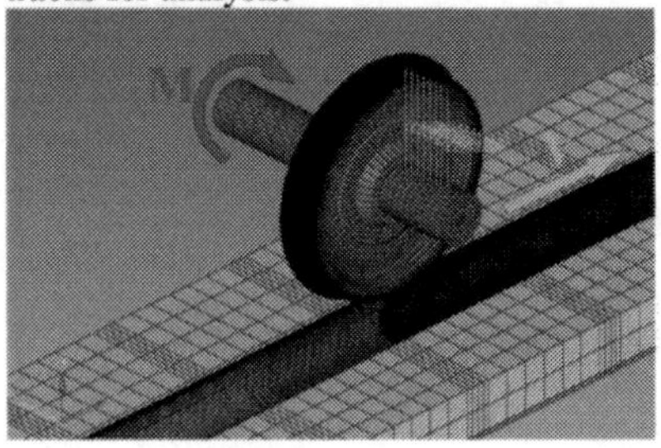

(a)

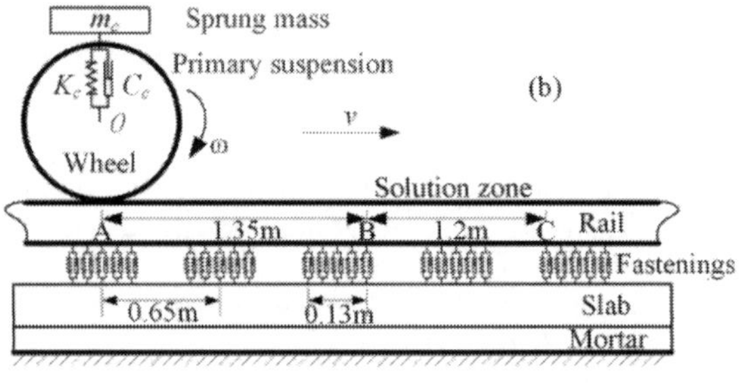

(b)

(a) Finite element mesh;(b) Dimension diagram
Fig. 2 Three-dimensional transient rolling contact finite element model

Wheel-rail vibration caused by rail corrugation grinding is a high-frequency vibration. The train above the primary suspension is less affected by high frequency vibration. Therefore the train above the primary suspension is simplified to a mass, and the value of the mass is determined by the axle weight of the train. The primary suspension consists of 8 parallel spring-damped structures for connecting the mass and the wheel axle. The contact between the wheel and the track uses a face-to-face contact criterion based on a penalty function. The friction between the contact faces is Coulomb's law with a coefficient of friction of 0.5.

From point A to point B is a dynamic relaxation zone with a distance of 1.35m to ensure that the wheel can roll smoothly. From point B to point C is a calculation zone, and a corrugation of 1.2 m in length is applied to the top surface of rail, in this section. The total length of the track is 13.5m, including 20 sleepers. In the dynamic relaxation zone and the calculation zone, the division of the rail elements is relatively fine, and the division away from the wheel rolling is relatively sparse. At the starting point of the wheel, point A, the minimum element size of the rail is 1 mm; The minimum element size in the dynamic relaxation zone is 2.5 mm; The minimum element size in the calculation zone is 1 mm;

The fasteners were simulated with 5 x 5 spring damping structures (5 in the lateral direction and 5 in the longitudinal direction of the track). The arrangement of the springs is evenly distributed according to the size of the rail pad, and each rail pad has a size of 0.13 m. The distance between the two sleepers is 0.65m. Both the slab and the mortar layer are simulated by elastic solid elements, and the material parameters are determined according to the measured values. Each slab is 6.5m long and the distance between the slabs is 50mm, which is filled with concrete.

The x-axis of the model is longitudinal, that is, the direction in which the wheel rolls; The y-axis of the model is horizontal and the y-axis is forward pointing to the inside of the track; The z-axis of the model is vertical. The boundary conditions of the model are as follows: The mass point is coupled to the axle to ensure that the mass point has only a vertical movement relative to the wheel axle. A lateral symmetry constraint is applied to the end of the wheel axle. A symmetrical constraint is applied to both ends of the rail. The fasteners, slabs and mortar layers are restrained in the horizontal and vertical directions to ensure that they can only move in the vertical direction. The bottom of the mortar layer is fixed.

The parameters used in the model are shown in Table 1[9].

Table 1 Values of parameters

Parameter		Value
Sprung mass, m_c(kg)		8000
Stiffness of primary suspension, K_c(kN/m)		880
Damping of primary suspension, C_c (kN·s/m)		4
Stiffness of fastenings, K_f (MN/m)		22
Damping of fastenings, C_f(kN·s/m)		200
Material of wheel & rail	Young's modulus, E (GPa)	205.9
	Poisson's ratio, υ	0.3
	Density, ρ (kg/m^3)	7790
Material of slab	Young's modulus, E (GPa)	34.5
	Poisson's ratio, υ	0.25
	Density, ρ (kg/m^3)	2400
Material filled in slab gaps	Young's modulus, E (GPa)	29.5
	Poisson's ratio, υ	0.25
	Density, ρ (kg/m^3)	2400
Mortar material	Young's modulus, E (GPa)	8
	Poisson's ratio, υ	0.2
	Density, ρ (kg/m^3)	1600

The calculation process of the model can be divided into three steps, (1) stress initialization, (2) dynamic relaxation, (3) computational solution. In the first step, the implicit solution method is used to calculate the deformation displacement of the wheel-rail system under the action of gravity when the wheel is placed on the track. In the second step, the displacement field obtained in the first step is applied to the corresponding node as the initial deformation. At the same time, the translational speed and the rotational speed are applied to the wheel at t=0 as an initial condition. The transient rolling contact and the dynamic behavior of the wheel-rail system is calculated by the explicit solver solution method. From point A to point B in Fig. 2(b) is the dynamic relaxation zone. The reason for setting this region is that the calculation methods of implicit solution and explicit solution are different. The conversion of an implicit solution to an explicit solution introduces numerical instability that can result in unreal high-frequency vibrations in the wheel-rail system. When the wheel rolls over the dynamic relaxation zone, the wheel and rail will reach an approximately steady-state rolling, and the unreal high-frequency vibration will also disappear under the action of damping, thus ensuring that the obtained result is effective and reliable when entering the calculation solution zone. The third step is to calculate the various response results when the wheel rolls over the rail corrugation.

Among them, the wheel-rail contact force, the acceleration of the wheel and the track are common indicators. The schematic diagram of the calculation process is shown in Fig.3.

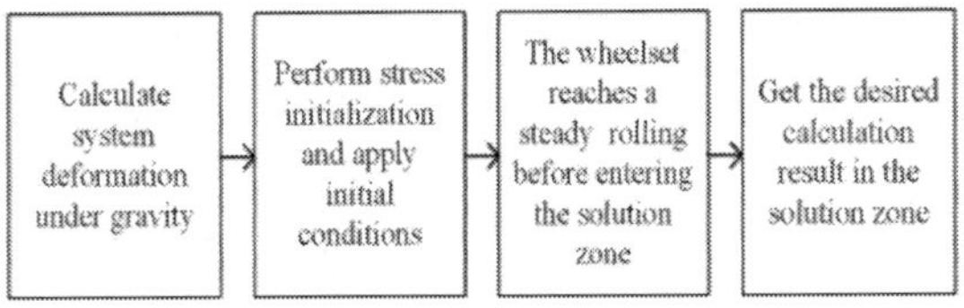

Fig.3 A schematic diagram of the calculation process.

Under the action of gravity, the stress cloud diagram of the rail when the wheel pair stays on the rail is shown in Fig. 4. There is an elliptical contact spot at the rail top position of the rail, and the maximum stress value in the contact patch is 1.325×10^9 Pa.

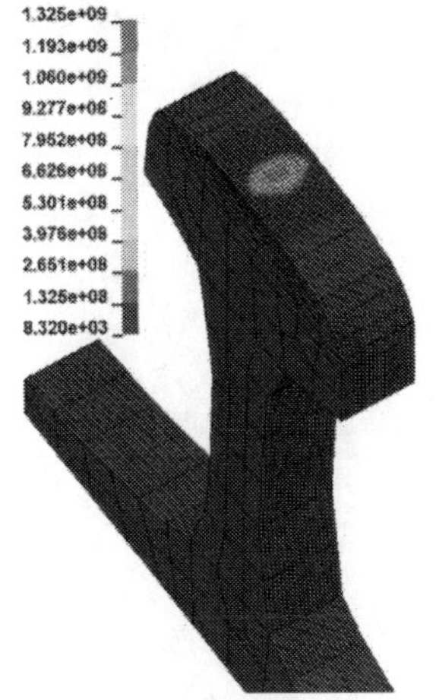

Fig. 4 The stress cloud diagram of rail

Subsequently, the wheel-rail contact force when the wheel rolled over the smooth rail was analyzed, as shown in Fig. 5. The curve in the figure shows the change of the vertical wheel-rail contact force with time, and the points A, B and C are the points in Fig. 2. It can be seen from the figure that there is a high-frequency vibration of the wheel-rail contact force from the point A to the point B, , which is due to the inevitable vibration of the wheelset from static to dynamic. As time increases, the amplitude of the vibration of the wheelset is gradually reduced by the damping. When point B is reached, the wheelset is basically in a steady rolling state. From point B to point C, the wheel-rail force is basically stable and coincides with the axle weight, which can be used for further analysis and calculation.

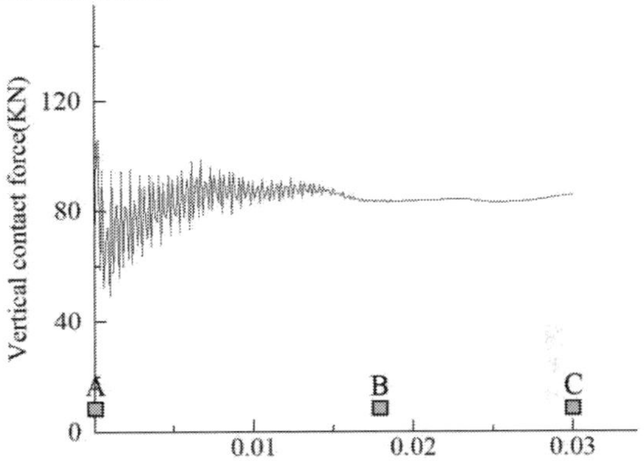

Fig. 5 Vertical wheel-rail contact force

3 Vibration response caused by rail corrugation

First, select a high-speed line containing the rail corrugation, measure the rail corrugation, and obtain the data such as the wavelength and depth, as shown in Fig. 6, the length of the rail corrugation is 1 m.

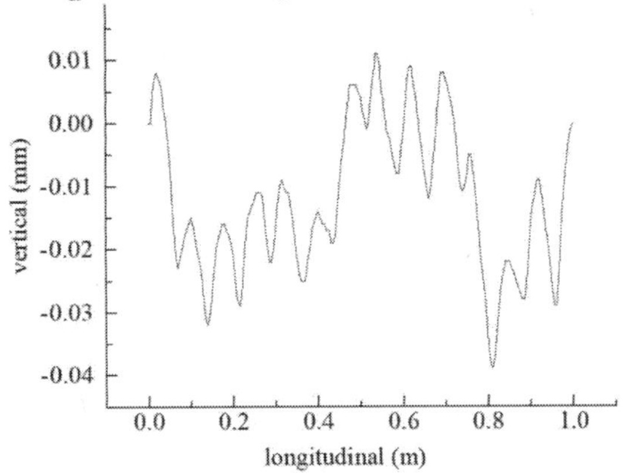

Fig. 6 corrugation

The wave curve is Fourier transformed to obtain a spectrogram, and the frequency characteristics of the curve are analysed, as shown in Fig. 7.

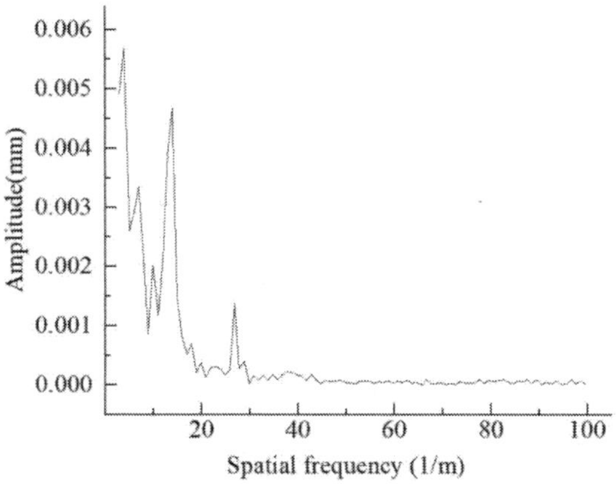

Fig. 7 Spatial frequency of the rail corrugation

The curve in Fig. 6 is applied to the top surface of the rail in the solution zone to obtain the wheel-rail contact force, as shown in Fig. 8. It can be seen from the figure that the wheelset enters the rail corrugation section at 0.017 s and exits the rail corrugation section at 0.03 s.

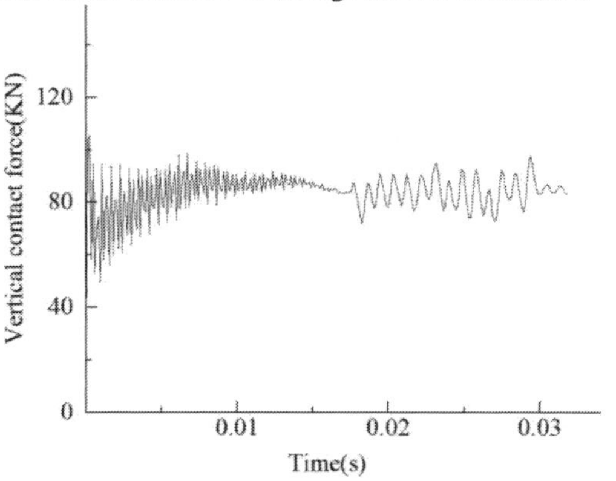

Fig. 8 Wheel-rail contact force

Fourier changes were made to the wheel-rail contact force of the rail corrugation section to analyze the frequency characteristics, as shown in Fig. 9. It can be seen from the figure that the frequency is mainly concentrated in 0~1220HZ and near 2220HZ, and the maximum amplitude occurs at the frequency of 1155HZ.

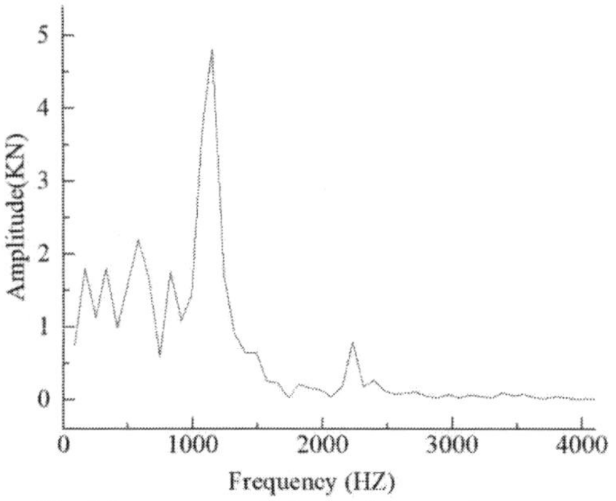

Fig. 9 Spectrogram of wheel-rail contact force.

In order to obtain the frequency characteristic relationship between the wheel-rail force and the rail corrugation, the two frequencies are put together for comparison, and the spatial frequency of the rail corrugation is multiplied by the speed of the wheelset to obtain the excitation frequency to the wheel, as shown in Fig. 10.

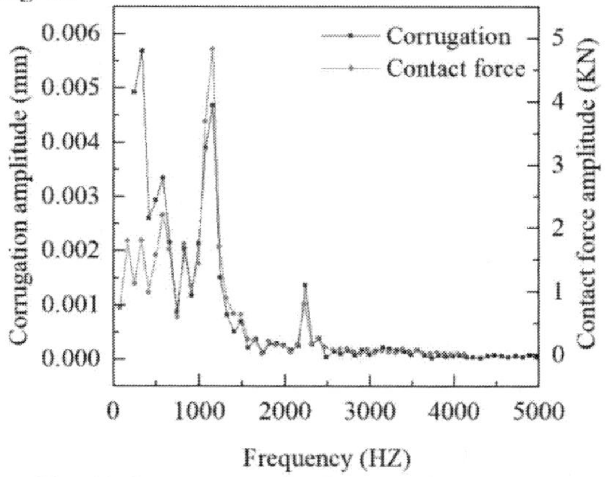

Fig. 10 Spectrogram of rail corrugation and wheel-rail contact forces.

Modal analysis is performed on the wheel and the track respectively. The track contains rails, fasteners, slabs and mortar layer, and the bottom of the mortar layer is fixed. In order to save space, only the first few modal values of the wheel and the track, and the modal values in the frequency range with larger amplitude in Fig. 10 are listed. The modes of the wheel and track are shown in Table 2.

Table 2 Vibration mode of wheel and track

Wheelset		Rail	
Mode number	Frequency	Mode number	Frequency
1	0	1	51.26
3	0	2	65.18
5	0	3	82.07
6	0	4	101.77
7	300	20	285.92
8	300	21	304.86
9	448	22	335.86
10	448	23	340.71
11	473.51	33	574.69
12	731.57	34	595.29
13	731.57	35	611.83
14	1114.1	49	1041.2
15	1114.1	50	1098.6
16	1146.2	51	1157.8
17	1439.5	52	1207.7

It can be seen from the Table 2 that in the vicinity of the frequency at which the amplitude of the wave is largest (333HZ), only the 22nd mode of the track

(335.86HZ) is relatively close, and the wheel has no close value. At a frequency of 581 Hz, the amplitude of the rail corrugation has reached a peak, but in this frequency range, only the 33rd mode of the track (574.69HZ) is relatively close, and the wheel has no close frequency. Near the frequency (1155HZ) where the wheel-rail contact force amplitude is the largest, the 16th mode of the wheel (1146.2HZ) and the 51st mode of the track (1157.8HZ) are close to this value. It can be concluded that the vibration response of the wheel-rail system is determined by the external excitation and the mode of the wheel-rail system itself. When the excitation frequency of the rail corrugation is close to the natural frequency of the wheel-rail system, the frequency amplitude of the wheel-rail contact force is maximized.

4 Conclusions

In order to analyze the influence of rail corrugation on the vibration of the wheel-rail system in high-speed railway, this paper adopts the explicit and implicit finite element method to establish the transient rolling contact model of the wheel rolling over the rail corrugation section, and calculate the wheel-rail contact force and vibration mode, and reached the following conclusions:

(1) The frequency of the measured wave grinding is in good agreement with the calculated frequency of the wheel-rail force. By analyzing the frequency of the wheel-rail contact force, the main frequency range of the rail corrugation can be obtained, which provides a certain reference for the detection of the rail corrugation.

(2) The vibration response of the wheel-rail system is determined by the excitation frequency of the rail corrugation and the natural frequency of the wheel-rail system itself. When the excitation frequency of the rail corrugation is close to the natural frequency of the wheel-rail system, the frequency amplitude of the wheel-rail contact force is maximized.

References

1. S. Li, Z. Li, A. Núñez and R. Dollevoet, Applied Sciences, **7** (2017)
2. L. Zhou and Z.-y. Shen, Journal of Zhejiang University SCIENCE A, **14** (2013)
3. S. Grassie and J. Kalousek, Proceedings of the Institution of Mechanical Engineers, Part F: Journal of Rail and Rapid Transit, **207** (1993)
4. K. Knothe and A. Groß-Thebing, Vehicle system dynamics, **46** (2008)
5. J. C. Nielsen, Wear, **265** (2008)
6. W. Zhai, H. Xia, C. Cai, M. Gao, X. Li, X. Guo, N. Zhang and K. Wang, International Journal of Rail Transportation, **1** (2013)
7. K. D. Vo, H. T. Zhu, A. K. Tieu and P. B. Kosasih, Wear, **322-323** (2015)
8. X. Zhao, J. Yang, B. An, C. Liu, Y. Cao, Z. Wen and X. Jin, Proceedings of the Institution of Mechanical Engineers, Part F: Journal of Rail and Rapid Transit, **232** (2016)
9. X. Zhao, Z.-f. Wen, H.-y. Wang, X.-s. Jin and M.-h. Zhu, Journal of Zhejiang University SCIENCE A, **15** (2014)

Khartoum sky trains network: a study on implementing caterpillar trains concept in Khartoum

Mohammed A. Mohammed[1], Omer Mohammed[1], Mohamed Ibrahim[1], and Huda G. Osman[1,a]

1 University of Khartoum, Civil Engineering Department, P.O. box 321. Khartoum, Sudan

Abstract. The experience of using public transportation in Khartoum is hectic due to over (1.5) Million seats in daily deficit. This research explores the daunting issues of public transport in developing countries like Sudan, taking Khartoum as a case study, and studies the possible alternatives to find an innovative solution that counters them. This paper reviewed the concept presented by MIT scholars in 2017 which was an innovative idea of elevated mini-trains that are intended to run over city streets called (Caterpillar Trains). The paper investigated the use of the C-Trains when integrated with LRT to form the Khartoum Sky Trains Network (STN). It was found that the STN is an optimum model for developing countries, as they were found able to meet the demands of public transport, in both short term and long term. The expected various environmental, social and economic impacts of such large infrastructure project were also studied. Finally, the research suggests that the Build-Operate-Transfer (BOT) contracting module can be utilized to overcome the construction costs, the concession period of which was found through a FORTRAN95 program.

1 Introduction

Public transportation in Khartoum has been a hot topic for quite some time now, which started after South Sudan gained its independence in 2011, causing a major economic disaster for the North [1].

This instantly affected the oil prices and caused currency devaluation, which in turn caused a spike in spare parts prices and availability, leading to severe scarcity in the only available mode of transport; buses [2]. The problem was briefly relieved by buses dubbed by the public "Al-waly Buses" that had generous seating capacity.

However, that didn't last long. The expensive spare parts due to sanctions, taxes, and customs caused Al-waly Buses' owners, who had then become private sector drivers after the privatization of the firm, to ditch their buses; causing a severe scarcity in the number of operational buses. With the buses going scarce, the people, government and country as a whole are now suffering a number of consequences that would not be present in an otherwise stable sector. Nearly millions of dollars of productivity hours are being wasted every day, jobs are harder to access; businesses are negatively affected, and social injustice is on the rise [2].

This research aims to suggest a system of transportation that is modern and comfortable for citizens in both the short and the long term. It studies transportation and economic feasibilities as well as the social impacts of creating a transit network that covers all areas in the state, using two types of trains; main trains to serve the more crowded lines, and the cheaper, much more flexible (C-Trains) to approximately all the roads in the city with relatively lighter traffic.

2 Literature review

In its Master Plan for Transport and Mobility [3], the ministry of transport and infrastructure in Khartoum states that the deficit in daily seating is over 1.5 million seats at the time of making the plan. As part of an ambitious attempt, the master plan presented several options that were considered viable for filling the deficit gap. They include Underground metro, local tram, regional train, and Bus Rapid Transit (BRT).

2.1 Caterpillar Trains (C-Trains)

Ashwani Kumar Upadhyaya and his colleague, Emil Jacob, both Ph.D. researchers who have interest in railway engineering, recently won an award at an MIT-held competition for their innovative idea of mini elevated C-Trains (Caterpillar trains), which are a network of lightweight elevated train cabins that will cover the whole city. [4]

3 Methodology

[a] Corresponding author: hosman@mail.bradley.edu

In this section, we will discuss how the design procedure was carried out, as well as the cost-benefit analysis and the assessment of the environmental, economic, and social impacts.

3.1 Design and appraisal procedure

This assessment was made according to the transport appraisal process which we adopted from the Transport Analysis Guidance (TAG), provided by the UK's Department for Transport, issued in January 2014 [5]. The guidance provides a very detailed, three stages procedure to make sure every proposed project is worth the investment. The first step of this procedure is called the option development. It is concerned with preliminary studies that aim to demonstrate that a specific project is feasible. The second and third stages are conducted after the initial approval of the project and are meant to provide specific, detailed information and estimations for construction and maintenance purposes [5].

The appraisal process has several steps of "elements" that flow logically to test the feasibility of any project. These processes were tailored to meet the Sudanese conditions while keeping in mind that the overall objectives are met. These processes, or elements, are:

3.1.1 "Understanding the current situation"

By means of studying past literature and careful observation, we developed an understating of the current transport sector. This included studying legislative policies, levels of service, and demand patterns.

3.1.2 "Understanding the Future Situation"

This step mainly focused on studying population growth based on the UN's population forecast and, consequently, the changes in the demand, LOS, and the transportation systems themselves. This allowed us to anticipate a range of possible scenarios that helped us greatly in assessing possible options.

3.1.3 "Establishing the Need for Intervention"

Drawing on the aforementioned information, we then made a sound body of arguments to show that the current and future situations indeed need an intervention. The legislative deficiencies at the ministry of transport in Khartoum state, and its interlined authority with the local state, the federal government, and traffic police were some of the main arguments that led to this result.

3.1.4 "Identifying Objectives"

Reflecting on the previously discussed issues, we arrived at a preliminary set of targeted objectives that we wish to accomplish: These objectives were identified based primarily on the six features of the modern public transport, as well as an examination of failed past projects and inspection of the criteria needed to avoid those failures. The objectives are High Accessibility, High Frequency, Acceptable Speed, Comfort, Low Costs, reduced emissions and noise pollution, and Clear administrative status. [4]

3.1.5 "Generating Options"

In accordance with TAG guidelines, we developed a wide range of possible substitutes considering all modes, substructure, guidelines, and pricing, that are likely to meet all the objectives mentioned above, and are applicable in Khartoum. These options included many transport options already considered by the ministry, as well as the C-Trains. These alternatives included the Metro, Light rail transit, regional rail transit, and Bus Rapid Transit (BRT) network.

3.1.6 "Initial Sifting"

With the help of the Early Assessment and Sifting Tool, and based on the evidence acquired from Khartoum transportation master plan, we arrived at the following:

1. While the light rail transit system and the C-Train network don't quite address all the issues separately, when combined, they became better capable of doing so. Therefore, they were branded as one potential option for further assessment.
2. In spite of the fact that it strongly meets all the objectives, the Metro network option was omitted due to issues concerning its high cost and the surface water table issue in Khartoum, caused by inadequate sanitary systems. Issues with the width of the streets and low land availability caused the elimination of the regional rail transit option as well.
3. The absence of basic road infrastructure makes the otherwise perfect option of BRT network very costly to deliver and is, therefore, rejected.

3.1.7 "Option Assessment Report"

In this step, we needed to shorten the list of options to only include the best-performing ones by gathering information regarding costs, advantages, and their impacts. Then we assessed them against the Transport Business Case criteria using the Option Assessment Framework provided by the TAG.[5]

Having collected the current and future demand in the previous stages and done all the desk-based work, the most important parts of which will be shown in the calculations section, we arrived at the optimum mode of transfer in Khartoum state to be, indeed, the STN network.

3.2 Environmental, Economic & Social Impacts

3.2.1 Environmental Impact

The Environmental Impact Assessment (EIA) was conducted in accordance with the transport analysis guide. At this early stage, this assessment was based on literature published by specialists or people with a far-reaching understanding of how the evaluation of this impact area is carried out.

3.2.1.1 Noise Impact

In a study on a newly proposed line of construction of the BLRT extension project [6], it was established that short-term elevated noise levels are expected and cannot be avoided for a project of this kind during the construction phase. However, they can be mitigated by barriers. [7]

3.2.1.2 Greenhouse Gases

From the Federal Transit Administration (FTA) Greenhouse Gas Emissions Programmatic Assessment, carried out to analyze the GHG emission rates produced by transit projects [8], results for light rail projects suggest that projects with high ridership are expected to lower the overall GHG emissions regardless of number of stations, length of trips, or alignment.

3.2.1.3 Air Quality

A report from the American Public Transit Association (APTA) presents evidence that every individual shifting from driving a private vehicle to riding a light rail transit mode reduces the hydrocarbon, nitrogen oxide, and carbon monoxide emissions drastically [9].

3.2.1.4 Landscape

The C-train was designed with the intention of making it far less visually intrusive than any other kind of elevated transit in mind. This is obvious in that the train is around half the height and width of any monorail system and the fact that the arches which hold the train are not any more invasive than a normal concrete pole [10].

3.2.2 Economic Impact

We assessed projected changes in consumer surplus, economic development, and Land and property values by Studying previous, similar economic impact assessments on rail transit projects to identify the direction on which the STN will steer the local economy.

3.2.2.1 Changes in Consumer Surplus

While the exact amount was not determined, it is established that the STN, when compared with the current transport system, will have no problem in saving more than enough time for riders which, in turn, will increase the consumer surplus. [11]

3.2.2.2 Economic Development

RIMS II is a set of regional multipliers maintained by the Bureau of Economic Analysis, used to predict the future impact of large infrastructure projects. Since Sudan is yet to develop an independent RIMS II Model, we adopted an already made model for Florida, U.S. [12], closest to represent the circumstance in Khartoum at the time. The analysis indicated that the impact of this kind of transit projects tends to give back generously.

3.2.2.3 Land and Property Value

Improved accessibility to urban areas can cause major development projects in them. Demonstrating this positive impact is a case study of property prices and sales conducted in the area of Atlanta [13], The study focused on the link between property values and mass transit stations. A model was presented and analyzed, and the results indicated a general rise of prices in these areas. The results cite a $62 rise for floors above the mean number of floors, determined at 5, among many other findings.

3.2.3 Social Impact

The Social Impact study is an indirect result of the previous two impact studies. To cite a compelling point supporting this argument is [14]: "The lower costs and increased accessibility due to transport improvements reduce the costs of the households' travel and, in turn, increase demand for goods and services. Such changes ripple through the market, increasing employment, output, and income in the short run. Over time, dynamic development effects derive from the mechanisms set in motion when transport service improvements activate a variety of interconnected economy-wide processes and yield a range of sectoral, spatial, and regional effects, that augment overall productivity."
It could be argued that these changes will result in boosted employment rates and in an expansion in transport-using firms, which is expected to result in increasing the social inclusion of the ignored parts of the society.

3.3 Calculations

3.3.1 Demand Calculation

3.3.1.1 Key Assumptions

1. C-Trains' Utility Function being a third of the demand for transportation.
2. Linear growth of demand.
3. Patterns of demand are not changing.

3.3.1.2 Basic Data

1. The C-Trains capacity is 50 passengers per carriage [15].
2. The C-Trains practical lower limit on headway is 1.5 minutes during rush hour [16].
3. Maximum capacity of Light rail trains is 300 per train [17].

3.3.1.3 Future demand of the network:

Peak demand percentage = Total population of the state in 2011÷Peak hourly demand for public transport [3].
= 2,000 ÷ 8,000,000 = 0.025%

Putting to use the UN Population forecast; the population of Khartoum state in their last forecasted year, 2100, was calculated and found to be 30.6 Million [18].

Demand = 70 % of Khartoum's population [3]
= 21,392,266 Passengers
Demand for STN = Total demand × STN Utility Function
= 7,273,370 Passengers
Peak hourly demand in 2100 = Khartoum population × peak demand percentage
= 7640 Passengers/ lane/ hour

C-Trains hourly capacity = Trains per hour × Capacity of train × No. of tracks (two; above and below)

= 4000 Passenger/hour/lane

When the peak demand exceeds 4000, the integration of an LRT system that covers high demand areas will be needed. Interpolation between the existing demands and the calculated demand in 2100 found that 2054 is the year when the integration of the LRT system into the STN will be needed.

Light Rail capacity = Trains per hour × Capacity of trains × No. of tracks (one) = 9000 Passenger/hour/lane

3.3.2 Costs Calculations

3.3.2.1 Initial construction costs

Basic Data:

1. Costs of building the C-Train network was estimated by the researchers to be $1.7M per kilometer. [19]
2. The length of the network that needs to be covered by C-Trains is approximately 470 Km, while the LRT will cover approximately only 130 Km. [20]
3. Cost of the 600 MWA Thermal plant to run the network = $850 Million. [21]

∴ Total Construction Costs = $3.2 Billion

3.3.2.2 Running Costs:

Annual running costs; which includes the operation and maintenance costs (O&M), personnel hiring costs, electricity costs, among others, were evaluated using the LRT model to perform a sensitivity analysis, introduced by Utpal Dutta of Detroit Mercy University [22]:

$$\text{O\&M} = 223462.7 \times (\text{Route Miles})^{0.744} \times (\text{Car-Miles})^{0.144}$$

Definition of System Variables:

Annual Route Miles (Route Miles): Stands for the miles taken in a directional route, average of 10 Km was used:
Route miles (D) = 2 × Route length (one-way)
Route miles (D) = 2 × (10 kilometers) = 12.42742 Miles

Annual Car Revenue-Miles (Car-Miles): It is the total number of miles served by all the trains: Car-Miles = Number of cycles during one day × Number of operating days in a year × Route miles per cycles × Number of cars per train
Cycles during peak hours = 40 cycles per hour
Cycles during off-peak hours = 6 cycles per hour

it is estimated that there are six peak hours and an additional 12 off-peak hours in a typical working day:
total number of cycles = 6 × 40 + 6 × 12 = 312 cycle/day

Adopting an estimate of 300 working days per year based on the assumptions, and lastly, the C-Trains prototype is dependent on one car per train.

∴ Car-Miles = 312 × 300 × 12.42742 × 1
= 939512.952 Miles

∴ Total annual network O&M cost using LRT model
= $10.6 Million/year

Total annual O&M cost of the plant
= $11.3 Million/year [21]

∴ Total annual O&M costs are approximated at $22 Million/year.

3.3.3 Cost-Benefit Analysis

For a project this big, the most suitable way to provide the necessary funding is by using a BOT contract. we conducted a cost-benefit analysis to decide whether a BOT contract is feasible. As we already calculated the costs of the project, we only need to calculate the profits and the payback plus profits period.

3.3.3.1 Total Revenues: Key Assumptions:

1. The number of passengers throughout a given year will be constant.
2. Only the daily ridership will be taken into consideration, the other viable side-investments (Like advertisement or small shop rentals) are ignored.
3. The initial Inflation rate will be taken as 55.6%, with an annual increase of 12%, as it stood in March 2018 at the time of this analysis. [23]
4. The initial ticket price of 10 SDG will be adopted, with the exchange rate to USD being 45 SDGs, as it stood in March 2018. [23]

The Costs vs Revenues chart was developed using a FORTRAN95 Software specifically programmed for this purpose using a function to adjust the ticket prices over the years to account for the inflation rates being at 12%.

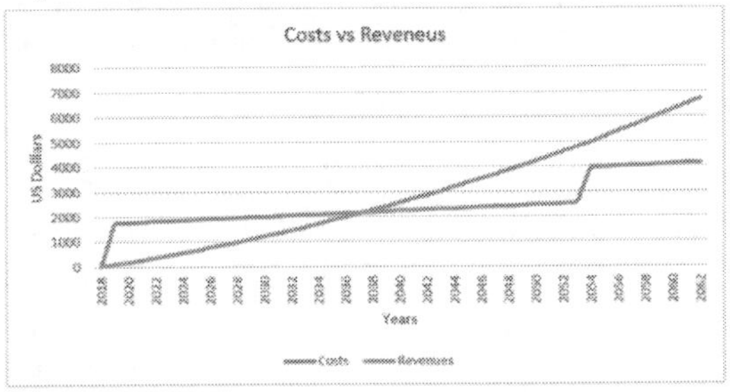

Figure 1. Costs vs Revenues of the STN.

As clearly seen from the Cost-benefit analysis, the breakeven point would take place somewhere around 2037, with the full capital invested retrieved in equivalent US dollars.

4 Discussion and Conclusion

Upon examination, we concluded that the suggestions offered by the government were poorly planned and do not fully satisfy the current much less the future demand. The concept of Caterpillar Trains, however, is both cheap and convenient for solving the transportation problem. In this research, we investigated the impacts of applying a rail network using this particular mode in Sudan.

Our findings indicated that the use of the C-train in constructing the STN as a solution is quite satisfactory from an engineering point of view. In fact, it will have a positive impact on traffic as there are major congestions in the city due to poor infrastructure.

These results also suggest that this mode of transport can be implemented in any developing country as it is economic, lightweight, fast, and can easily run over the narrowest of roads. Also, the C-Train is expected to have minimal visual invasiveness, little noise impact, and reduce Greenhouse Gases (GHG).

We also found that the most convenient option to fund this project is through a BOT contract. The Cost-Benefit-Analysis of the STN then was calculated using a programmed FORTRAN95 software.

References

1. A. Maasho, "Sudan, South Sudan reach oil deal, will hold border talks.".
2. Intibaha Newspaper, "Alwaly buses - manufacturing problems awaiting a solution", 2016.
3. Ministry of Transport and Infrastructures, "Khartoum Transport and Mobility Master Plan," 2010.
4. Jacob Innovations, "Jacob Innovations' MINI ELEVATED C-Trains (caterpillar train)," 2016. [Online]. Available: http://jacob-innovations.com/cTrain.html.
5. Department for Transport, "Transport Appraisal Guide," 2014, pp. 1–35.
6. Donald W. MacGlashan, "Analysis: Impact of Noise from the Purple Line Project Prepared," Federal Transit Administration, pp. 1–12.
7. I. (CSA) the metropolitan council in cooperation with Cross-Spectrum Acoustics, "Noise and Vibration Technical Report."
8. G. Filosa, C. Poe, and M. Sarna, "Greenhouse Gas Emissions from Transit Projects: Programmatic Assessment," Federal Transit Administrative Report No. 0097, no. 0097, 2017.
9. American Public Transit Association: Transit Fact Book. Washington, D.C. 104, 1993.
10. E. Jacob, "Elevated Trains Made Feasible - Station design challenge," Jacob Innov.
11. D. J. Forkenbrock and S. Benshoff, "Assessing the Social and Economic Effects of Transportation Projects," National Cooperation of Highways Research Program, vol. 31, no. February 2001.
12. T. Lynch, "Analyzing the Economic Impact of Transportation Projects Using RIMS II, IMPLAN, and REMI", vol. 22181, no. 703. 2000.
13. A. Nelson, "Transit stations and commercial property values: a case study with policy and land-use implications," Journal of Public Transport, vol. 2, no. 3, pp. 77–93, 1991.
14. T. R. Lakshmanan, "The broader economic consequences of transport infrastructure investments," Journal of Transport and Geography, vol. 19, no. 1, pp. 1–12, 2011.
15. K. Kumar, "Gurgaon considers caterpillar train for improved connectivity," Hindustan Times, 2016. [Online]. Available: https://www.hindustantimes.com/gurgaon/gurgaon-considers-caterpillar-train-for-improved-connectivity/story-uFdOlWX0yHr2BbQTGMnyFO.html.
16. Transit Information, "Transit Technology Choice Report," 2009.
17. "South Florida East Coast Corridor Transit Analysis Study: Environmental Impact Statement," 2006.
18. "United Nations Population Division DESA," 2019.
19. "Mini Elevated GreenTrains - complete transit solution - powered by green energy," 2016. [Online]. Available: https://www.climatecolab.org/contests/2016/transportation/c/proposal/1329506.
20. Researchers work; using Satellite Imagery. 2018.
21. A. A. Sulieman Nada, O Osama, "Construction of Infrastructure Projects in Sudan Using BOT," University of Khartoum, 2009.
22. D. Zhong, "An LRT Operation & Maintenance (O & M) Cost Model to Perform Sensitivity Analysis," Transp. Res. Board, 92nd Annu. Meet., vol. 820, 2013.
23. Central Bureau of Statistics, Republic of Sudan, "Inflation Rate in Sudan from April 2017 to March 2018", 2018.

Modelling Mode Choice at Sharjah University City, United Arab Emirates

Lubna Obaid[1,a] and Khaled Hamad[1]

[1]*Department of Civil and Environmental Engineering, University of Sharjah, Sharjah, U.A.E.*

Abstract. A growing interest in the behaviour of travelers to university campuses has recently emerged whether by university administrators or transport officials. Understanding the modal choice determinants of university travellers increases the opportunity for finding appropriate policies and solutions to reduce traffic congestion and parking needs as well as to encourage active transportation hence achieving more sustainable mobility. This research study investigates the differences in mode choice habits among the various groups of travellers to Sharjah University City (SUC) in the United Arab Emirates (UAE), including students, staff, faculty, and university visitors. A revealed preference survey was distributed randomly throughout SUC. Using information collected from this survey, multinomial discrete logit choice models were developed to evaluate the SUC travellers' mode-choice likelihood for the following modes: car, private bus, public bus, taxi, and active transport (walking and biking). It was found that travel time, travel distance, trip makers' characteristics (gender, citizenship, car ownership, car sharing, and the number of cars per household), and other contributing factors (Weathers conditions, Infrastructure adequacy, and bus services quality) are the main factors that affect significantly the mode choice at SUC. Further, a sensitivity analysis was conducted to study how the considered factors influence the mode choice. The developed model can be used in future studies to predict travel demand at SUC in response to new policies and solutions set by university administrators or transport officials.

1 Introduction

Many university campuses worldwide experience increasing travel demands leading to excessive traffic congestion and parking shortage. University campuses have complex, unique travel and usually characterized by their pro-active environment [1]. To solve transport problems at their university campuses, various researchers have studied this problem in recent years to understand university travelers' characteristics, travel patterns, and modal choice factors. Understanding the travel behavior of university travelers, and specifically their dependence on the private auto for commuting, can help universities and other stakeholders work together toward improvements to policies, programs, and infrastructure that encourage students' use of public transport or non-motorized modes of travel [2]. This is critical particularly in the case of large universities since student travel behavior directly impacts congestion levels in adjacent streets, the well-being of students and staff, as well as that of residents and businesses in the university neighborhood.

Like many other university campuses around the world, Sharjah University City (SUC)—which is the tertiary-education district in the City of Sharjah, United Arab Emirates)—suffers from severe traffic congestion on daily basis. With the difficulty of expanding the existing roadway network's capacity, the interest of the SUC's multi-stakeholders has turned into exploring any kind of transportation system, policy, or technique that reduces vehicular traffic. Towards this, a research project was awarded to develop a travel demand forecasting model to help plan for future sustainable transport improvements and policies at SUC.

The study's goal is to investigate SUC travelers' mode choice. Towards achieving this goal, the following four objectives were set: (1) collect socio-demographic and trip data using an online-survey containing both qualitative and quantitative questions, (2) develop discrete mode choice models, namely multinomial logit models (MNL), to investigate statistically the factors that are most likely to influence drivers utility and likelihood for choosing a specific transportation mode over other alternative modes and replicates the SUC travelers mode choice behavior, (3) conduct a sensitivity analysis to review the influence of multiple factors on travelers mode choice patterns, and (4) suggest solutions to encourage active and public transportation modes. While this paper focuses on SUC, it still sets an example of how universities can analyze the students' modal choice determinants to address their sustainability goals. Further, the study contributes to the emerging literature on understanding student behavior at university campuses in the region and worldwide.

[a] Corresponding author: lobaid@sharjah.ac.ae

2 Previous studies

Mode choice at university campuses has been an active research area for over two decades now.

Delmelle & Delmelle explored the spatial, temporal, and gender differences in transportation commuters mode choice within Idaho University students [3]. The main finding of this research is that male students are more likely to switch their driving mode through the year whereas female students are more to keep using cars. Also, noticeable commuters' preferences were found between graduate and undergraduate students.

Longo et al conducted a study to analyze mobility issues at the campus of an Italian university [4]. A web-based survey was carried out to identify solutions to reduce traffic congestion and improve accessibility to their campus. The survey participants were mostly residents and commuters and prefer to travel using public transportation as busses or trams. The Analytical Hierarchy Process was applied to prioritize these solutions from the perspectives of different user groups. The prioritizing criteria included several factors as travel time, cost, comfort, safety, flexibility, and environment.

Shannon et al assessed the commuting habits of students and staff at an Australian university campus [2]. The results of an online survey highlighted the transport modes being used, factors affecting transport decisions, and the potential for changing the transit mode. The survey results indicated that around 20% of the staff and 46% of the students used active modes of transport such as walking, cycling, and transit. Furthermore, potentially 30% of staff and students would like to switch to active transport modes; therefore, the study suggested reducing barriers to achieving this shift. Suggested polices include implementation of a subsidized public transport pass, increased student housing on or near campus, increased the cost of parking, and improved bus services and cycle networks.

Miralles-Guasch studied the transport patterns at the Autonomous University of Barcelona, including motivations, barriers and user preferences, and explores the main transport challenges faced by the their campus [5]. Over 5500 members of the university community were surveyed and results showed that the primary challenges of switching from private to public transportation are the lack of adequate infrastructure, the marginal role of walking and cycling as means of transportation, and the longer trip time when using public transportation.

Balsas explored how college campuses have encouraged a modal shift from cars to bicycling and walking [6]. The purpose of the study was to identify some of the best practices undertaken by several universities to make their campuses more sustainable communities for bicyclists and pedestrians. A survey of eight bicycle and pedestrian friendly campuses across the USA was conducted. A key finding of the survey was that college campuses are de-marketing automobile commuting and promoting alternative transportation modes. A final recommendation of the research was to promote the shift from private to public transportation through universities since they are a proactive educational environment.

Abubakar et al. explored students' perspectives towards the sustainability of their campus based on a survey of 152 students of the College of Architecture and Planning at the University of Dammam in the Kingdom of Saudi Arabia [7]. The study results indicated that students showed a great deal of awareness and concern about campus environmental sustainability; nevertheless, they lacked interest and willingness to participate in initiatives towards achieving sustainability. An interesting point found in this study is that transportation initiatives were fewer compared to sustainable landscaping and waste conservation. Their research concluded by highlighting the importance of incorporating sustainability into campus travel operations and training university students in promoting environmental sustainability in the KSA and the MENA region.

Aoun et al. conducted a study at the American University of Beirut (AUB) in Lebanon to explore potential practices for better sustainable transportation practices at AUB [8]. Given the high-income population and high car-ownership at AUB, conventional strategies might not be effective. Their main learned lessons from reviewing the experience of five American university campuses can be summarized as follow: First, campus transport services do not always stem from existing public transport services. Second, designing a campus transport service catered to the target population can be extremely effective with high-income clients. Finally, mass transport ridership is partly dependent on driving disincentives such as limited and priced parking. The study proposed to have a new dynamic mobility solution, i.e. dynamic taxi sharing. Their preliminary mode-shift survey demonstrated the potential market and possible positive reductions in-vehicle trips.

Villanueva et al measured pedometer steps for university students who used transit and passenger vehicles to travel to and from the campus of a University of Western Australia [9]. A field experiment over university students where about 100 undergraduate students wore a pedometer .for five consecutive days and completed a travel activity diary. The results of the experiment were and then compared to motor vehicle users. The comparison indicated that public transport users made more steps daily compared to private vehicle users. The study recommended encouraging public transport use could help increase and maintain community physical activity levels.

Table 1 summarizes and compares several related research studies in terms of the statistical analysis used, the modes assessed, as well as the variables considered. As can be seen in Table 1, various statistical approaches were used to assess the general public modal choice behavior in many countries in the Middle East and North Africa (MENA) region; yet only few studies focused on mode choice for university travelers [10].

Table 1: Summary of the variables applied in statistical models for university students

Ref.	University	Modes	Models	Variables/Attributes
[11]	Seven campuses in Greater Toronto	Active modes, transit, Car as a driver, and Car as passenger	Multinomial logit models (MNL)	- Socio-economic: Gender, household size, number of cars in H.H., building type, age, number of members in the residence, income level. - Travel cost: Trip distance, trip frequency. - TDM: car sharing, transit pass ownership, bike ownership. - Residential self-selection factors: The primary factor of house location is transit, walking, bike. - Built environment: Density of pedestrian networks, land use mix
[12]	University of Michigan	Single occupant vehicle (SOV), Car pool, Walk, Bicycle, and Bus	Binary Logit Model	- Socio-economic: Gender, Occupation. - Mode specific: *Bicycle*: Adverse weather, bicycle theft, general crime, lack of bicycle lanes, No bicycle, distance, lack of interest. *Walking:* Police presence, lighting, safe intersections, traffic enforcement, emergency phones, fuel cost, parking cost.
[3]	Los Angeles	Transit (Bus), Bike, Walk, Drive alone, Carpool, and Telecommuting	MNL	- Trip makers': Gender, residence choice, age, occupation. - Trip: Travel time, time of travel, commute distance. - Travel demand management: Parking permit. - Mode specific: Transit pass, direct bus lane.
[13]	University of Coimbra	Car, and Public transportation	MNL	Occupation, income per capita, residence distance, gender, Public transportation usage frequency in the last month.
[14]	North Dakota State University	Automobile, Bus, Carpool, motorcycle, Non-motorized	MNL	In-vehicle travel time (IVTT), out-of-vehicle travel time (OVTT), Parking demand factor (PKD)
[15]	University of Texas at El Paso	N.A.	MNL	The distance of the parking facility i from the desired destination zone of the driver, Distance of the parking facility i from the current vehicle position of driver, and Parking fee of the parking facility i.
[10]	University of Beirut	Car, Bus, and Jitney	Nested Logit model (NLM)	- For students: Cost, travel time, monthly H.H. income, Gender, Number of cars per H.H. - For general public: Cost, travel time, monthly H.H. income, trip frequency per week
[16]	University of Cantabria	Parking within the Campus, Park-and-ride, Parking in the surrounding area, Bus, and Bike-sharing	Mixed logit model	Travel time, waiting time for the bus, time to find a parking space, access time to the Campus from the parking spot, Cost of parking the car or travelling by bus or bike (in the case of the bike-sharing system)
[17]	University of Calabria	Walk, and Electric minibus	MNL	Travel time (walking by foot/ transit), transit waiting time, bus crowding, bus comfort, age, patience, laziness
[18]	Istanbul Technical University	Private car, Service vehicle, and Public transport	MNL	Mode specific variables: *Private car*: H.H. income, trip running cost, parking fee. *Service vehicle*: Age, vehicle time, OVTT. *Public transport*: IVTT, OVTT, ticket fee.
[19]	McMaster University, Hamilton	Car, Bicycle, Walk, and Bus	MNL	Street and sidewalk density, travel time, access time, vehicle ownership, age, gender, living arrangements, vehicle comfort, shelter, parking (TDM)

3 Methods and materials

3.1 Study area

Sharjah University City is the tertiary-education district in the City of Sharjah with a total area of over 1,500 hectares. Located southeast of the City of Sharjah's downtown (**Figure 1**), SUC has a beautifully crafted landscape and distinguished Islamic architecture for most of the buildings. It encompasses over 15 tertiary education and research institutions including: University of Sharjah (with its men, women, medical, and fine arts campuses), American University of Sharjah, Higher Colleges of Technology (with its both men and women campuses), the Arabic Language Academy in Sharjah, Al-Qasimia University, and Sharjah Police Academy. Over 20,000 students are enrolled in the various institutions inside SUC. These students are served by over 2,000 faculty and staff members. Out of these, around 6,000 students and faculty members live in the various dormitories and houses on their respective campuses.

3.2 Data Collection

A well-prepared questionnaire was distributed among travelers on campus, including students, faculty, staff, and visitors. A web-based platform was used for developing and administering the questionnaire. The survey asked participants to report their trips that had at least one end at SUC to obtain comprehensive information about the activity-travel patterns of the university population. In addition, the survey collected data on intra-campus trips where both trips ends are located on the same campus to fully measure all travel demand that occurs in and around the university campus. The survey was randomly distributed over SUC road commuters, and over 4000 responds were collected.

The questionnaire survey consisted of two sections focusing on:

- Demographic characteristics: Gender, nationality, status, vehicle ownership, etc.
- Commuting characteristics: Trip purpose, mode of travel, trip duration and length, origin and destination, etc.

Five transportation modes widely used by SUC travelers were assessed in this study: private car, taxis, private bus, public bus, and active transportation (i.e., walk and bike). For this research, three types of variables affecting mode choice were considered:

1. Trip makers' characteristics: Gender, citizenship, status, car ownership, and number of cars per household.
2. Trip characteristics: In-vehicle travel time (IVTT) in minutes, out-of-vehicle travel time (OVTT) in minutes, and trip distance in kilometers.
3. Other contributing factors: Car sharing, weather conditions, bus services, and status of transport infrastructure.

These variables were selected following previous studies. Only the trip duration is considered for active modes. The OVTT for cars and taxis was calculated using **Eq. 1** whereas for public and private busses using **Eq. 2.**

$$OVTT\ for\ Car = Parking\ time + Walking\ time\ to\ destination \quad (1)$$

$$OVTT\ for\ Bus = Walking\ time\ to\ bus\ station + Waiting\ time\ for\ bus \quad (2)$$

3.3 Descriptive analysis of the selected variables

Descriptive statistics of the selected variables are summarized in **Table 2** and **Table 3**. Out of all the travelers who participated in the survey, 84% were university students while the remaining 16% were staff, faculty, and visitors. Around 55% of them were males. The majority (over 70%) were expatriates while the remaining were UAE nationals. Almost two-thirds of the respondents (66%) prefer to drive alone rather than car sharing and 64% owned their own cars. In contrast, a low percentage (almost 13%) travel from and to the university with their colleagues/family members.

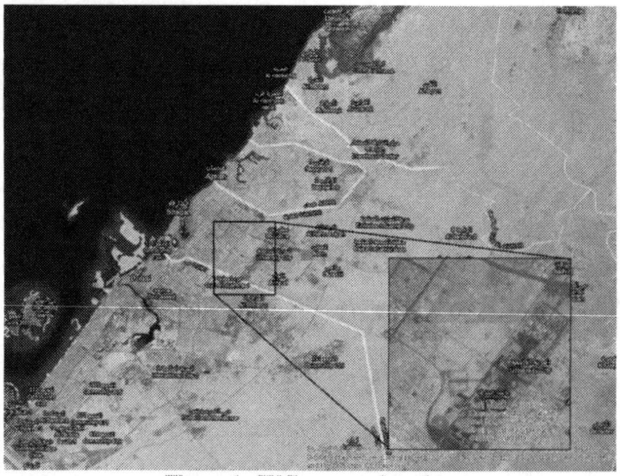

Figure 1: SUC campus location

Table 2: Descriptive statistics of continuous variables

	Variables	Mean	Median	Std. dev.	IQR	Min	Max
Trip related factors	IVTT (min)	26.41	25	19.88	25	1.000	180
	OVTT (min)	16.14	15	11.61	17	0	45
	Trip distance (km)	10.04	8.507	13.37	11.17	0.1	169

Table 3: Descriptive statistics of categorical variables

n= 4300 respondents

	Variables	**Code**	**% of total**
Trip maker's characteristics	Gender	Male	56.21
		Female	43.8
	Citizenship	Local	25.8
		Expatriate	74.21
	Academic Standing	Student	84.13
		Faculty	6.64
		Staff	7.05
		Visitor	2.18
	Car Ownership	Yes	64.3
		No	35.7
	Number of Cars per Household	One car or less	6.85
		Multiple cars	93.15
Other contributing factors	Car Sharing	Yes	12.63
		No	66.44
	Inconvenient Bus service	Yes	34.7
		No	65.3
	Harsh Weather Conditions	Yes	38.78
		No	61.22
	Inadequate Sustainable Transportation Infrastructure	Yes	10.2
		No	89.8

As for the other contributing factors, weather conditions had a slightly higher impact on the travelers' attitudes toward shifting for more sustainable transportation modes while the adequacy of sustainable transport infrastructure (sidewalks and bike lanes) had a minor impact.

Travelers' trips ranged from a minimum of 100m for in campus trips to almost 170km for out of campus intra emirates trips. Their in-vehicle trips' duration ranged from 1 to 180 minutes with an average of around 26 minutes, while the out-of-vehicle trip duration ranged from 0 minutes to 45 minutes with an average of 15 minutes.

3.4 Modelling approach

To help analyze the relationships between mode choice behavior of SUC travelers and its determinants a multinomial logit (MNL) model was developed using the aforementioned data collected. As shown in **Table 1**, the MNL approach has been widely used to model mode choice around the world. Basically, the probability the traveler i would choose the mode of transport j among the set of available modes k is given by **Eq.3.:**:

$$P_{ij} = \Pr\left(U_{ij} > U_{ik}\right), \forall\, k \neq j \tag{3}$$

where k represents variable transportation means, and U_{ik} is the maximum utility required for a traveler i if the traveler chooses the mode j. The utility of traveler i for choosing mode j is given by **Eq.4.:**

$$U_{ij} = \beta_j X_{ij} + \varepsilon_{ij} \tag{4}$$

where β_j is a vector of the unknown coefficients for each of the explanatory variables, X_{ij} is the vector of the known variables, and ε_{ij} is the random component of the utility. If the stochastic terms ε_{ij} are assumed to be independent, the logit choice probability can be expressed as the following **Eq.5.:**:

$$P_{ij} = \frac{e^{\beta_j X_{ij}}}{\sum e^{\beta_j X_{ij}}} \tag{5}$$

The widely used statistical package IBM SPSS Version 24 was utilized to estimate our model. It is also worth mentioning that the β_s are estimated by maximizing a log likelihood function.

https://doi.org/10.1051/matecconf/202030802004

Table 4:Multinomial logit model estimation results

Variable	Cases (Case Code)	Car (1)	Taxi (2)	Private Bus (3)	Public Bus (4)	Active Transport (5)
		N= 2598	N= 73	N= 135	N= 707	N= 787
	Intercept	14.034	0^b	4.718	3.639	5.347
*Gender	Male (0)	0.251	0^b	-1.334	-0.256	0.658
	Female (1)	0^b	0^b	0^b	0^b	0^b
**Citizenship	Expat (0)	-1.628	0^b	-1.484	-1.214	-1.489
	Local (1)	0^b	0^b	0^b	0^b	0^b
*Academic Status	Student (0)	-1.088	0^b	-0.37	0.204	0.781
	Not a student (1)	0^b	0^b	0^b	0^b	0^b
*Car Sharing	No (0)	0^b	0^b	0^b	0^b	0^b
	Yes (1)	6.487	0^b	2.943	1.415	-0.45
***No of Cars per Household	One or less (0)	0.059	0^b	0.387	0.626	0.522
	Multiple Cars (1)	0^b	0^b	0^b	0^b	0^b
*Car Ownership	No (0)	0^b	0^b	0^b	0^b	0^b
	Yes (1)	8.229	0^b	-0.171	-0.02	1.426
***Inconvenient Bus Service	No (0)	0^b	0^b	0^b	0^b	0^b
	Yes (1)	3.182	0^b	3.273	2.531	1.654
***Harsh Weather Conditions	No (0)	0^b	0^b	0^b	0^b	0^b
	Yes (1)	2.068	0^b	4.193	2.4	2.911
***Inadequate Infrastructure	No (0)	0^b	0^b	0^b	0^b	0^b
	Yes (1)	1.293	0^b	3.223	-5.11	30.207
*IVTT	Continuous	0.013	0^b	0.028	0.037	-0.077
*OVTT	Continuous	-0.311	0^b	0.435	-0.21	N.A.
*Trip Distance	Continuous	0.08	0^b	-0.036	-0.033	-0.273

Notes: Taxi is the reference mode, Nagelkerke $R^2 = 0.841$, df = 4, the reference (base) categories are set to zero because it is redundant, *Significant at 1% level, ** Significant at 5% level, * Significant at 10% level.**

4 Results and discussion

4.1. MNL approach results

Table 4 summarizes the results obtained from analyzing the developed MNL mode-choice models for the five modes of transportation considered in this study. The base (reference) mode, denoted as 0^b, was chosen to be the taxi because it has the smallest number of observations in order to minimize the number of estimated coefficients for the corresponding utility functions [19]. Not all the variables considered in this study are statistically significant; thus, the model was refined to include only significant factors. Furthermore, the variables reported varying levels of significance. For instance, the number of cars per household is significant at a significance level of $\alpha = 10\%$ whereas the citizenship variable is significant at a significance level of $\alpha = 5\%$; the remaining study variables were significant at a significance level of $\alpha = 1\%$. **Table 4** reports the value of the log-likelihood of the null model (i.e., intercept only) and the full model. After the significance of the factors is tested, the model was reduced so the final utility function for traveler *i* with respect to mode *j* is defined as the following **Eq.6.**::

$$U_i(j) = \alpha_j + \beta_{1j} \times \textit{Trip Duration} + \beta_{2j} \times \textit{Trip Distance} + \beta_{3j} \times \textit{Gender} + \beta_{4j} \times \textit{Citizenship} + \beta_{5j} \times \textit{Status} + \beta_{6j} \times \textit{Car sharing} + \beta_{7j} \times \textit{No. of cars per household} + \beta_{8j} \times \textit{Car ownership} + \beta_{9j} \times \textit{Inconvenient Bus Service} + \beta_{10j} \times \textit{Harsh Weather Conditions} + \beta_{11j} \times \textit{Inadequate Infrastructure} \quad (6)$$

It is concluded that the private cars have the highest utility, followed by active transportation, private and public buses, and taxis respectively. This conclusion is made based on the following: (1) The ordinal nature of the utility functions, where an alternative mode j will be chosen by a traveler i over another alternative when it has a higher utility, regardless to the magnitude difference between both modes, and (2) the MNL model results obtained.

Regarding the traveler sociodemographic dichotomous factors, six of these factors showed high significance in affecting the mode choice. Regarding the trip maker's gender, male travelers showed higher positive utility for using active transportation while female users are more likely to use both private and public buses. As clearly shown in **Table 4**, students are the highest category that tends to use active transportation and public transportation in comparison to staff, faculty, and visitors who mostly depend on driving alone their cars. Unsurprisingly, travelers that do not own car tend to use busses and taxis. This is actually related to the number of cars per household factor; that is, the more cars in the household, the higher the odds for travelers to use their own cars. It is interesting to notice that the difference between UAE citizens and non-citizens. The latter group (i.e., expatriates) mostly prefer using taxi and public transportation whereas citizens prefer to use cars to reach the university campus.

An interesting finding is that active transportation is actually the most attractive mode for the following groups: female travelers, students, travelers that do not own a personal private car, travelers with one car or less in their households, and among non-citizens. Car sharing indicated a positive utility for active transportation; that is travelers who share their ride with a family member or colleague usually tend to walk for their internal trips within the SUC campus. The weather condition, lack of adequate infrastructure, and quality of offered bus service affect travelers' mode choice and increase the odds for choosing more sustainable transportation modes.

As expected, factors related to trip cost (namely trip length and duration) affect the mode choice at SUC. The results showed that the longer the trip length and the IVTT required to complete the trip, the more travelers tend to use modes other than active transportation as indicated by the negative utility sign for both factors. There is also a negative utility for choosing private cars and public transit relative to the OVTT. Increasing the times to wait for bus, find parking space for cars, or the walking to final destination or bus stop, clearly decrease the utility for choosing the travel mode.

4.2. Sensitivity analysis

The preliminary results indicated that the trip distance affects the travelers' likelihood to choose their transportation mode more than the trip duration. In order to evaluate the magnitude of the utility of trip duration and distance on the mode choice, the probability for using cars, public and private busses, and active transportation were calculated using **Eq. 7** while that for the reference mode (i.e. taxi in our study) was calculated using **Eq. 8.**

$$P_{r(j=K-1)} = \frac{e^{aj-1+\beta_{j-1}X_{ij}}}{1+\sum_{k=1}^{k-1} e^{aj-1+\beta_j X_{ij}}} \quad (7)$$

$$P_{r(j=k)} = 1 - \sum_{k=1}^{k-1} P_j(j) = \frac{1}{1+\sum_{k=1}^{k-1} e^{\beta_j X_{ij}}} \quad (8)$$

The estimated probabilities of the five considered modes were plotted with varying trip duration and distance while keeping the remaining variables as constant. The following assumption was made: An average value for continuous variables was used while the most frequent category was used for the qualitative binary variables. Hence, the following values were utilized: trip average distance = 10.04 km, in-vehicle trip average duration= 26.4 minutes, out-of-vehicle tip average duration = 16.14, Gender = Female, Status = Student, Car ownership = Yes, Citizenship = No, Number of cars per household = Multiple, Car sharing = No, Inconvenient bus service= No, Harsh weather conditions = No, and Inadequate Infrastructure= No). A graphical summary of the obtained results of this analysis are illustrated in **Figure 2** which shows the probability of choosing the five modes when the in-vehicle trip duration is varied and **Figure 3** which shows the probability of choosing the five modes when the trip distance is varied.

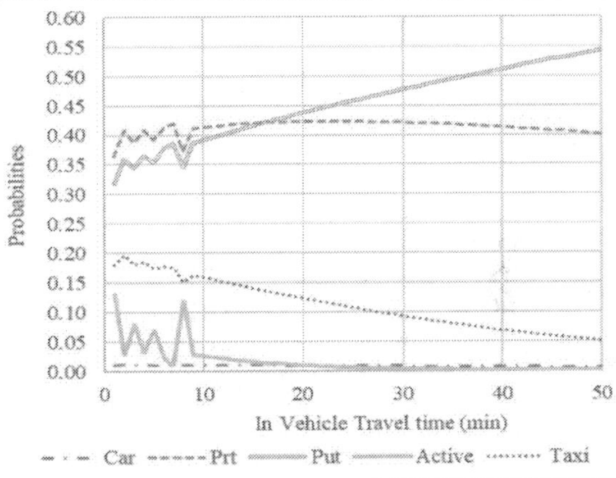

Figure 2: Estimated Probabilities of travelling by different travel modes as a function of in-vehicle travel time.

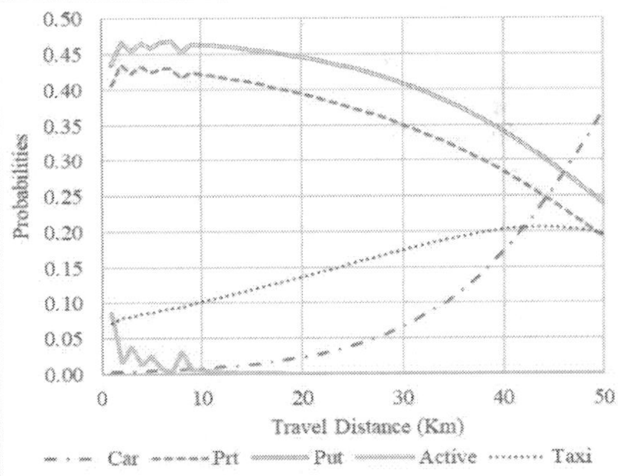

Figure 3: Estimated Probabilities of travelling by different travel modes as a function of Trip Distance.

The probability of choosing active transportation declines with the increase of distance and duration reaching almost 0% after 10 km or 30 mins of a trip. This is unsurprising given the fact that traveller's tendency to walk/bike for long distances is the least among all modes. In contrast, the probability of using the car increases with the increase of the trip distance until it becomes the dominant mode at trips that are longer than 45 km since it offers the shortest travel time. It is important to point out that the trends showed a low utility for the cars over long duration trips. This can be justified by the fact that the survey respondents' trips for long distances took less time using cars in contrast to public and private transportation.

It is also observed that both the public and private transportation have the highest utilities among all other travel alternatives. Lastly, there was no noticeable change of the utility of taxi with the change of trip duration due to the fact that only a small proportion of SUC road commuters travel use taxi to travel to campus (a declining utility is observed for the active transportation).

To validate the model developed to predict the travellers' utility of choosing their transportation modes at SUC, the results of the model were compared to those reported by other studies for other university campuses. According to the SUC MNL model, a direct relation between travel time and the mode choice where observed whereas the longer the trip the smaller the utility for the active transportation becomes. This was also indicated by a previous study conducted by Zhou, which showed that with increasing travel time, the drive-alone and public transit become the most dominant modes whereas the active transportation mode becomes more dominant for trips with small duration [3].

Furthermore, this study showed that undergraduate students are the user group that has the highest tendency for choosing active transportation modes. Similar studies, such as a study by Moniruzzaman and Farb, showed comparable results where positive utilities for active transportation were reported for both female users as well as those who have one car or less in their household [11]. In addition, this study clearly showed a negative utility for all transportation modes relative to the car as a driver with the increase of travel distance.

Lastly, our model showed that females, non-citizens, and travellers who do not own a private car, have higher utility for using active transportation and therefore can be targeted for future modal shift solutions. This finding was also stated by Nguyen-Phuoc et al who studied different factors including the travellers' gender, status, car ownership, as well as travel distance and time. The analysis results revealed that females, travellers who do not own a personal car, short travel distances, and short trip duration are all factors with a positive utility for prioritizing active transportation over other modes [20].

4.3 Suggested polices and solutions

The following are some solutions to encourage sustainable transportation modes and to reduce daily traffic congestion at SUC: (1) apply parking restrictions to limit the maximum free hours to 2 or 3 hours per day; (2) build a park-and-ride facility close to the campus to allow travellers to park their cars then take the shuttle bus to their final destination; (3) Launch a pay-to-ride bike and scooter to supply students with easy to use and cheap alternatives to travel inside campus; (4) Develop an application where students can register their names for more reliable private and public transport modes that suit their schedules; and (5) Offer mobility on demand private door-to-door services.

5 Conclusion

Modelling transportation mode-choice behaviour is an important step to accurately predict future travel demand. Sustainable transportation helps in decreasing the vehicular traffic demand and improving the tendency for using public and active transportation. Though many researchers analysed the modal choice determinants in the MENA region, only a few of these studies focused on university campuses even though university campuses have special travel patterns. This study analysed the most significant determinants that affect travellers' mode choice at Sharjah University City in the UAE. In addition, the study developed discrete mode choice (i.e., multinomial regression) models to examine the mode choice of SUC's travellers. Several sociodemographic factors of the trip makers and trip-specific variables were assessed in terms of their contribution and significance to the final mode choice decision of these travellers.

Although the models developed in this study can not necessarily be generalized to all university campuses; yet they provide insights for other higher-educational institutes when investigating their travel patterns and mode choice determinants. This study has several limitations that hopefully will be overcome in future research. The sample size could be expanded to eliminate any biasness for a specific group of respondents. Furthermore, additional variables believed to affect the mode-choice models at Sharjah University City could be added.

6 References

[1] T. Limanond, T. Butsingkorn, and C. Chermkhunthod, "Travel behavior of university students who live on campus: A case study of a rural university in Asia," *Transp. Policy*, vol. 18, no. 1, pp. 163–171, 2011.

[2] T. Shannon, B. Giles-Corti, T. Pikora, M. Bulsara, T. Shilton, and F. Bull, "Active commuting in a university setting: Assessing commuting habits and potential for modal change," *Transp. Policy*, vol. 13, no. 3, pp. 240–253, May 2006.

[3] J. Zhou, "Sustainable commute in a car-dominant city: Factors affecting alternative mode choices among university students," *Transp. Res. Part A Policy Pract.*, vol. 46, no. 7, pp. 1013–1029, 2012.

[4] G. Longo, G. Medeossi, and E. Padoano, "Multi-criteria analysis to support mobility management at a university campus," in *Transportation Research Procedia*, 2015, vol. 5, pp. 175–185.

[5] C. Miralles-Guasch and E. Domene, "Sustainable transport challenges in a suburban university: The case of the Autonomous University of Barcelona," *Transp. Policy*, vol. 17, no. 6, pp. 454–463, 2010.

[6] C. J. . Balsas, "Sustainable transportation planning on college campuses," *Transp. Policy*, vol. 10, no. 1, pp. 35–49, Jan. 2003.

[7] I. R. Abubakar, F. S. Al-Shihri, and S. M. Ahmed,

"Students' assessment of campus sustainability at the University of Dammam, Saudi Arabia," *Sustainability*, vol. 8, no. 1, pp. 1–14, 2016.

[8] A. Aoun, M. Abou-Zeid, I. Kaysi, and C. Myntti, "Reducing parking demand and traffic congestion at the American University of Beirut," *Transp. Policy*, vol. 25, pp. 52–60, Jan. 2013.

[9] K. Villanueva, B. Giles-Corti, and G. McCormack, "Achieving 10,000 steps: A comparison of public transport users and drivers in a University setting," *Prev. Med. (Baltim).*, vol. 47, no. 3, pp. 338–341, 2008.

[10] M. Danaf, M. Abou-Zeid, and I. Kaysi, "Modeling travel choices of students at a private, urban university: Insights and policy implications," *Case Stud. Transp. Policy*, vol. 2, no. 3, pp. 142–152, Dec. 2014.

[11] M. Moniruzzaman and S. Farber, "What drives sustainable student travel? Mode choice determinants in the Greater Toronto Area," *Int. J. Sustain. Transp.*, vol. 12, no. 5, pp. 367–379, 2018.

[12] G. Rybarczyk and L. Gallagher, "Measuring the potential for bicycling and walking at a metropolitan commuter university," *J. Transp. Geogr.*, vol. 39, pp. 1–10, 2014.

[13] E. Barata, L. Cruz, and J. P. Ferreira, "Parking at the UC campus: Problems and solutions," *Cities*, vol. 28, no. 5, pp. 406–413, Oct. 2011.

[14] R. Bridgelall, "Campus parking supply impacts on transportation mode choice," *Transp. Plan. Technol.*, vol. 37, no. 8, pp. 711–737, 2014.

[15] B. I. Bustillos, J. Shelton, Y.-C. C. Chiu, B. I. Bustillosa, J. Shelton, and Y.-C. C. Chiu, "Urban university campus transportation and parking planning through a dynamic traffic simulation and assignment approach.," *Transp. Plan. Technol.*, vol. 34, no. 2, pp. 177–197, 2011.

[16] L. Dell'Olio, M. Bordagaray, R. Barreda, and A. Ibeas, "A methodology to promote sustainable mobility in college campuses," *Transp. Res. Procedia*, vol. 3, pp. 838–847, Jan. 2014.

[17] L. Eboli, G. Mazzulla, and A. Salandria, "Sustainable Mobility at a University Campus : Walking Preferences and the Use of Electric Minibus," *Int. J. Transp.*, vol. 1, no. 1, pp. 21–34, 2013.

[18] H. Tezcan, "Using parking pricing as a travel demand management tool at a university campus: an example for Istanbul Technical University," *Transp. Lett.*, vol. 4, no. 3, pp. 181–192, 2012.

[19] K. E. Whalen, A. Páez, and J. A. Carrasco, "Mode choice of university students commuting to school and the role of active travel," *J. Transp. Geogr.*, vol. 31, pp. 132–142, Jul. 2013.

[20] D. Q. Nguyen-Phuoc, R. Amoh-Gyimah, A. T. P. Tran, and C. T. Phan, "Mode choice among university students to school in Danang, Vietnam," *Travel Behav. Soc.*, vol. 13, pp. 1–10, Oct. 2018.

Research on Expressway Travel Time Prediction Based on Exclusive Disjunctive Soft Set

Qingqing Wang[1], Huamin Li[1,a] and Weixin Xiong[2]

[1] *School of Automation, Chongqing University, Chongqing, 400044, China*
[2] *School of Mechanical Engineering, Chongqing University, Chongqing, 400044, China*

Abstract. In order to study the prediction problem of expressway travel time, due to the ambiguity and uncertainty in the road traffic system, the travel time prediction model is established based on the exclusive disjunctive soft set theory. Through the parameter reduction theory of soft set, the main influence factors are extracted, and the mapping relationship between the influence factors and the travel time is obtained through the exclusive disjunctive soft set decision system. The travel time model is established based on the soft set theory, and the travel time is calculated through the mapping relationship. The experimental results show that, compared with the BPR function model, the travel time model based on the exclusive disjunctive soft set theory reduces the prediction error and effectively improves the calculation accuracy of the travel time.

1 Introduction

With the rapid development of intelligent transportation system, the traffic information that can be obtained and provided is more and more abundant, and the travel time prediction is an indispensable part of the advanced highway travel information system.Facing the increasingly prominent traffic problems, dynamic traffic control and traveller behavior decision-making require highway travel time prediction. It is important to improve the accuracy of the estimation when considering the estimated value, it is of great practical significance to quantitatively describe the uncertainty of the predicted value.

The researches on travel time mainly focus on the construction of prediction models and travel time function forms, which have achieved rich research results. More Rohan applied artificial neural networks to road traffic flow prediction, which reduced the difficulty of establishing accurate mathematical models, while ensuring high prediction accuracy [1]. Mori et al. introduced the literature on travel time model and thoroughly classified existing technologies, most of the attention was focused on travel time estimation and travel time prediction [2]. Allstr M et al. combined parametric and nonparametric traffic state prediction techniques through assimilation in aggregating Kalman filtering. For nonparametric prediction, the neural network method was adopted; the parameter prediction was performed by using the cell transmission model with velocity as the state [3]. Celikoglu et al. investigated the performance of the flow model in providing efficient travel time estimation for varying flow patterns of freeway traffic by adopting the two-phase fundamental diagram [4]. Zhong et.al. proposed an online travel time prediction model with emphasis on capturing the effects caused by anomalies. A Functional Principal Component Analysis (FPCA) framework was adopted to forecast link travel times based on historical data and real-time measurements [5]. Sehyun Tak et.al. proposed the new algorithm called multi-level k-Nearest Neighbor (k-NN), which was designed for predicting travel time with higher computational efficiency and prediction accuracy, and provided the data fusion method to overcome the problems of data errors that combined the traffic data from ILDs and DSRC [6]. Lu et al. proposed a piecewise function based on traffic flow and vehicle type in order to estimate the travel time function of heterogeneous flow, and used the VISSIM model of actual traffic data calibration to generate traffic data for estimating parameters [7]. Chen et al. used the data source from highway charge system to establish the prediction model based on Bootstrap to improve the reliability of point prediction in travel time [8]. Wang et al. used the K nearest neighbour algorithm to predict the short-term travel time of the expressway based on the location and time information of the vehicle entering and leaving the toll station in the highway toll data [9]. The experimental results showed that the algorithm could effectively improve the prediction accuracy of the model, but the impact of weather conditions on traffic conditions was not considered, and the historical database wasn't classified according to different weather conditions.

However, when studying the travel time, the following problems are rarely considered: the judgment of the traffic participants on the traffic state is inherently vague, not precise. Under the premise of ensuring the effective provision of information, traffic participants choose

[a] Corresponding author: huaminli@cqu.edu.cn

different paths by judging road congestion conditions, etc., but their judgments on road congestion conditions are qualitative, not quantitative. That is to say, the ambiguity should be fully considered for the estimation and prediction of the travel time. Soft sets are powerful theories for dealing with ambiguity and uncertainty, its advantages are not only for quantitative analysis, but also for qualitative analysis [10-12]. However, the existing soft set is mainly the aspect of complete information, but the aspect of incomplete information system needs to be studied. From the point of view of data mining, this paper introduces the exclusive disjunctive soft set to describe the incomplete information system, and quantifies the data in the incomplete information system by redefining the characteristic functions in the exclusive disjunctive soft set.

The rest of the paper is organized as follows. Section 2 introduces the basic concepts. Section 3 describes the prediction model and algorithms used in the study. Section 4 uses the numerical examples to verify the validity of the model, and the model is compared with the prediction result of the BPR function model. Finally, conclusions are discussed in Section 5.

2 Preliminary

2.1. Soft set

Molodtsov proposed the new uncertainty mathematics tool of soft set theory [13]. For the first time, he systematically introduced some basic theories and applications of soft sets, and used soft set as a general mathematical tool for dealing with uncertain, fuzzy and inaccurate objects. Soft set is a new method for solving uncertain problems that can't be solved by traditional methods, the biggest difference between the soft set and the traditional data method is that it can set different forms of parameters according to its own preferences, which can solve the problem of ambiguity and uncertainty well.

Definition 2.1 (F,E) is a soft set on the domain U, if and only if F is a mapping of E into the set of all subsets of the set U[13].

Definition 2.2 Let $P(U)$be the fuzzy set of the domain U, $A_i \subset E$, and the two-group $(F_i, A_i)$is called fuzzy soft set defined on the universe U, where F_i is a mapping[14]:

$$F_i: A_i \to P(U) \tag{1}$$

2.2 Exclusive disjunctive soft set

Definition 2.3 (F,E) is a soft set on the domain U, x is an element of the domain U, e is an element of parameter set $E.\xi_x(e)$ represents the e-element feature function of soft set (F,E), defined as

$$\xi_x(e) = \begin{cases} 1, x \in F(e) \\ 0, x \notin F(e) \end{cases} \tag{2}$$

Definition 2.4 $(F,E)$is a soft set on the domain U,where F is the mapping, $F: B \to P(U)$, B is a non-empty parameter set, (F,B) is the exclusive disjunctive soft set, , if and only if (F,B) meets

(i) $\bigcup_{e \in B} F(e) = U$;

(ii)Supposing x is an element of the domain, if $\sum_{e \in B} \xi_x(e) > 1$, $\xi_x(e)$ can be interpreted as :

a)For each x is unknown, but $\sum_{e \in B} \xi_x(e = 1)$;

b)And x is called an indeterminate element of soft set(F,B)[15].

Both the bijective soft set and the exclusive disjunctive soft set are developed on the basis of soft set, the bijective soft set is a special exclusive disjunctive soft set. The bijective soft set doesn't contain uncertain elements, and exclusive disjunctive soft set can be used to describe an incomplete information system.

2.3 Exclusive disjunctive soft set decision system

Soft set theory as a new mathematical method to solve the problem of uncertainty obtains fast development, the paper introduces the exclusive disjunctive soft set, and proposes to systemize the incomplete information into the exclusive disjunctive soft set decision system, so that the exclusive disjunctive soft set can process the decision analysis based on the incomplete information system. In the process of quantification, it can make up for the information that may be missing, and comprehensively reflect the overall information of the system.

Definition 2.5 Supposing (F_i, E_i) $(i = 1,2,3,\dots,n)$ is n exclusive disjunctive soft sets defined on the domain U, where $E_i \cap E_j = \emptyset, (i = 1,2,3,\dots,n; j = 1,2,3,\dots n, i \neq j)$, $(G,B)$is the bijective soft set defined on the domain U, and$B \cap E_i = \emptyset(i = 1,2,3,\dots n)$, (G,B) are the soft set of decisions. Supposing$(F,E) = (\tilde{\cup}_{i=1}^{n}(F_i, E_i))$, the triplet $((F,E),(G,B),U)$ is called the exclusive disjunctive soft set decision system defined in the domain U. The dependence of the exclusive disjunctive soft set decision system is denoted by κ, which is defined as[15]:

$$\kappa = \gamma(\wedge_{i=1}^{n}(F_i, E_i), (G,B)) \tag{3}$$

3 Travel time model based on exclusive disjunctive soft set

3.1. Problem formulations

In the case of the same road network conditions of the transportation system, the intuitive factors affecting the travel time are mainly road capacity, traffic flow, vehicle speed and occupancy, including: time factors (general time, peak hours, etc.), weather conditions (sunny, cloudy, rainy, etc.), traffic events (road construction, traffic failure, traffic control, etc.), etc.. The travel time is often caused by one or more of these factors, because different factors have different degrees of influence on the travel time, and the same factor has different influences on the travel time on different occasions, the more important factors are extracted and the relatively minor factors are removed. Combined with actual traffic data and road traffic conditions, according to expert experience, data and natural language are transformed into a fuzzy vector map of the domain of travel time. The parameter reduction theory of soft set is used to eliminate the redundant information and preserve the mapping features that can't be removed [16].

According to the definition of soft set, firstly establishing a mapping relationship between the influence factors of road traffic (including qualitative and quantitative factors) and the travel time, so as to quantify the fuzzy factors [17]. A fuzzy soft set (F, X) on the domain $S = (S_1, S_2, \dots, S_n)$, S_i represents the power set of travel time under different traffic conditions. For the parameter set X, $X = X_1 \cup X_2 \cup \dots \cup X_i$, X_i represents the ith factor that affects travel time.

3.2 Construction of the travel time prediction model based on exclusive disjunctive soft set

On the basis of the above, giving the following definitions:

Definition 3.1 Universe $U = \{T_1, T_2, T_3\}$ indicates the formula mode of travel time under different traffic conditions. There is currently no uniform regulation on the division of road traffic states, but we think it is possible to satisfy the following possibility: when the number of vehicles is less than or equal to the maximum value to ensure that the vehicle is free to travel, the maximum value indicates the actual traffic capacity of the road section, the travel time is generally stable and small, and the travel time is denoted as T_1; when the number of vehicles exceeds the threshold, the threshold represents the traffic jam of the road section, the travel time gradually changes and increases, and the travel time is denoted as T_2; when there is severe congestion on the road, few vehicles can enter the network, and the travel time required is longer, the travel time is denoted as T_3. E indicates the parameter set, $E = \{E_1, E_2, \dots, E_i\}$, E_i represents the ith factor of affecting travel time. Here, the soft set (F, E) describes the prediction of the travel time.

The specific steps to establish the travel time prediction model are as follows:

Step 1 Considering the uncertain and fuzzy factors, according to the parameter reduction theory of soft set, and retaining the important influence factors, and eliminating the secondary factors.

Step 2 Constructing the exclusive disjunctive soft set decision system $\cup_{i=1}^{\sim n}(F_i, E_i), (G, B), U), (\mathrm{G}, \mathrm{B}) = (F_j, E_j)$.

Step 3 Establishing the membership function and the mapping relationship between the influence factors and travel time according to the decision rules, and then quantifying the fuzzy factors to calculate the dependence κ, $\kappa = \gamma(\wedge_{i=1}^{n}(F_i, E_i), (F_j, E_j))$ of the exclusive disjunctive soft set decision system $(\cup_{i=1}^{\sim n}(F_i, E_i), (F_j, F_j), U)$.

Step 4 Establishing the parameter form of the soft set between the travel time and the influence factors.

Step 5 According to the mapping relationship, predicting the travel time.

According to the mapping relationship, the following expression of the travel time is obtained, $P(U) = \left\{T_a^0, T_a^0\left(\frac{Q_a - C_a}{Z_a - C_a} + e^{\alpha\kappa^{\beta}}\right), T_a^0\left(\frac{Q_a - Z_{a+\varepsilon}}{\varepsilon} + e^{\alpha(\kappa+\delta)^{\beta}}\right)\right\}$, corresponding to T_1, T_2, T_3 of the domain U, which respectively indicate the prediction model of the travel time. For any of the above traffic conditions, the corresponding mapping relationship can be found in the power set.

According to the congestion state of the road, it is divided into three sections according to the probability of occurrence. Traffic states serves as a classification criteria. Because different influence factors can cause different traffic conditions, which decide how long the travel time will take. In general, when $\kappa \epsilon [0,0.3)$, the prediction model of the travel time is T_1, when $\kappa \epsilon [0.3,0.7)$, the prediction model of the travel time is T_2, when $\kappa \epsilon [0.7,1.0]$, the prediction model of the travel time is T_3.

Among them, T_a^0 indicates the time of travel on the free flow on the road section a, Q_a indicates the traffic flow on the road section a. C_a indicates the actual traffic capacity of the road section a, and the threshold value Z_a indicates the traffic jam of the road section a, these two parameters can be obtained by consulting relevant experts or by calculating the arithmetic mean value of the historical record of the section. κ denotes the dependence of the exclusive disjunctive soft set decision system. Usually, $\alpha = 0.15, \beta = 4.0, \varepsilon > 0, \delta > 0$, they are the constants.

4 Numerical example

4.1. Experimental data source

The data in this paper is derived from UK Highway Network, which can help find and use open government data developed by data.gov.uk, using a variety of sources to estimate traffic flow parameters such as travel time and speed, including automatic license plate recognition (ANPR) cameras, on-board global positioning systems (GPS), and induction circuits built into the road, including average traffic time, vehicle speed, road occupancy, traffic flow, etc. In order to facilitate data selection, the data is first preprocessed. This paper chooses to use the road segment (A1 northbound at a minor junction between A167 near Washington Birtley and A692 (120014301)) as the research object of this paper, and calculates the average travel time of the two stations. Collecting 10 consecutive days of data, the collection interval was 15 minutes, and a total of 500 sets of data were collected. Taking a certain section as the predicted section, the data collected in the first 9 days of the section is taken as the learning sample, and the data on the 10th day of the section is the test sample, and the newly established prediction model is used to predict the travel time.

According to the relevant information and expert experience, and parameter reduction theory of soft set [16-17]. This paper mainly considers objective factors (traffic flow, vehicle speed), weather conditions and traffic event as the main factors affecting travel time. E represents the factor parameter set that affects the travel time, $E = E_1 \cup E_2 \cup E_3 \cup E_4$, E_1 indicates the quality of the weather condition, E_1 = {sunny, cloudy, rainy}; E_2 indicates the magnitude of the traffic flow per unit lane, the parameter set of traffic flow are defined as large, general and small; E_3 indicates the magnitude of the average speed, the parameter set of average speed are defined as fast, general and slow; E_4 indicates the emergency of road traffic, E_4 = {road construction ,traffic failure, traffic control, normal}, among them, the traffic control refers to the control measures for vehicles and personnel in some or all traffic sections for certain safety reasons, such as road and bridge construction, etc.. The membership functions of each set are formulated as shown in Table 1, Table 2, Table 3, Table 4 [18].

Table 1. Membership function of weather condition.

Weather factor	Rainy	Cloudy	Sunny
Degree of membership	1.0	0.5	0

Table 2. Membership function of traffic flow.

Language value	Traffic flow (pcu)				
	240	480	720	960	1700
Large		0.1	0.7	0.9	1.0
General		0.6	1.0	0.6	
Small	1.0	0.8	0.2		

Table 3. Membership function of average speed.

Language value	Average speed (km/h)		
	60	75	90
Fast		0.8	1.0
General	0.5	0.7	0.4
Slow	0.7	0.2	

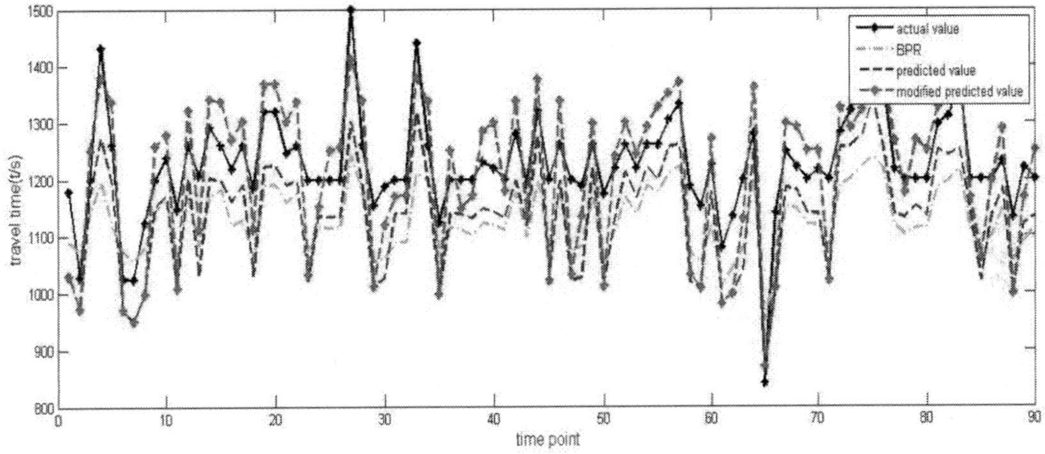

Figure 1. Comparison of results of three calculation models for travel time of road section.

Table 4. Membership function of emergency.

Emergency	Road construction	Traffic failure	Traffic control	Normal
Degree of membership	0.7	0.6	0.5	0

In summary, the dependence of the exclusive disjunctive soft set decision system was calculated. Through the mapping relationship [19] and the decision rule, the corresponding travel time prediction model was obtained from the interval range of the dependency degree. In order to verify the validity of the travel time prediction model established based on the exclusive disjunctive soft set theory, it was compared with the calculated value of the BPR function model. When only considering factors such as traffic flow and speed, comparing the prediction results of the new model and BPR function, and the result at this time was called the predicted value. When adding weather factors and emergencies of road traffic, comparing the prediction results of the prediction model, the result at this time was called the modified predicted value, and the results of the three calculation models of the road section travel time were compared as shown in Fig.1, in the figure, X-axis represented the number of samples tested and Y-axis represented the travel time, the unit was t/s.

4.2. Analysis of results

Three sets of prediction experiment results were output, and the evaluation indicators were used for data analysis. The indicators for the prediction results are selected as follows:

$$MAPE = \frac{\sum_{i=1}^{N} |\frac{T_p - T_a}{T_a}|}{N} \quad (4)$$

$$\mathrm{RMSPE} = \sqrt{\frac{\sum_{i=1}^{N} (\frac{T_p - T_a}{T_a})^2}{N}} \quad (5)$$

The two evaluation indicators are: MAPE (Mean Absolute Percent Error) is the average of the absolute error between the true value and the predicted value, which can reflect the actual situation of the prediction error; RMSPE (Root- Mean Squarer Phase Error) is the measure of the deviation between the predicted value and the observed value. Where T_p is the predicted value, T_a is the actual value, and N is the predicted number. Table 5 shows the results of the comparison of the indicators of the two optimization methods.

Table 5. Comparison of prediction error results.

	Modified predicted value	Predicted value	BPR
MAPE	0.0573	0.0704	0.0811
RMSPE	0.0046	0.0065	0.0075

From the comparison curve of travel time, the calculation results of the two models (BPR function model and travel time model based on exclusive disjunctive soft set) can better estimate the actual travel time. However, it is easy to know through the error analysis table: Under the same conditions, predicting the travel time based on the exclusive disjunctive soft set theory, MAPE and RMSPE of the experiment reduce by 0.0107 and 0.001 respectively. Taking into account the weather and the sudden situation of the actual traffic, the modified predicted model predicts better results than the original model, MAPE and RMSPE of the experiment reduce by 0.0131 and 0.0019, it can be seen that the accuracy of the prediction is obviously improved.

5 Conclusions

In this paper, the influence of ambiguity and uncertainty on travel time is studied based on traffic flow data, and using the exclusive disjunctive soft set decision system and soft set theory to establish the corresponding prediction model and verify it with actual data. The experimental results show that, compared with the BPR function, the model established in this paper can improve the accuracy of prediction. Certainly, the results of this paper are only a relatively preliminary study, and there are still many aspects that need to be discussed in depth. For example, when dealing with fuzzy factors, there is inevitably some subjectivity, which leads to some deviation of the prediction results, this requires multiple observations, consultation with multiple experts, and repeated correction of relevant data to improve the prediction process and results. For the influx of massive data, whether the calculation speed can meet the requirements of real-time applications in real projects, and so on.

Acknowledgements

The project is supported by the National Key R&D Program, China (Grant No. 2018YFB1600600).

Reference

1. R. More, A. Mugal, S. Rajgure, et al. Road traffic prediction and congestion control using artificial neural networks. International Conference on Computing, Analytics and Security. IEEE. 52-57, (2016).
2. U. Mori, A. Mendiburu, M. álvarez, et al. A review of travel time estimation and forecasting for Advanced Traveller Information Systems. Transportmetrica A: Transport Science. 11(2):119-157, (2015).
3. A. AllstrM, M.J. Ekstr, R.D. Gundleg, et al. Hybrid Approach for Short-Term Traffic State and Travel Time Prediction on Highways. Transportation Research Record: Journal of the Transportation Research Board. 2554:60-68, (2016).
4. H.B. Celikoglu. Flow-based freeway travel-time estimation: A comparative evaluation within dynamic path loading. IEEE Transactions on Intelligent Transportation Systems. 14(2):772-781, (2013).
5. R.X. Zhong, J.C. Luo, et al. An Online Travel Time Prediction Model Based on Functional Data Analysis. Transportation Research Board 96th Annual Meeting. (2017).
6. S. Tak, S. Kim, K. Jang, et al. Real-Time Travel Time Prediction Using Multi-Level k-Nearest Neighbor Algorithm and Data Fusion Method. 2014 International Conference on Computing in Civil and Building Engineering. (2014).
7. Z.Y. Lu, Q. Meng, G. Gomes, et al. Estimating link travel time functions for heterogeneous traffic flows on freeways. Journal of Advanced Transportation. 50(8), 1683-1698, (2016).
8. J.N. Chen, X. Zhang, S.R. Zhang. Analysis and empirical study of highway travel time interval prediction based on Bootstrap-KNN. Control and Decision. 33(11):2080-2086, (2018).
9. X. Wang, X.H. Chen, X.M. Yang. Short-term travel time prediction of expressway based on K nearest neighbor algorithm. China Journal of Highway and Transport. 28(01), 102-111, (2015).
10. X. Zhi, W.J. Chen, L.L. Li. A method based on interval-valued fuzzy soft set for multi-attribute group decision-making problems under uncertain environment. Knowledge and Information Systems. 34(03), 653–669, (2013).
11. X. Zhi, K. Gong, Y. Zou. A combined forecasting approach based on fuzzy soft sets. Journal of Computational and Applied Mathematics. 228 (1):326–333, (2009).
12. Y.C. Shao. An application of fuzzy rough sets in predicting on urban traffic congestion. 14th International Symposium on Distributed Computing & Applications for Business, Engineering and Science. IEEE. 208-211, (2016).
13. D. Molodtsov. Soft set theory–first results. Computers and Mathematics with Applications. 37(4-5), 19–31, (1999).
14. P.K. Maji, R. Biswas, A.R. Roy. Soft set theory. Computers and Mathematics with Applications. 45(4-5), 555-562, (2003).
15. X. Zhi, K. Gong, et al. Exclusive disjunctive soft sets. Computers & Mathematics with Applications. 59(6), 2128-2137, (2010).
16. S. Danjuma, M.A. Ismail, T. Herawan. An alternative approach to normal parameter reduction algorithm for soft set theory. IEEE Access.5, 4732-4746, (2017).
17. K. Gong, P.P. Wang, X. Zhi. Bijective soft set decision system based parameters reduction under fuzzy environments. Applied Mathematical Modelling. 37(6), 4474-4485, (2013).
18. R. Li, C. Jiang, F. Zhu, et al. Traffic Flow Data Forecasting Based on Interval Type-2 Fuzzy Sets Theory. IEEE/CAA Journal of Automatica Sinica. 3(2), 141-148, (2016).
19. H.L. Yang, Z.L. Guo. Kernels and closures of soft set relations, and soft set relation mappings. Computers and Mathematics with Applications. 61(3), 651-662, (2011).

MATEC Web of Conferences **308**, 03001 (2020) https://doi.org/10.1051/matecconf/202030803001
ICTTE 2019

Research on Liquefied Natural Gas(LNG) inland river transportation security risk assessment and security technology based on risk management

Hu Yuchang[1,a] and Xu Liansheng[1]

[1] *China Waterborne Transport Research Institute, Beijing, China*

Abstract. Using the Analytic Hierarchy Process and taking a certain section of the lower reaches of the Yangtze River as an example, this paper analyzed and identified the risk of inland river transportation of LNG, established the risk assessment criteria. According to the assessment results, it can be concluded that the current security risk situation of inland river LNG transportation is between "general" and "higher", and puts forward corresponding suggestions to improve security.

1 Introduction

In recent years, with the accelerated implementation of China's new strategy for supporting the Yangtze River's economic belt and the coordinated development of Beijing, Tianjin and Hebei, China has increased the spatial layout of Reform and Opening-up and accelerated the coordinated development of regional economy. As the coordinated development zone of China's East-West interaction and cooperation and the pioneer demonstration zone of ecological civilization construction, the new support belt of the Yangtze River economy, led by ecological priority and green development, is an inland river economic belt with global influence. General Secretary Xi Jinping pointed out that the development of the Yangtze River Economic Zone is a major strategy related to the overall development of the country and is of great significance to the realization of the goal of "two hundred years" and the realization of the Chinese dream of the great rejuvenation of the Chinese nation.

The Yangtze River covers nine provinces and two cities, with the population and total economic volume exceeding 40% of the whole country. In recent years, with the rapid economic development, the demand for energy is also growing rapidly. Under the high pressure of ecological priority, green development requirements and environmental protection, the demand for LNG and other clean energy in the central and Western regions of China is growing rapidly. LNG is a clean energy source. It is of great significance to optimize the energy structure, improve the ecological environment and achieve sustainable development by using LNG.

Inland waterway operation is a sustainable mode of transportation and an important part of China's modern comprehensive transport system, accounting for nearly 40% of China's water transport. It is a convenient way to transport LNG from coastal receiving stations to demand terminals in central and western regions by inland waterway transportation. Because of its high risk of fire and explosion, the domestic understanding of its inland waterway transport safety is insufficient, and the lack of

[a] Corresponding author: hycray@163.com

relevant safety management standards for LNG from inland waterway transport, the current standard of inland waterway transport LNG also has many conditions, which restricts the development of LNG inland waterway transport and affects the promotion and application of LNG clean energy.

Previously, for the safety risk assessment of LNG transportation, domestic research mostly focused on navigation conditions, LNG ship operation and risk identification. Gan Haoliang [1] considered that the safety zone of LNG ships in actual navigation should vary according to the characteristics of LNG ships, the conditions of port waters and the changes of traffic flow density. The safety zone of LNG ships should be divided into delimited safety zone and variable safety zone. Wen Yuanqiao et al. [2] proposed a quantitative method to calculate the width of LNG vessel's berthing safety zone. The results show that the width of LNG vessel's berthing safety zone is closely related to the size of LNG vessel, the size of vessel in adjacent waters and the speed of navigation. Wang Chaomin [3] quantitative studied the Adaptability of LNG Inbound and Outbound Channels Using Fuzzy Comprehensive Evaluation Model of Aggregation Method. Guo Weizhi [4] uses ship maneuvering simulator to obtain track distribution records of large LNG ships, and designs a saving channel width algorithm for large LNG ships. The evaluation results show that the scheme not only meets the safety requirements, but also saves engineering funds. Hu Huiguo et al. [5] constructed an optimization method of LNG ship entry and exit organization scheme based on lattice order decision theory. Zhang Xiaodong et al. [6] proposed a queuing model considering non-simultaneous and simultaneous interference types to simulate the passage status of waterway vessels and effectively evaluate the safety and efficiency of LNG ships entering and leaving the port. Li Hongliang et al. [7] calculated the influence of LNG ships on the main channel capacity of Gaolan Port by using the navigation impact rate model and the channel occupancy rate model respectively. Wang Hairong et al. [8] used the method of fuzzy analytic hierarchy process to determine the weight vector and the fuzzy evaluation matrix, calculated the decision vector, and realized the quantitative risk identification of LNG receiving station. Wang Qingfeng et al. [9] quantitative calculated and analyzed of typical leakage and diffusion of inland river thin film LNG carrier and the consequences of fire and explosion accidents by using PHAST risk analysis software, considering the actual situation of inland river waters in China. Dou Xu et al [10] carried out numerical study on structural damage caused by collision of small LNG ships during inland navigation. Wang Kan et al. [11] proposed an optimization fireball model by introducing the atmospheric transmission rate τ into the original TNO dynamic model.

MOLEN et al.[12] verified the correctness of TERMISM model in 6-DOF model by measuring actual Mooring Loads of LNG ships under different conditions. At the same time, calculate and analyze the swaying motion and mooring force of ships under surge action. HIGHTOWER et al. [13] analyzed the risk of LNG overflow and diffusion caused by ship accidental damage and intentional damage. The results show that within the range of 500m overflow, it will have a greater impact on the safety of personnel and public property, but little impact when it is larger than 1600m. JOBNSON et al. [14] used the LNG leakage consequence analysis model proposed by the Federal Energy Regulatory Commission to calculate the dangerous area by estimating the leakage rate, diffusion rate, evaporation rate in open water and thermal radiation. HISSONG et al. [15] pointed out that the leak of LNG in water is more complicated than that of storage tanks on land. The key is to estimate the size of liquid pools. They used the method of fuel and energy balance in establishing evaporation and LNG liquid pool models, taking into account multiple LNG characteristic parameters and their changes, as well as the influence of mixing with water. It makes the model more real and effective. KIM et al. [16] analyzed the potential hazards causing the leakage of thin film LNG vessels at the LNG receiving station in Pingze Port, Korea, and quantitatively evaluated the overall leakage accident rate using fault tree analysis method. ELSAYED [17] used Mamdani and Sueno fuzzy reasoning methods to evaluate the risk of LNG ship during loading and unloading. BUBBICO et al. [18] preliminarily assessed the risk of LNG vessel receiving terminal in Panigaglia, Italy. The results show that LNG spillover caused by accident can be basically eliminated, and intentional damage will lead to a large number of LNG leaks causing pool fires, but the impact on the residential population can be neglected. NWAOHA et al. [19] used risk matrix method to identify the main hazards in LNG ship maneuvering process, in which the transmission arm LNG overflow and control system failure have the highest hazard level. Further, through fault tree analysis of its causes and fault logic, the method of fuzzy logic reasoning was used to determine its safety level. Kim [20] analyzed the failure behavior of LNG sealed cargo hold under impact.

2 LNG transportation risk analysis and identification

2.1 Risk analysis of LNG ships

Chinese "Statistical Measures for Marine Traffic Accidents" will be included in the statistics of water traffic accidents into collision accidents, grounding accidents, reef-striking accidents, damage accidents, wave damage accidents, fire and explosion accidents, wind accidents, self-sinking accidents and other water traffic accidents causing casualties, direct economic losses and so on.

Inland river waters are characterized by complex channel conditions, small channel width, more shoals, more bends, dramatic changes in water flow, more bridges, and high traffic density. There are a large number of ships on inland waterways, with dense wharf layout and frequent operation of ships entering and leaving the wharf, which results in the risk of ship traffic accidents. The risk of inland waterway transportation of LNG ships depends on many factors, among which the most important are: ship type, physical and chemical characteristics of dangerous substances, characteristics of waterway, etc. Natural factors such as wind, cyclone and fog are also important factors causing ship collision, self-sinking, grounding and other traffic accidents. At the same time, LNG is a flammable and explosive dangerous cargo, which has occurred. Accident risks such as fire and explosion of live ships.

Therefore, the main types of accidents that may occur in inland LNG ship transportation are grounding, collision between ship and ship, collision between ship and wharf or bridge, self-sinking of ship, fire and explosion of ship, etc.

2.2 Risk identification of LNG ships

Explosion or leakage is common in LNG waterway transportation. Explosion or leakage has a great destructive effect, and accompanied by air and water pollution, it brings great economic losses to society. For the risk factors of goods in LNG transportation, the hazards of goods and the possible environmental hazards caused by accidents should be considered.

The factors affecting the navigation environment of inland LNG vessels can be roughly divided into four aspects: natural environmental factors, channel factors, traffic factors and management factors. These factors interact with each other and are complex and changeable. Therefore, it is necessary to accurately grasp the changing law of each factor, accurately and timely locate the position and draw up the correct route in order to ensure the safety of ship navigation system. When analyzing the navigation environment of inland LNG ships, the following aspects should be considered. Channel characteristics include channel curvature, channel intersection angle scale, channel obstruction, etc. Meteorological and hydrological characteristics including wind, current, visibility, etc. Navigation aids including natural and manual navigation aids. Ship trends in waterways include ship types, sailing time, dynamic characteristics, rules of motion, etc. Meteorological characteristics include the characteristics of meteorological changes in navigational waters and their impacts on navigation, etc.

In the actual evaluation, the number of evaluation factors and the accuracy of evaluation are not directly related, so we should pay attention to the role of evaluation factors in the evaluation. The basic principle of determining evaluation factors is to reasonably compose evaluation factor set with as few main factors as possible.

3 Establishment of risk assessment indicators for LNG transport accidents

3.1 Establishment of evaluation index system

After investigation and statistics, the evaluation factors of navigation environment of inland LNG ships are divided hierarchically, and the indexes affecting navigation safety risk assessment of inland LNG ships are determined as follows: wind, visibility, flow, channel width, channel curvature, channel intersection angle, obstacles, traffic flow, special operations, traffic management and navigation assistance. Set the first level of evaluation, evaluation set **U**={μ_1, μ_2, μ_3, …, μ_m}, to construct the first level risk factors set, **U**={hydrological conditions μ_1, channel μ_2, traffic μ_3, management μ_4}. Similarly, the second level risk factors set μ_1= {flow, visibility, wind}, μ_2= {channel width, channel curvature, channel crossing angle, obstacles}, μ_3= {traffic volumes, special operations}, μ_4= {traffic service, aids of navigation}.

In the analysis of related factors, the evaluation criteria are divided into five levels according to the attributes of the target layer. The order from low to high is low, lower, general, higher and high. On the basis of determining the index system of inland LNG ship transportation safety risk assessment, here takes a certain section of the lower reaches of the Yangtze River as an example to determine the safety evaluation criteria by field analysis and consultation with relevant experts. The following evaluation criteria are obtained, as shown in Table 1.

Table 1. Risk classification standard of navigation safety assessment indicators for LNG ships

Risk	Low	Lower	General	Higher	High
Days of poor visibility (days/years)	0-15	15-20	20-30	30-40	>40
Wind speed (m/s)	0-5	5-10	10-15	15-25	>25
Velocity of flow (m/s)	0-0.4	0.4-1.2	1.2-2.2	2.2-3.2	>3.2
Width (m)	0-0.3	0.3-0.5	0.5-0.8	0.8-1.0	>1.0
Maximum turning angle (degree)	<15	15-30	30-45	45-60	>60
Crossing angle (degree)	<20	20-45	45-60	60-70	>70
Distance from Obstacles (m)	>200	100-200	50-100	20-50	<20
Traffic volume (ship/day)	0-10	10-15	15-25	25-70	>70
Special operations (%)	0-10	10-16	16-30	30-60	>60
Transportation services (%)	95-100	95-90	80-90	70-80	<70
Improvement rate of navigation aids (%)	90-100	80-90	65-80	50-65	<50

3.2 Determining the weight of risk factors

The results of expert consultation collected are sorted out and the weights of factors at different levels are calculated respectively. Following is an example of the first level risk factors set. By calculating the first level of risk factors, the second level of risk factors can be obtained, as shown in Table 2.

Table 2. Comparison and judgment of the first level risk factors

	Hydrological conditions	**Channel**	**Traffic**	**Management**
Hydrological conditions	1	0.922	0.983	2.793
Channel	1.085	1	0.864	2.148
Traffic	1.017	1.157	1	3.113
Management	0.358	0.466	0.321	1

From the Table 2, we can get the judgment matrix **A**.

$$\mathbf{A}=\begin{pmatrix} 1 & 0.922 & 0.983 & 2.793 \\ 1.085 & 1 & 0.864 & 2.148 \\ 1.017 & 1.157 & 1 & 3.113 \\ 0.358 & 0.466 & 0.321 & 1 \end{pmatrix}$$

Processing the above matrix, the weights of $\mathbf{u}_1$, $\mathbf{u}_2$ and $\mathbf{u}_3$ are obtained respectively. $\mathbf{W}=(\mathbf{W}_1\ \mathbf{W}_2\ \mathbf{W}_3\ \mathbf{W}_4)=(0.292\ 0.276\ \ 0.320\ \ 0.112)$

Consistency check CR=0.0052<0.90

Therefore, the consistency checks of the judgment matrix of the first level risk factors set can be judged, and the weights of the factors u_1, u_2 and u_3 can be obtained, as shown in Table 3.

Table 3. Weight of the second level risk factors

The first level factors	**The second level factors**	**Weights**
Hydrological conditions $\mathbf{u}_1$	Visibility (u_{11})	0.333
	Wind (u_{12})	0.356
	Flow (u_{13})	0.311
Channel $\mathbf{u}_2$	Width (u_{21})	0.303
	Curvature (u_{22})	0.248
	Crossing angle (u_{23})	0.236
	Obstacles (u_{24})	0.213
Traffic $\mathbf{u}_3$	Traffic volume (u_{31})	0.599
	Special operations (u_{32})	0.401
Management $\mathbf{u}_4$	Transportation services (u_{41})	0.599
	Navigation aids (u_{42})	0.401

3.3 Recognition of calculation results

Taking the navigation safety index system of LNG ships in a certain section of the lower reaches of the Yangtze River as an object of study, consulting relevant experts and professionals to determine the value of the evaluation index, the evaluation index affecting the navigation safety of LNG ships in this section can be obtained, as shown in Table 4.

Table 4. Measurement and estimation of navigation safety assessment indicators for LNG ships

Evaluation index	Definition	Value
Visibility	Days of poor annual visibility (days/years)	29
Wind	Wind speed (m/s)	14.8
Flow	Maximum velocity of flow (m/s)	1.73
Width	Average length of vessel allowed to navigate/narrowest channel width (%)	1.21
Curvature	Maximum channel bending angle (degree)	29
Crossing	Maximum crossing angle (degree) between channel and sub channel	50
Obstacles	Distance (m) from the nearest obstruction to the channel	180
Traffic Volume	Average annual throughput (weighted traffic volume) (number of vessels/day)	41.7
Special operations	Annual vessel traffic volume/total traffic volume (non-weighted) (%) for special operations	8
Transportation services	Including VTS, VHF, cruise boats and other equipment rate (%)	75
Navigation aids	Number of channels with perfect navigation aid information/total channel number (%)	85

The detailed steps for the second-level fuzzy comprehensive evaluation of LNG ships in a certain section of the lower reaches of the Yangtze River are as follows:

$$\mathbf{B}_1 = \mathbf{W}_1^{(2)} \bullet \mathbf{R}_1 = (w_{11}, w_{12}, w_{13}) \bullet \begin{pmatrix} r_{11} & r_{12} & r_{13} & r_{14} & r_{15} \\ r_{21} & r_{22} & r_{23} & r_{24} & r_{25} \\ r_{31} & r_{32} & r_{33} & r_{34} & r_{35} \end{pmatrix}$$

$$= (0.317 \quad 0.428 \quad 0.255) \bullet \begin{pmatrix} 0 & 0 & 0.567 & 0.433 & 0 \\ 0 & 0 & 0.54 & 0.46 & 0 \\ 0 & 0 & 0.97 & 0.03 & 0 \end{pmatrix}$$

$$= (0 \quad 0 \quad 0.6582 \quad 0.3418 \quad 0)$$

Similarly, the evaluation vectors **B**2、**B**3、**B**4 can also be obtained for the channel condition.

$$\mathbf{B}_2 = (0 \quad 0 \quad 0.1677 \quad 0.5946 \quad 0.2376)$$
$$\mathbf{B}_3 = (0.1013 \quad 0.2057 \quad 0 \quad 0.6078 \quad 0.0852)$$
$$\mathbf{B}_4 = (0 \quad 0 \quad 0.2868 \quad 0.7132 \quad 0)$$

Therefore, the final evaluation **B** of navigation environment safety can be obtained by the results of the second level evaluation.

$$\mathbf{B} = \mathbf{W} \bullet \mathbf{R} = (w_1 \quad w_2 \quad w_3 \quad w_4) \bullet \begin{pmatrix} \mathbf{B}_1 \\ \mathbf{B}_2 \\ \mathbf{B}_3 \\ \mathbf{B}_4 \end{pmatrix}$$

$$= (0.0324 \quad 0.0658 \quad 0.2706 \quad 0.5383 \quad 0.0928)$$

3.4 Anti-fuzzy of final evaluation vector

$$M = \frac{\sum_i^n b(u_i) * u_i}{\sum_i^n b(u_i)}$$

$$= (0.0324 \quad 0.0658 \ 0.2706 \ 0.5383 \ 0.0928) \bullet (-2 \ -1 \ 0 \ 1 \ 2)^T / (0.0324 + 0.0658 + 0.2706 + 0.5383 + 0.0928)$$

$$= 0.5933$$

According to the preceding contents, we can know that the membership degree of the first level risk factors is the evaluation vector calculated for the second level risk factors. Based on this, we can get the navigation safety risk status of inland LNG ships. According to the results of anti-fuzzy, it can be concluded that the current inland LNG shipping safety risk situation is between "general" and "higher".

4 Safety guarantee measures of LNG inland river transportation based on risk management

4.1 Legal safeguard

Firstly, we should speed up the formulation of LNG shipping and operation regulations by referring to the relevant LPG specifications. Secondly, we should speed up the research and formulation of LNG inland river standard ship form, and strive to include it in the National Outline for the Development of Inland River Standardized Ship Form.

4.2 Personnel safeguard

Workers involved in LNG loading and unloading operations should be trained in LNG professional knowledge. They should be qualified by the competent transport authorities (crew members are qualified by the maritime administration), and only after obtaining the qualification certificate, can they work on the job. Operating units must organize employees to learn laws and regulations and raise their legal awareness. At the same time, enterprises should organize practitioners to learn according to the specific conditions of their departments, so that they can master the knowledge of LNG risks and emergency measures when accidents occur, and specific requirements of safe handling, so as to improve the professional quality of practitioners.

4.3 Ships safeguard

According to the "IGC Rules" and specifications, the structure, equipment, accessories, devices and materials of LNG carriers shall be subject to initial inspection, periodic inspection, interim inspection, annual inspection and additional inspection. At the same time, it is necessary to establish LNG ship fault database as soon as possible. Ship Inspection Departments and Maritime Departments should strengthen LNG Ship Inspection. Shipping companies should scientifically manage LNG ships under their jurisdiction. The crew on board should have a good grasp of the safety situation of the equipment on board and the maintenance work. Especially they should focus on the inspection and maintenance of the equipment.

4.4 Security management

In view of the unsafe factors existing in port LNG operation, it is necessary to speed up the training of professional personnel of maritime authorities, formulate safety measures for LNG ships entering and leaving the port, strengthen ship safety inspection, and promote port and shipping enterprises to establish a standardized system of safety and quality. At the same time, the owners, operators or managers of LNG ships and the personnel on board should attach importance to the safety management of LNG ships and strictly implement ISM rules. It is necessary to draw up corresponding safety operation rules according to the regulations of LNG ship management in western countries and the safety operation guidelines provided by equipment manufacturers. Establishing a scientific, systematic and procedural safety management system to ensure the institutionalization and standardization of maintenance, maintenance and troubleshooting.

4.5 Emergency plans

In accordance with the requirements of the guidelines for the preparation of emergency plans for production security accidents in production and operation units, corresponding emergency rescue plans are formulated to form an emergency plan system, which is connected with the emergency plans of local people's governments and relevant departments. Emergency plan can be divided into three levels: general plan, special plan and on-site disposal plan. The general plan expounds the emergency policy, emergency organization structure and corresponding responsibilities of the plan as a whole, as well as the general idea of emergency action. Special plans are aimed at specific and specific types of emergencies, such as emergency plans for flood control and prevention, emergency plans for environmental emergencies, etc. The on-site plan is based on the special plan and is compiled according to the specific situation of the site. It is a plan for a specific site (i.e. the site is the target), usually for a place, device or important protective area where the accident risk is high.

Examples of on-site plans to be developed are as follows:

Emergency disposal plan for ship leakage, fire and explosion. Emergency Plan for Major Safety Accidents of Special Equipment. Emergency Disposal Plan for Personnel Asphyxia Accident.

4.6 Exercise of emergency plans

According to the actual situation, regular drills of accident emergency plan will be held. Through the drill, we can find the defect of the plan procedure, improve the coordination among the emergency departments, agencies and personnel, enhance the public's confidence and emergency awareness in dealing with emergency rescue, improve the proficiency and technical level of emergency personnel, further clarify their positions and responsibilities, and improve the coordination among departments. Improving the overall emergency response capability.

5 Conclusion

By analyzing and evaluating the risk of LNG transportation, the risk assessment index of LNG accident is established, and the safety safeguard measures of LNG inland waterway transportation are put forwarded. The main conclusions are as follows.

The factors affecting the navigation environment of inland LNG vessels can be divided into four aspects: natural environment factors, channel factors, traffic factors and management factors. These factors are interrelated, interactional, complex and changeable. Only by accurately grasping the changing law of each factor, accurately locating in time and drawing up the correct route can the navigation system be guaranteed.

Accidents such as collision, collision, damage, wave damage, grounding, fire, explosion and sinking may occur during inland navigation of small LNG vessels. Relatively speaking, LNG leakage accident, fire and explosion accident and ship damage accident are the main dangers of LNG inland navigation. The main causes of the above-mentioned accidents are unsafe human behavior, quality defects or faults of equipment and facilities, other external adverse factors, improper operation or mismatch of ships, and poor hydrological and meteorological conditions.

Indicators affecting navigation safety risk assessment of inland LNG vessels are wind, visibility, flow, channel width, channel curvature, channel intersection angle, obstacles, traffic flow, special operations, traffic management and navigation information perfection. Applying the analytic hierarchy process to determine the evaluation index system, the risk assessment index of LNG transport accident is established. Taking a certain section of the lower reaches of the Yangtze River as an example, the navigation safety risk situation of inland LNG ships is evaluated and analyzed.

Risk identification, risk assessment and ship operation simulation experiments are the main technical means of LNG ship safety management. Based on risk identification and risk assessment of LNG ship transportation, here puts forward safety assurance measures of LNG ship transportation based on risk management from aspects of personnel, ships, safety management, laws and regulations, and emergency plans.

References

1. H. L. Gan, WHUT (2012)
2. Y. Q. Wen, L. Wang, X. Yang, C. S. Xiao, C. H. Zhou, CSSJ, **24**, 5 (2014)
3. C. M. Wang, WHUT (2011)
4. W. Z, Gou, WHUT (2011)
5. H. G. Wu, C. J. Liu, Y. J. Xiao, Journal of Shanghai Maritime University, **33**, 2 (2012)
6. X. D. Zhang, H. Zhang, W. J. Chen, Y. J. Xiao. Journal of Shanghai Maritime University, **35**, 1 (2012)
7. H. L. Li, C. B. Yang, Port & Waterway Engineering, 6, 85-88+113 (2013)
8. H. R. Wang, X. Q. Ma, CSSJ, 3, 131-135+177, (2007)
9. Q. F. Wang, X. Y. Cui, Journal of Jiangsu University of Science and Technology (Natural Science Edition), **33**, 1 (2019)
10. X. Dou, W. Q. Wu, M. Song, Navigation of China, **40**, 2 (2017)
11. K. Wang, Y. R. He, Z. Y. Liu, X. M. Qian, J. Hazard. Mater, 366 (2018)
12. W. Molen, H. Ligteringen, J. C. Lem, et al. J. Waterw. Port Coast. Ocean Eng, **129**, 1 (2003)
13. M. Hightower, L. Gritzo, A. Luketa-Hanlin, Process. Saf. Prog. **24**, 3 (2005)
14. D. W. Johnson, J. B. Cornwell, J. Hazard. Mater.**140**, 3 (2007)
15. D. W. Hissong, J. Hazard. Mater. **140**, 3 (2007)
16. H. Kim, J. S. Koh, Y. Kim, THEOFANOUS T G, Korean J. Chem. Eng. **22**, 1 (2005)
17. T. Elsayed, Appl. Ocean Res. **31**, 3 (2009)
18. R. Bubbico, S. D. Cave, B. Mazzarotta, J. Loss Prev. Process Ind. **22**, 5 (2009)
19. T. C. Nwaoha, Z. Yang, J. Wang, et al, Ocean Eng. **72**, 11 (2013)
20. M. S. Kim, S. B. Kwon, S. K. Kim, Constr. Steel. Res. 156 (2019)

 https://doi.org/10.1051/matecconf/202030803002

Identification of interventions for better child road safety in Jordan

Khair Jadaan[1,a], Chika Sakashita[2], and Ruba Hikmat *[3]

[1] *Professor, Civil Engineering Department, University of Jordan, Amman, Jordan*

[2] *International Consultant, Global Road Safety Solutions Pty Ltd, Sydney, Australia.*

[3] *Child Protection Specialist, UNICEF, Amman, Jordan*

Abstract. Children of Jordan below 18 years are particularly at risk and their proportion of all casualties is rising reaching about 28% in 2016, yet limited research has been carried out to address this problem. This study investigates the magnitude and nature of the child road safety problem with the final aim of identifying the necessary interventions to improve child road safety in Jordan. Road crash death data were provided by the Jordan Traffic Institute. Additional data were sought through interviews with the Jordanian Government and other non-government stakeholders such as the UNICEF Makani Centres (community centres for children). A seat belt observational survey was also conducted. All the data were synthesised to provide a much better understanding of the magnitude and nature of the child road safety issue in Jordan. Potential interventions for Jordan to improve its child road safety situation are provided based on the present data analyse.

1 Background

In the most recent World Health Organization (WHO) Global Status Report on Road Safety [1] the rate of road deaths per 100,000 population in Jordan was estimated at 24.4, making it one of the worst road safety performing countries in the world. Pedestrians and passengers make up the majority (62.3%) of the road deaths in Jordan with most of those pedestrian and passenger deaths suffered by children under the age 18 [1].

WHO reports traffic injuries are the leading cause of death for children between 5-14 years and young adults between 15-29 [1]. The problem is more acute in developing than developed countries. The results of a study in Bangladesh reported that one third of the pedestrians dying in road crashes were under the age of 16 years, leading to the development of a child road safety strategy to reduce child injuries and fatalities [2]. In India, research revealed that road crashes accounted for 55% of all child accidental death and in almost all of these, the unsafe behavior of child was considered to be at fault [3]. In Malaysia, road traffic injuries among children aged 0–19 years account for about 35 % of total admission to public hospitals [4].

Jordan is a lower middle-income country that has been experiencing persistent growth in population, economy, and motorization level. The annual population growth rate over the past few years is about 6% and children below 18 years of age constitute about 45% of the total population of about 9.9 million [5].

Along this growth, Jordan has also experienced a high level of traffic crashes. There were about 16,000 fatalities from all ages between the years 1989 and 2016, while children below the age of 18years constitute about 26% of total casualties [6] with this proportion found to be increasing in recent years [7]. The lack of reliable data is a serious problem in most of developing countries, including Jordan. Therefore, it is quite reasonable to assume that the incidence of crashes is much larger than actually reported. Clearly, urgent actions are needed to improve the child road safety situation in Jordan.

A review of the literature identified many of the road safety studies in Jordan were general analyses of road

[a] Corresponding author: kjadaan@gmail.com

crash data reported by the Jordan Traffic Institute (JTI). Past studies on pedestrian crashes have highlighted the overrepresentation of children in pedestrian casualties [8,9]. A hospital-based study on childhood injuries in Jordan suggested that school aged children suffered the greatest number of injuries and sustained the injuries most commonly at school followed by on roads and playgrounds [10]. Overall, research literature on road crash casualties specifically involving children in Jordan is limited.

This comprehensive study, the first of its type in Jordan, aims to determine the magnitude of and potential factors contributing to road related causalities among children and identify key opportunities to improve child road safety in Jordan.

2 Methods

This study was concerned with children under the age of 18. It was conducted in 2017. Road crash death data were obtained from the Jordan Traffic Institute. Data were also sought through interviews with the Jordanian Government and other non-government stakeholders such as parents and teachers at select public schools and UNICEF Makani Centres (community centres for children). Approval of interviews with parents and teachers at select public schools was granted from the Ministry of Education. All those interviewed provided consent to participate. The government stakeholders included the Jordan Traffic Institute, Central Traffic Directorate, Highway Patrol Directorate, Ministry of Public Works and Housing, Ministry of Municipalities, Ministry of Interior, Ministry of Education, Ministry of Transport and Greater Amman Municipality. A brief observational survey of seat belt wearing was also conducted. All the data were synthesised to provide a better understanding of the magnitude and possible risk factors of child road deaths and injuries in Jordan.

3 Results

3.1 Magnitude of child road deaths in Jordan

The crash data from the Jordan Traffic Institute provide a picture of magnitude of the child road safety problem in Jordan. In the 10 years between 2007 and 2016, Jordan lost 2,060

children under 18. The 10-year data were analysed to provide annual average estimates of deaths by age group and Governate (Table 1) and by age group and road user group (Table 2).

The death data by age and Governate show that the three most populous Governate in Jordan – Amman, Irbid and Zarqa – have suffered the greatest number of child road deaths between 2007-2015 (Table 1). The data also show that all age groups under 18 are at risk of road deaths in Jordan, but younger children suffer a greater proportion of all deaths under 18 – in increasing order from ages 12-17 (24%), 6-11 (30%) and 0-5 (46%).

Table 1. Annual average (rounded to closest whole number) number of deaths by age group and Governate in Jordan

Age Group	0 to 2	3 to 5	6 to 8	9 to 11	12 to 14	15 to 17	Total under 18
Amman	18	22	13	13	10	11	87
Irbid	8	10	6	6	4	5	39
Zarqa	6	7	4	4	3	4	28
Balqa	2	3	2	1	1	1	10
Karak	1	2	1	1	1	1	7
Mafraq	3	3	2	2	1	1	12
Ma'an	0	1	0	0	0	0	1
Aqaba	1	1	1	1	0	0	4
Madaba	1	1	1	1	0	0	4
Tafilah	0	1	0	0	0	0	1
Jerash	1	1	1	1	0	1	5
Ajloun	1	1	1	0	0	0	3
Total	42	53	32	30	20	24	201

The child death data by age and road users (pedestrians, drivers and passengers) show that children under 18 are at particularly significant risk as pedestrians (55% of all crash deaths under 18) and vehicle passengers (40% of all crash deaths under 18), making up 95% of the road crash deaths suffered by children under 18 in the last 10 years (Table 2). This is in stark contrast to pedestrians (25%) and passengers (31%) making up 56% of deaths in adults aged 18 and over in Jordan and pedestrians (38%) as well as passengers (36%) making up 74% of child road deaths worldwide [11]. Again, younger children suffer a greater proportion of pedestrian and vehicle passenger deaths – in increasing order from ages 12-17 (21%), 6-11 (30%) and 0-5 (48%).

Table 2. Annual average (rounded to closest whole number) number of deaths among children under 18 by age and road user group

Age Group	Pedestrians	Drivers	Passengers	Total
0 to 2	19	2	25	46
3 to 5	37	0	14	51
6 to 8	21	0	10	31
9 to 11	16	1	12	29
12 to 14	11	2	11	24
15 to 17	10	6	11	27
Total under 18	114	11	83	208
Percentage by road users	55%	5%	40%	100%

It is also notable that children aged 0-2 and 9-17 were also killed as drivers in the past 10 years even though the legal licensing age is 18. When consulted, the JTI explained that some children, especially between the ages 15 - 17 take the car without the knowledge of the parents and some children under the age 15 have been found in the driver seat at the time of the crash. The JTI suspect that the child moved a parked car. It is also not uncommon observation of children being nursed by an adult driver in Jordan, this may also have led to a child being killed in the driver's seat and being mis-recorded as the driver.

3.2 Possible factors contributing to child road deaths and injuries

The seat belt observational survey and the interviews with the Government stakeholders, parents and teachers of school children at select public schools and UNICEF Makani centres identified possible factors contributing to road crash deaths and injuries of children in Jordan.

The causes of casualty deaths identified by the Jordan Traffic Institute (JTI) were all related to human factors (e.g. Tail Gating, Not Taking the Necessary Precautions While Driving, Using Incorrect Lane, Priorities false, Reversing Incorrect, Failing to Comply with Obligatory signs, Speed Limit Exceeding, Loss of Control While Driving, Incorrect Bending and Turning, Disregarding A Traffic Light Signal, Driving Opposite To Traffic Direction, Priority to Disallowing Pedestrian, Others). A few of these causes relate to excessive speed(e.g. Tailgating, Exceeding Speed Limit, Loss of Control). High travel speeds, speeding and non-compliance with traffic lights and other road rules were also commonly reported as a problem in Jordan in the interviews with Government and non-government stakeholders. However, non-use of seat belts was not included in the JTI data records. In addition, all the causes identified by the JTI were attributed to human factors and no causes were attributed to non-human factors such as infrastructure (e.g. Non-functional Street Lights; Hit Fixed Object on Roadside; Inappropriate Speed Limit) or vehicle (e.g. No Seat Belt Installed).

The seat belt observational survey revealed that seat belt wearing rates were very low in Jordan (Table 3). When compared with wearing rates in other countries, it is apparent that Jordan has the lowest wearing rate and especially lower wearing rates for children (Table 4).

Table 3 . Seat belt wearing rates in Jordan

Adults	
Position	% Wearing a seat belt
Driver	23.5 %
Front seat passenger	0 %
Rear seat passenger (and cargo area of trucks or vans)	0 %
Weighted mean for sample	15.8 %
Children aged 10-18	
Position	% Wearing a seat belt
Driver	0 % (but children seen driving are quite rare with only one child seen driving and so the sample is too small to be reliable)
Front seat passenger	8.3 %
Rear seat passenger (and cargo area of trucks or vans)	0 %
Weighted mean for sample	2.7 %
Children under age 10	
Position	% Wearing a seat belt
Front seat passenger	4.8%
Rear seat passenger (and cargo area of trucks or vans)	0 %
Weighted mean for sample	1.2%

Table 4. Comparison of seat belt wearing rates in different countries

Occupant position	Jordan	Australia	Sweden	Georgia	Kosovo	Serbia	Macedonia
Driver	23.5%	(97%)	(98%)	71%	32.1%	(66%)	38.6%
Passenger – front	0	(97%)	(98%)	58%	30.9%	(66%)	25.4%
Passenger – rear	0	(96%)	(84%)	Not enough data	3.8%	(3%)	3.4%
Children	2.4 %	-	-	33%	54.5%	-	21.4%
Total	11%	(97%)	(93%)		28.6%	(45%)	32.5%

Source: Global Road Safety Solutions Pty Ltd. Figures in brackets are based on WHO [12]

While seat belt legislation exists, child restraint legislation does not exist in Jordan. However, children under 10 are required by law to sit in the back seats and yet the seat belt law only requires drivers and front seat passengers to wear seat belts. The usage of child restraints, whilst not nil, was extremely rare in Jordan.

It was also reported in the interviews that teenage boys even if unlicensed drive in Jordan and show speeding and stunt behaviours. High risks of young drivers especially males and unlicensed drivers are well established in the literature [13] and they are likely to in part account for the road deaths and injuries of children in Jordan. While data on the levels of enforcement could not be obtained, the abovementioned risk behaviours which are against existing laws in Jordan (e.g. seat belt laws, speed limits, licensing age of 18 and over) suggest the enforcement is generally weak.

Data on the total number of schools in Jordan and how many of those are placed on major roads could not be obtained. However, the present interviews revealed that schools along major roads with high traffic volume and schools with the main gate facing high volume traffic roads, at complex intersections or on curved roads are common. High risk locations of schools in Jordan are likely to contribute to a high risk of death and injury in children commuting to school.

In addition, travel speeds do not appear to be well controlled around schools in Jordan. When teachers and parents were asked in the interviews about the speed limit around schools, they did not know. However, they reported to see cars pass the schools at speeds greater than 60 km/h. The interviews further revealed that clear and enforced school zones are not implemented in Jordan. All schools are advised to have a school sign and a 40km/h speed limit sign, but they are guidelines and not legal requirements.

The interview with government stakeholders also uncovered the lack of funding as a major barrier to improving road safety in Jordan, including child road safety. The interview also highlighted the lack of a national road safety management body and no particular agency was responsible for managing child road safety in Jordan.

4 Discussion

This study identified the magnitude of the child road safety problem and the possible factors contributing to the problem through a multi-method approach. The interviews identified some interventions are under development in Jordan. These included school zones, public transport, improving the walkability of Amman city and licence penalty point system. The present findings suggest certain other interventions have potential to improve the child road safety problem in Jordan. These are discussed in turn.

Jordan would benefit from making improvements to the current crash data collection and analysis mechanisms which will be able to produce more reliable fine-level crash data and allow in-depth analysis. For example, regular seat belt surveys can provide Jordan with data that can inform where and when enforcement is needed. Such data also provide Jordan with the capacity to monitor and evaluate the benefits of seat belt initiatives such as increased seat belt enforcement. Evaluation data not only have the benefit of improving practice but also sustaining political and public support for the initiatives.

A comprehensive seat belt law applying to all vehicle occupants anywhere and anytime is vital to improve seat belt wearing by children and therefore reduce the number of road deaths and injuries in children of Jordan. Mandating seat belts for all vehicle occupants including children in the back seats is critically needed in Jordan. Countries with seat belt laws that apply to all front and back seat occupants (e.g. Australia, Sweden) have high seat belt wearing rates for both front seat occupants (close to 100%) and back seat occupants (well over 80%). Any exclusion clauses such as only for front seat occupants and only on highways can interfere with effective implementation of the law including enforcement. In addition, implementation of child restraint legislation mandating the use of child restraints based on age, height or weight will further help prevent deaths and injuries of children as vehicle passengers in Jordan.

The legal licensing age is 18 in Jordan yet younger drivers were reported in road crash death data. Interviews with parents and teachers confirmed that it is not uncommon to find underage drivers not only driving but also speeding and performing stunts in Jordan. A strict licensing scheme such as the Graduated Licensing Schemes (GLS) which require drivers to progress through incremental licensing stages with gradual lifting of restrictions is needed in Jordan. Under such schemes non-compliance with the restrictions result in heavy penalties. These schemes are shown to be effective in reducing risk taking and casualty crashes involving young drivers [14].

Jordan's plan to implement school zones should be fully supported. This study found high travel speed around schools is of particular concern. School zones must be clearly marked through highly visible speed limit signs of 30km/h or lower on every road the school property sits. Self-enforcing infrastructure measures to reduce travel speed such as speed humps at regular intervals throughout the school zones and marked and raised pedestrian crossing [15]

Leadership and political commitment backed with sustainable funding for road safety are clearly needed to urgently and effectively bring down the road crash deaths and injuries suffered by the children of Jordan. Jordan's child road safety problem can improve dramatically with the top decision-making positions of Jordan giving clear recognition of the road safety problem and embarking with appropriate policy, institutional and financial changes.

It is hopeful that a National Road Safety Strategy (NRSS) 2018 -2027 has been developed in Jordan.

The strategy is primarily based on two main safety approaches namely Safe System and Vision Zero [16]. Safe System thinking has been at the core of success for the best performing road safety countries and jurisdictions globally. Within a safe system the road transport system must be designed, built, maintained, operated and used to accommodate inevitable human errors and vulnerability of human bodies. That is, a safe system ensures that crash forces are not beyond the forces the human body can withstand, in the event of a crash.

The Safe System thinking also drives a moral demand that people should not die or suffer debilitating injury on our roads. It accepts that human error is inevitable and thus that crashes are inevitable, but does not accept that death and serious injury are inevitable consequences of these crashes.

Vision Zero policy requires that fatalities and serious injuries are reduced to zero by a specific target year. The national road safety work in Jordan should be based on a refusal to accept human deaths or lifelong suffering as a result of road crashes. A successful implementation of the National Road Safety Strategy 2018 -2027 in Jordan may bring reductions in child road deaths and injuries in Jordan.

Acknowledgements

This study was funded and supported by UNICEF Jordan. The authors also acknowledged the cooperation and collaboration of the government stakeholders, non-government stakeholders, and UNICEF Makani Centres in Jordan.

References

1.WHO, *Global status report on road safety. World Health Organisation*: Geneva (2018).

2.M. Hoque, H. Mahmud, A. Azad, S. Sohel , S. Sarkar. The Risk of Children in Road Traffic Accidents in Bangladesh. WEB international conference, 20th Jan. (2009)

3.H. Singh, A. Aggarwal D. Fatal Road Traffic Accidents among Young Children. J Indian Acad Forensic Med, **32**(4). (2010).

4.N. Mohamed N., W. Voon. *An Overview of Road Traffic Injuries Among Children in Malaysia and Its Implication on Road Traffic Injury Prevention Strategy*, Research Report, Malaysian Institute of Road Safety Research(MIROS). (2011).

5.Department of Statistics, *Jordan Statistical Yearbook*, Amman Jordan. (2018)

6.k. Jadaan, M.Al-Fayyad, and H. Gammoh, Prediction of Road Traffic Accidents in Jordan using Artificial Neural Network. Journal of Traffic and Logistics Engineering, **2: 92-94**. (2014)

7.K. Jadaan, H. Gammoh, S. Alrafayah, E. Braizat. Road Traffic Injuries and Fatalities Among Children in Jordan. The International Conference on Advances in Software, Control and Mechanical Engineering (ICSCME'2015). Antalya, Turkey. Sept. 7-8. (2015)

8.G. Al-Khateeb. Analysis of Accident Data and Evaluation of Leading Causes for Traffic Accidents in Jordan. Jordan Journal of Civil Engineering, **4(2):76-94.** (2010)

9.H. Al Masaeid. Traffic Accidents in Jordan. Jordan Journal of Civil Engineering, **4(2):331-343.** (2009)

10.M. Fa'ouri and F. Najdawi. Patterns of childhood injuries. Journal of the Royal Medical Services. **9(2):35-39.** (2002)

11.WHO, *Ten strategies for keeping children safe on the road.* World Health Organization: Geneva (2015) http://apps.who.int/iris/bitstream/10665/162176/1/WHO_NMH_NVI_15.3_eng.pdf

12.WHO, *Global status report on road safety. World Health Organisation*: Geneva. (2015)

13.C. Sakashita, A. Graham, M. de Roos, S. Croft and M. Elliot. "Comparing provisional and unrestricted licence holders on speeding offences and crash rates." Paper in Proceedings of the Australasian Road Safety Research Policing and Education Conference, Melbourne, Australia, 17-19 October 2007. (2007)

14.L. Hartling, N. Wiebe, K. Russell, J. Petruk, C. Spinola, & T. Klassen. Graduated driver licensing for reducing motor vehicle crashes among young drivers. Cochrane Database Syst Rev, **2(2).** (2004)

15.M. de Roos. and F. Marsh, Speed limits: Getting the limit right – the first step in effective Speed Management. Journal of the Australasian College of Road Safety, **28(2), 55-59**. (2017)

16.K. Jadaan, *National Road safety strategy 2018 – 2027,* a report submitted to the Minister of Transport, Jordan. (2018)

Modelling Pedestrian Crossing Behaviour based on Human Factor.

Naasah Nasrudin[1,a], Marlyana Azyyati Marzukhi[1] , Oliver Ling Hoon Leh[1,] Zaharah Mohd Yusoff[2] and Nurul Shakila Khalid[1]

[1] Centre of Town and Regional Planning Studies, Universiti Teknologi MARA, Malaysia
[2] Centre of Surveying Science and Geomatics, Universiti Teknologi MARA, Malaysia

Abstract. Research on pedestrian crossing behaviour in urban areas is extensive and has contributed to very useful insights into the role of road, traffic and pedestrian characteristics on the crossing decisions of pedestrians, their compliance with traffic rules and the related safety. However, human factors are rarely incorporated in pedestrian crossing behaviour research. The objective of this research is to analyse the development of pedestrian crossing choice models on the basis of road traffic and human factors. For that purpose, a questionnaire was distributed to 663 respondents among pedestrians in the Shah Alam district. The respondents were asked to fill in a questionnaire about their travel motivations, ability characteristics, risk perceptions and preferences with respect to walking and road crossing, as well as their opinion on drivers, etc. From the modelling analysis, the results showed that there is a significant relationship between Human Factor and Crossing Behaviour; there were two components of Human Factor that influenced pedestrian crossing behaviour to emerge, namely a "risk-taker" and a "rule-follower". Based on 'path coefficient' of Human Factor analysis, this study concludes that a 'risk-taker' component contributed more to Crossing Behaviour. The findings of this research can be used to evaluate the implementation of new pedestrian crossings and a redesign of existing pedestrian crossing environments.

1. Introduction

Unsafe pedestrian behaviour is a major factor in the injuries and fatalities of pedestrians. In the year 2013, the number of road accidents in Malaysia was 135,624 but increased to 141,808 in 2014 and 148,302 in 2015. Involvement in accidents in Malaysia was also alarming at 71% road death with approximately 11% being pedestrians. Pedestrian death rate per 100,000 population in Malaysia can be considered among the highest in the Southeast Asia region. Existing research on pedestrian crossing behaviour in urban areas is extensive and has contributed useful insight into the role of road, traffic and pedestrian characteristics on the crossing decisions of pedestrians, their compliance with traffic rules and the related safety.

2. Literature Review

The highest incidence of pedestrian accidents occurs mainly when the pedestrian crosses a road in non-authorized places [1]. However, the analysis is complex given that pedestrian crossing behaviour is influenced by multiple factors. They include aspects such as the road and traffic environment, socio economic and psychological characteristics of the individual, and the travelling conditions [2-4].The road infrastructure (i.e. number and width of lanes, the presence of signals for crosswalks, zebras, and bridges) and the purpose of the trip are two decisive factors for the pedestrians behaviour when they need to cross a road [5]. Other factors such as the composition of traffic flow passing through a pathway and the design of phases in the case of signalized intersections, can also affect the behaviour of pedestrians [6]. In sum, the literature shows that pedestrian crossing behaviour is affected by socioeconomics (e.g. age, gender), traffic characteristics (e.g. speed and flow), the road environment (e.g. crossing walking distance, presence of traffic/pedestrian control devices and signals, presence of crossing facilities), and subjective factors (e.g. perceptions and attitudes). All of the above, except the subjective factors, are commonly found in reported pedestrian crossing studies.

However, human factors are rarely incorporated in pedestrian behaviour and safety models; therefore, the explanatory power of these factors on the observed behaviour should be tested. The objective of this research is to highlight the development of pedestrian crossing choice models whilst considering road, traffic and human factors. More specifically, to develop choice models in order to estimate the probability of crossing at each location along a pedestrian's trip in relation to roadway design, traffic flow, traffic control, and human factors. The analysis of pedestrian crossing behaviour in urban areas may assist in understanding the way pedestrians interact with the road and traffic environment, as well as with other pedestrians. It may also help to understand the way they balance the need for comfort and safety at the cost of delays, within the framework of existing traffic rules. This study therefore, expects to meet the government's strategy in reducing road accidents and creating more sustainable mobility environments in our cities.

[a] Corresponding author: naasa717@uitm.edu.my

3. Data Collection & Questionnaire Design

For the questionnaire survey, a total of 663 respondents were selected randomly from among the residents of Shah Alam. Selected of the sample were calculated based on the total population of 336590 people, with a 99% degree of confidence, and 5% margin of error. For the development of the questionnaire, several questionnaires from the existing literature were studied. The questions were designed to be rated based on Likert Scales such as always/never or agree/disagree scales. The behavioural questionnaire of Papadimitriou et al, 2016 was used as a basis. The questionnaire was designed based on related crossing behaviour elements; for example, perception attitudes, beliefs, motivation etc. The questionnaire includes 4 sections, and the detail of section B and section C are shown in Table 1:

- Section A: Demographics
- Section B: Risk Perception and Value of Time (Human Factors)
- Section C: Pedestrian Crossing Behaviour
- Section D: Pedestrian Perception of Drivers

Table 1.Questionnaire Design

B	**Risk Perception and Value of Time (Human Factor)** As a pedestrian, how much would you agree with each one of the following statements:*
B_1	Crossing roads is easy.
B_2	Crossing roads at designated locations reduces the risk of accident.
B_3	Crossing roads outside designated locations is wrong.
B_4	Prefer routes with signalized crosswalks.
B_5	Trial to make as few road crossings as possible.
B_6	Trial to take the most direct route to the destination.
B_7	Willingness to make a detour to find a protected crossing.
B_8	Willingness to take any opportunity to cross.
B_9	Crossing roads outside designated locations saves time.
B_10	Crossing roads outside designated locations is acceptable because other people do it.
B_11	Willingness to make dangerous actions as a pedestrian to save time.
C	**Pedestrian Crossing Behaviour** As a pedestrian, how often do you adopt each one of the Following behaviours:**
C_1	Cross at designated crosswalk when there is no oncoming traffic.
C_2	Cross at designated crosswalk when in a hurry.
C_3	Cross at designated crosswalk when there is a shop on the other side.
C_4	Cross even though the pedestrian light is red.
C_5	Cross between vehicles stopped on the roadway during traffic jams.
C_6	Cross without paying attention to traffic.
C_7	Absent-minded while walking.
C_8	Cross while talking on cell phone.
C_9	Cross while listening to music with headphones.
C_10	Cross even though obstacles (parked vehicles, buildings, trees, etc.) obstruct visibility.
C_11	Cross despite seeing oncoming vehicles.
C_12	Cross at designated crosswalk when observing other people doing the same.
C_13	Cross at designated crosswalk when company prompts to do it.
C_14	Prompt company to cross at designated crosswalk.

* (1: strongly disagree, 2: disagree, 3: agree, 4: strongly agree)
** (1: never, 2: rarely, 3: sometimes, 4: often, 5: almost always)

The field survey design consisted of three walking conditions and several places were identified as a survey area according to these three crossing conditions.

• Crossing a main urban road with signal-controlled and uncontrolled crosswalks.
For this particular crossing condition, Section 7, Shah Alam was identified as a survey area which involved students of UiTM crossing the road in order to access commercial facilities near the campus. As well as this, crossing facilities near Shah Alam Hospital were also surveyed in order to measure the effectiveness of crossing pedestrian provided and its relation to crossing behaviour.
• Crossing a minor (residential) road with or without marked crosswalks.
Several schools located near residential areas were chosen as a survey area. For example, Section 6, Section 7, Section 9, Section 15 and Section 19. In addition, an area that facilitates public transport was also chosen as a survey area, such as Section 15 (Padang Jawa) and Section 19.
• Crossing a major urban arterial with signal-controlled crosswalks.
For this particular crossing condition, a high capacity urban road was chosen as an area of study. This included the crossing pedestrian needed to access the bus station at Section 13, close to the Federal Highway. This crossing pedestrian is also used to access AEON Mall.

The model's development consists of three steps, each serving a different purpose, namely (i) to determine the road and traffic variables that effect pedestrian crossing choice, (ii) to estimate human factors as latent variables, and (iii) to test whether these variables reflecting human factors, add to the explanatory power of the models. This study will introduce two components of pedestrian crossing behaviour as explanatory variables, namely a ''risk--taker'' component reflecting the tendency to cross at mid-block to save time, etc., and a ''rule- follower'' component, which represents a very rule-obedient pedestrian. The introduction of these components as explanatory variables into the choice models are expected to improve the modelling results and will help to prove that human factors have additional explanatory power over the road and traffic factors of a pedestrian.

4. The Result

4.1 Confirmatory Factor Analysis (CFA) in Measurement Model

Table 2 shows the summary of confirmatory factor analysis (CFA) for every construct in the measurement model. Based on Table 2, the value of factor loading for each item was greater than 0.60. Item B6 was deleted due to low factor loading of less than 0.60. The requirement for unidimensionality. was achieved through the item deletion procedure for low factor loading items.

The value of AVE obtained from every construct was greater than 0.50. Thus, the Convergent Validity for the measurement model was achieved since all the values for AVE were greater than 0.50 as suggested by Fornell and Larcker [7].

Table 2. Summary for Confirmatory Factor Analysis (CFA) in Measurement Model

Construct	**Component**	**Item**	**Factor Loading**	**CR**	**AVE**
Human Factor	Component 1	B2	0.74	0.89	0.68
		B3	0.87		
		B4	0.81		
		B5	0.86		
	Component 2	B8	0.87	0.93	0.81
		B9	0.86		
		B10	0.96		
Crossing Behaviour	Component 1	C1	0.61	0.84	0.65
		C2	0.94		
		C3	0.83		
	Component 2	C6	0.93	0.95	0.87
		C7	0.91		
		C8	0.96		
	Component 3	C9	0.90	0.93	0.80
		C10	0.88		
		C11	0.91		

Based on Table 3, when the three Fitness Indexes; Absolute Fit, Incremental Fit and Parsimonious Fit, achieved the requirements, the Construct Validity is achieved. CFI is equal to 0.90 or higher, RMSEA is equal to 0.08 or lower, and the ratio of Chisq/df is less than 5.0. [8]

Table 3. Summary for the Assessment of Fitness Indexes

Category	**Fit statistics**	**Recom-mended**	**Obtain**	**Comment**
Absolute Fit	RMSEA	<0.08	0.085	Satisfied
Incremental Fit	CFI	>0.90	0.954	Achieved
Parsimonious Fit	Chisq/df	<3.0	5.722	Satisfied

4.2 Relationship between Human Factor and Crossing Behaviour

Table 4 shows the standardized regression weight for the structural model. Based on the table, the path coefficient of Human Factor to Crossing Behaviour is 0.49. This value indicates that for every one-unit increase in Human Factor, its effects will contribute to a 0.49 unit increase in Crossing Behaviour. Since the p-value is less than 0.05 ($p=0.0001 < 0.05$), we can conclude that there is a significant relationship between Human Factor and Crossing Behaviour.

Table 4. Relationship between Human Factor and Crossing Behaviour.

Path Coefficient			Estimate	P value	Comment
Human Factor	to	Crossing Behaviour	0.49	0.0001	Significant

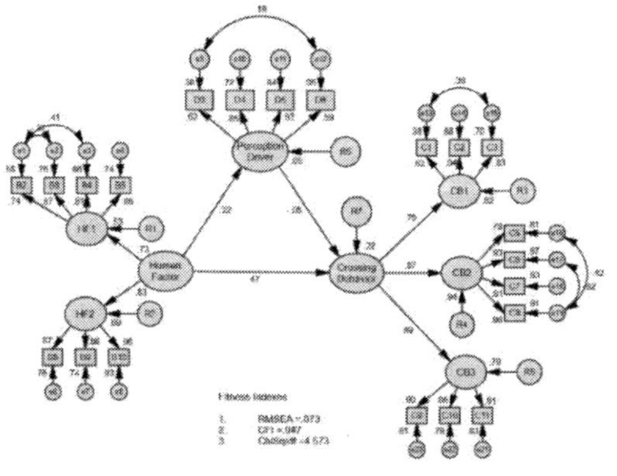

Figure 1. Structural Model

4.3 Contribution Dimension Human Factor To Crossing Behaviour.

From the modelling analysis, the results showed that there were two components of Human Factor that influenced pedestrian crossing behaviour, namely a "risk-taker" and "rule-follower". Based on the 'path coefficient' of Human Factor analysis, this study concludes that a 'risk-taker' component contributed more to Crossing Behaviour.

Table 5. Contribution Dimension Human Factor To Crossing Behaviour

Based on Table 5 above, the path coefficient of Human Factor -Component 1 (HFC1) to Crossing Behaviour was 0.11 and the path coefficient of Human Factor - Component 2 (HFC2) to Crossing Behaviour was 0.25. The value of beta estimate for HFC2 to Crossing Behaviour was higher than the value of beta estimate for HFC2 to Crossing Behaviour which was 0.25 > 0.11. Therefore, we can conclude that HFC2 contributes more to Crossing Behavior.

Path			Estimate	P value	Comment
HFC 1 (Rule-Follower)	to	Crossing Behaviour	0.11	0.029	Significant
HFC2 (Risk-Taker)	to	Crossing Behaviour	0.25	0.0001	Significant

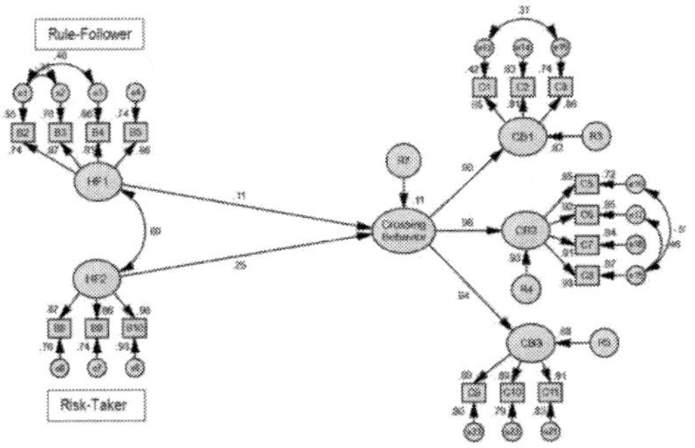

Figure 2. Contribution Dimension Human Factor to Crossing Behaviour

The two group components of the Human Factor that influenced pedestrian crossing behaviour can be described as follows:

- "Rule-Follower" pedestrians in this group have rather positive attitudes, perceptions and behaviour, as they have low scores for risk-taking (i.e. 'crossing roads at designated locations reduces the risk of accident', crossing roads outside designated locations is wrong', 'prefer routes with signalized crosswalks', and 'trial to make as few road crossings as possible')
- "Risk-Taker" pedestrians in this group have negative attitudes, perceptions and behaviour, as they have high scores for risk-taking behaviour ('willingness to take any opportunity to cross', 'crossing road outside designated locations saves time', and 'crossing roads outside designated locations is acceptable because other people do it').

5. Conclusion

The analysis of pedestrian crossing behaviour in urban areas may assist in understanding the way pedestrians interact with road and traffic environment, as well as with other pedestrians. It may also help to understand the way they balance the need for comfort and safety at the cost of delays, within the framework of existing traffic rules [6]. Incorporating attitudinal and behavioural data (human factors) in describing motivations, preferences and perceptions of pedestrians emerges as an effective approach to model crossing behaviour. This study introduced two components of pedestrian crossing behaviour as explanatory variables, namely a 'rule- follower' and a 'risk taker' component. This was applied to reflect the urban pedestrian in Shah Alam, Malaysia. The introduction of these components as explanatory variables into the choice models resulted in improvement of the modelling results, which indicates that human factors have additional explanatory power over traffic and road factors of pedestrian behaviour.

5. Acknowledgments

The authors would like to thank the Ministry of Higher Education, Malaysia for funding this research through the FRGS grant 600-IRMI/FRGS 5/3 (46/2016) and Universiti Teknologi MARA (UiTM) for supporting the research

6. References

[1] X. Zhuang and C. Wu. *Accident Analysis and Prevention*. **43(6):** 1927-1936 (2011).
[2] M. Hamed M. *Safety Science,* **38 (1)** 63-82 (2001)
[3] E. Papadimitriou. *Transportation Research Part F: Traffic Psychology and Behaviour,* **15 (1)** 75-94 (2012)
[4] O. Taubman-Ben-Ari & E. Shay. *Transportation Research Part F: Traffic Psychology and Behaviour,* **15 (2)**188-195 (2012).
[5] L. Cambon de, C. B Tijus, A. Poitrenaud, C. Leproux, Bergeron J., & JP Thouez. *Safety Science,* **47 (9)** 1248-1253 (2009).
[6] S Manski Das, CF & MD. Manuszak, Journal *of Applied Econometrics,* **20 (4)** 529-548 (2005).
[7] C. Fornell, and D. F. Larcker, *Journal of Marketing Research* **(18:1)** 39-50 (1981).
8] A Zainudin. *Research Methodology and Data Analysis*. 2nd Edition. Shah Alam: UiTM Press (2012).

MATEC Web of Conferences **308**, 03004 (2020) https://doi.org/10.1051/matecconf/202030803004
ICTTE 2019

Study on the effectiveness of navigation safety measures for Nanjing Yangtze River Bridge

Feilong Zhao[1,a], Guoping Guo[2]

[1]*School of Shipping, Wuhan University of Technology, Wuhan, China*
[2]*Key Laboratory of Inland Navigation and Hubei Province, Wuhan, China*

Abstract. Combined with navigation safety measures such as navigation mark adjustment, synchronous flash distribution, and real ship test adopted by Nanjing Yangtze River Bridge, theoretical analysis methods and data analysis are used to determine the safety evaluation risk index system, and an unascertained measure evaluation model based on entropy weight is constructed. Through the comparison of the effectiveness of the security measures analyzed by the questionnaire, the effectiveness of the bridge safety guarantee measures was evaluated; further suggestions on the optimization of relevant safety protection measures were proposed. The research results are of great significance for maintaining the safety of the bridge, the navigation order of the waters in the bridge area, protecting the navigation resources of the bridge area, and ensuring the navigation safety of the ship.

1 Introduction

Bridges play an irreplaceable role in promoting economic and social development. However, in the event of a bridge safety incident, the consequences will be unimaginable. Nowadays, more and more bridges are built in the same jurisdiction. It is especially important to carry out the necessary safety guarantees for the safety of the bridge and how to properly solve the problem of water traffic safety in the bridge section. Safety evaluation of the safety measures implemented by the bridge is carried out, and corresponding countermeasures are proposed in time. Through the analysis and evaluation of effectiveness, we can better understand and master the effectiveness of the safety and security measures implemented by the bridge, and provide scientific decision-making basis for the maritime management department to improve the navigation environment and traffic safety of the bridge section and the navigation management and control. And reference, thus having a certain practical value [1].

2 Brief Introduction to Safety Measures for Navigation in Bridge Area

In order to ensure the safety of the bridge, the water management department has adopted the following navigation safety guarantee measures, as shown in Table 1 below.

Table 1. Bridge navigation safety protection measures list

Measure	Effect
Adjusting the navigation area of the bridge area	Guide the ship's driver to accurately identify the navigation bridge hole and reduce the drift of the ship. As shown in Figure 1
Navigation bridge hole setting buoy synchronous flash	The navigation mark of the bridge area is more obvious, and the ship driver can correctly follow the navigation mark to guide the safe navigation through the bridge area. As shown in Figure 2
Lighting the navigation bridge hole	Solved the problem that the crew "can't see clearly and unclear" about the bridge hole. As shown in Figure 3.
Intelligent early warning monitoring	Improve the safe navigation ability of ships at night and effectively prevent ships from entering the non-navigate waters
Carry out the 6-hole real ship test of the bridge	Through the actual ship test data, the ship's navigation characteristics are analyzed, the navigation sign setting is optimized, the ship navigation method is simulated and simulated, navigation suggestions are provided, and the bridge navigation safety is guaranteed.

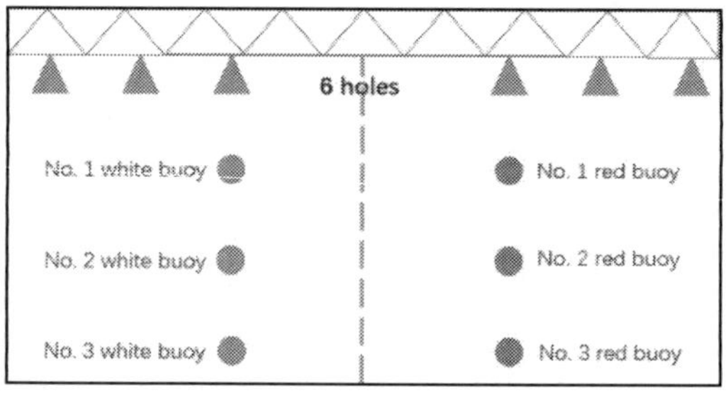

[a] Corresponding author: 15171456122@163.com

Figure 1. Schematic diagram of the adjustment scheme for the navigation mark hole navigation mark

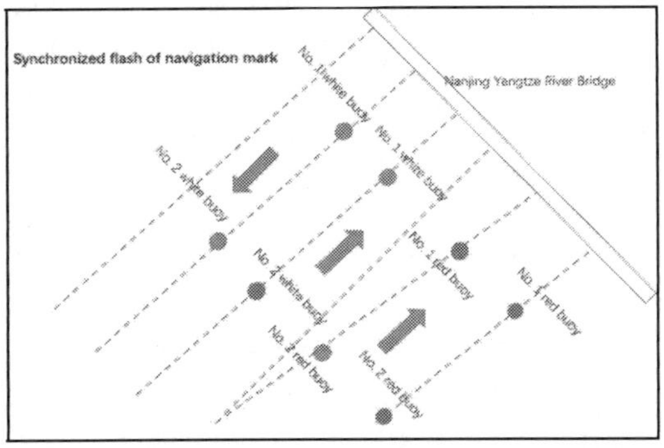

Figure 2. Schematic diagram of synchronous flash distribution scheme.

Figure 3. Schematic diagram of the distribution scheme of the navigation bridge hole lighting.

3 Safety Benefit Evaluation of Bridge Safety Protection Measures

3.1. Safety Benefit Evaluation of Bridge Segment

Establish a safety assessment risk indicator system for this jurisdiction, including people, ships, bridges, environment, and management, with a total of 26 secondary evaluation indicators. The expert judgment method is used to construct the comparative judgment matrix of the evaluation index. The brainstorming method is used to determine the membership degree of each index. Finally, the center of gravity method is used to defuzzify and obtain the comprehensive evaluation score of each index [2]. Five fuzzy comments were expressed for the safety benefits, and finally the effectiveness of the bridge safety guarantee measures was judged.

3.2. Unascertained measure evaluation model based on entropy weight

3.2.1 Single indicator not determined calculation

The calculation of the single-index unascertained measure matrix uses the method of constructing the unascertained measure function, and is determined by the expert scoring method. According to the characteristics of the specific problem to be solved and the severity of the state change, the undetermined measure function can be connected by different curves such as a straight line and a quadratic curve [3]. Among them, the linear unascertained measure function is currently the most widely used. Its calculation formula is as follows:

$$\mu(x \in c_{k+1}) = \frac{-x}{a_i + 1} + \frac{a_{i+1}}{a_{i+1} - a_i} a_i < x < a_{i+1} \tag{1}$$

$$\mu(x \in c_{k+1}) = x > a_{i+1} \tag{2}$$

Among them, a_i is the classification standard of each indicator, and C_k is the evaluation level. From the calculation formula (1) (2), it can be concluded that the observation value x_i of the index I_j with respect to the index x_{ij} causes the object to be in the single-index unsure degree matrix of each evaluation level C_k as follows:

$$\left(u_{ijk}\right)_{m\times k} = \begin{bmatrix} u_{i12} & u_{i12} & \cdots & u_{i1k} \\ u_{i21} & u_{i22} & \cdots & u_{i2k} \\ \cdots & \cdots & \cdots & \cdots \\ u_{im1} & u_{im2} & \cdots & u_{imk} \end{bmatrix} \tag{3}$$

$$i = 1, 2, \cdots, n; j = 1, 2, \cdots m; k = 1, 2, \cdots K$$

3.2.2. Entropy weight method weight calculation

Entropy is a measure of the uncertainty of the state of the system. The index weight assignment method based on entropy weight is a combination of quantitative and qualitative methods, which expresses and processes human subjective judgments in quantitative form [4]. Calculate the measure value of the evaluation index according to the corresponding membership function to construct $(\mu_{i1}, \mu_{i2}, \text{L}\ , \mu_{il})$, The entropy is now:

$$H = -\frac{1}{\log_{10} k} \sum_{k=1}^{k} u_{ijk} \log_{10} u_{ijk} \tag{4}$$

$$v_{ij} = 1 + \frac{1}{\log_{10} k} \sum_{k=1}^{k} u_{ijk} \log_{10} u_{ijk} \tag{5}$$

Obviously, $0 \leqslant u_{ijk} \leqslant 1$ let:

$$w_{ij} = v_{ij} / \sum_{j=1}^{m} v_{ij} \tag{6}$$

The weight $w_i = \left(w_{i1}, w_{i2}, \cdots, w_{im}\right)^T$ is used as the weight vector of the attribute set U, that is, its entropy weight. Where H is the entropy value of the evaluation index; v_{ij} is the proportion of the j-th evaluation object in the i-th indicator.

3.2.3 Multi-indicator comprehensive measure calculation

According to the single-index unascertained degree u_{ijk} and the index weight vector ω ,the unascertained degree vector x_{ij} of the evaluation object can be calculated, and the calculation formula is as follows:

$$u_{ik} = \sum_{j=1}^{m} w_j u_{ijk} \, (i = 1,2,\cdots n; k = 1,2,\cdots K) \quad (7)$$

Then we call the vector $u_i = (u_{i1}, u_{i2}, \cdots u_{ik})$, T is a multi-indicator synthesis Measure vector.

3.2.4. Confidence recognition

Because the evaluation grades are ordered, it is not possible to adopt the maximum membership recognition criterion for the out-of-order division, and instead use the confidence recognition criterion. Pre-set the confidence threshold λ , then:

$$k = \min_k (k : \sum_{k=1}^{K} uik \geq \lambda, 1 \leq k \leq K) \quad (8)$$

From this formula, it is judged that the object x_i belongs to the k_0 -th evaluation level and the confidence is higher than λ .

3.3. Establishment of evaluation index system

According to the regulations of the ship's alignment system, combined with the natural environment and navigation environment of the waters of the bridge area, comprehensive consideration of various factors affecting the establishment of water efficiency indicators of the bridge area, and selection of ships, environment and management [5] as evaluation Indicators; as shown in Table 2.

Table 2. Bridge area water area safety risk assessment index

<table>
<tr><th>Primary evaluation factor</th><th colspan="2">Secondary evaluation factor</th></tr>
<tr><td rowspan="4">Ship factor</td><td colspan="2">Improvement effect of dangerous goods ship accident</td></tr>
<tr><td colspan="2">Ship sailing freely to improve the effect</td></tr>
<tr><td colspan="2">Improve the effect of crossing the traffic lane</td></tr>
<tr><td colspan="2">Improved navigational safety index</td></tr>
<tr><td rowspan="7">envirnmental factor</td><td rowspan="3">Natural environment (impact effect)</td><td>Wind direction</td></tr>
<tr><td>Flow rate</td></tr>
<tr><td>Visibility</td></tr>
<tr><td rowspan="4">Traffic environment (impact effect)</td><td>water depth</td></tr>
<tr><td>Navigation aid sign</td></tr>
<tr><td>Obstruction</td></tr>
<tr><td>traffic flow</td></tr>
<tr><td rowspan="2">Management factor</td><td colspan="2">Bridge area water safety protection facilities</td></tr>
<tr><td colspan="2">First aid to improve the effect</td></tr>
</table>

The effective grades of each evaluation index are divided into five levels: no effect, some effects, general effects, large effects, and large functions. They are represented by C1-C5, and the evaluation criteria corresponding to each level are shown in Table 3.

Table 3. Bridge area water area safety risk assessment index

Not working at all	Some role	General role	Greater effect	Great effect
(0,60)	[60,70)	[70,80)	[80,90)	[90,10)

3.4. Determine evaluation indicator weights

The score of the evaluation is given to each indicator by the expert. The evaluation scores of the evaluation factors were counted, and the statistical results are shown in Table 4

Table 4. Bridge area water area safety risk assessment index

<table>
<tr><th rowspan="2">Primary factor</th><th rowspan="2">Secondary factor</th><th colspan="5">Survey results</th></tr>
<tr><th>C1</th><th>C2</th><th>C3</th><th>C4</th><th>C5</th></tr>
<tr><td rowspan="4">Ship</td><td>Improvement effect of dangerous goods ship accident</td><td>0</td><td>0</td><td>0</td><td>14</td><td>66</td></tr>
<tr><td>Ship sailing freely to improve the effect</td><td>0</td><td>0</td><td>0</td><td>12</td><td>68</td></tr>
<tr><td>Improve the effect of crossing the traffic lane</td><td>0</td><td>0</td><td>0</td><td>14</td><td>66</td></tr>
<tr><td>Improved navigational safety index</td><td>0</td><td>8</td><td>10</td><td>16</td><td>46</td></tr>
<tr><td rowspan="2">Envirn-mental</td><td>Natural environment (impact effect)</td><td>0</td><td>0</td><td>0</td><td>8</td><td>72</td></tr>
<tr><td>Traffic environment (impact effect)</td><td>0</td><td>0</td><td>4</td><td>10</td><td>66</td></tr>
<tr><td rowspan="2">Manage-ment</td><td>Bridge area water safety protection facilities</td><td>0</td><td>0</td><td>4</td><td>14</td><td>62</td></tr>
<tr><td>First aid to improve the effect</td><td>0</td><td>0</td><td>6</td><td>10</td><td>64</td></tr>
</table>

The fuzzy evaluation is obtained by using the single index measure calculation method to determine the membership degree of the index, and the entropy weight of the second evaluation factor is calculated. The membership function is constructed as follows: membership degree = factor to get the corresponding questionnaire valid questionnaire/total effective questionnaire.
The measure values of the respective evaluation indexes can be calculated according to the constructed membership function, and the calculation results are shown in Table 5 below.

Table 5. Indicator value vector

Measure vector				
C1	C2	C3	C4	C5
0	0	0	0.175	0.825
0	0	0	0.150	0.850
0	0	0	0.175	0.825
0	0.100	0.125	0.200	0.575
0	0	0	0.100	0.900
0	0	0.050	0.125	0.825
0	0	0.050	0.175	0.775
0	0	0.075	0.125	0.800

According to the entropy weight calculation formula (4) described above, the calculated values of the entropy weights of each evaluation index can be obtained as follows:

$$H_{V11}=-\frac{1}{\lg 5}(0+0+0+0.175*0.0825)=0.288$$

In the same way, the calculation is:
H_{V12}=0263,H_{V13}=0.288,H_{V14}=702,H_{V21}=0.202;
H_{V31}=405,H_{V3}2=393;
therefore,H_{V1}=(0.288,0.263,0.288,0.702),
H_{V2}=(0.202,0.353),H_{V3}=(0.405,0.393);
According to formula (5), it can be calculated:
V_{U1}=(0.712,0.737,0.712， 0.298);In the same way,
V_{U2}= （0.798,0.647）;V_U3= （0.595,0.607）;
Calculate the entropy weight of each indicator according to formula (6):
w_{U1}= （0.289,0.300,0.289,0.121）
In summary, the calculation results of the entropy weight of each of the secondary indicators and the primary indicators are shown in Table 6.

Table 6. Index entropy weight calculation result

Primary factor	Weights	Secondary evaluation factor	Weights
Ship	0.300	Improvement effect of dangerous goods ship accident	0.289
		Ship sailing freely to improve the effect	0.300
		Improve the effect of crossing the traffic lane	0.289
		Improved navigational safety index	0.121
envirnmental factor	0.380	Natural environment	0.5522
		Traffic environment	0.448
Management factor	0.320	Bridge area water safety protection facilities	0.495
		First aid to improve the effect	0.505

3.4. Bridge safety evaluation

3.4.1 Single indicator unascertained degree calculation

The evaluation results were processed by the mean value analysis method to obtain the scores of the respective indexes, as shown in Table 7.

Table 7. Effectiveness evaluation indicator score

Primary evaluation factor	Secondary evaluation factor		Score
Ship factor	Improvement effect of dangerous goods ship accident		93.25
	Ship sailing freely to improve the effect		93.50
	Improve the effect of crossing the traffic lane		93.25
	Improved navigational safety index		87.5
envirnmental factor	Natural environment	Wind direction	94.00
		Flow rate	
		Visibility	
		water depth	
	Traffic environment	Navigation aid sign	92.75
		Obstruction	
		traffic flow	
		Wind direction	
Management factor	Bridge area water safety protection facilities		92.25
	First aid to improve the effect		92.25

From the definition of the single-index measure function, an image of the single-index unascertained function of each index risk assessment can be obtained, as shown in Figure 4.

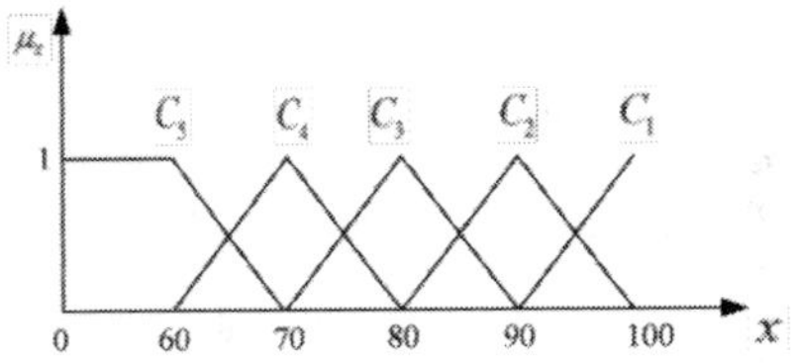

Figure 4. The unascertained measure function

The values of the indicators in Table 7 are respectively taken into the single-index measure function, and the single-index evaluation matrix for the safety and effectiveness evaluation of the safeguard measures is calculated [6].
Taking the "improvement effect of dangerous goods ship accident" in the ship factor as an example, according to the score value of Table 7, Figure 5 and formula (1). (2) and the "normality" of the unascertained measure, it can be concluded that The single-index measure vector that separates the opposite traffic flow to reduce the effect on the situation is (0 0 0 0.675 0.325). The same method can be used to find the unascertained measure of other indicators.

According to the calculation, the unascertained measure evaluation matrix of the ship safety factor for safety and safety measures is:

$$\mu_{15}=\begin{bmatrix}0&0&0&0.675&0.325\\0&0&0&0.650&0.350\\0&0&0&0.675&0.325\\0&0&0.25&0.750&0\end{bmatrix}$$

The unascertained measure evaluation matrix for environmental factors is:

$$\mu_{25}=\begin{bmatrix}0&0&0&0.600&0.400\\0&0&0&0.725&0.275\end{bmatrix}$$

The unascertained measure evaluation matrix for management factors is:

$$\mu_{35}=\begin{bmatrix}0&0&0&0.675&0.325\\0&0&0&0.675&0.325\end{bmatrix}$$

3.4.2 Multi-indicator unascertained degree calculationcalculation

The single indicator measure vector of the ship factor in the safety effectiveness assessment of the bridge navigation safety measures is:

$$\mu_{11}=\omega_1\times\mu_{15} \tag{8}$$

$$=(0.289\ 0.300\ 0.289\ 0.121)\times\begin{bmatrix}0&0&0&0.675&0.325\\0&0&0&0.650&0.350\\0&0&0&0.675&0.325\\0&0&0.25&0.750&0\end{bmatrix}$$

$$=(0\ 0\ 0.302\ 0.3759\ 0.2929)$$

By repeating the above calculations in the same way, the single-index unascertained measure vectors of the ship, environment and management factors in the safety and effectiveness assessment of the bridge navigation safety measures can be obtained as follows:

$$\mu_{22}=(0\ 0\ 0\ 0.6560\ 0.3440)$$

$$\mu_{33}=(0\ 0\ 0\ 0.6750\ 0.3250)$$

Therefore, a new unascertained matrix " μ " can be constructed. According to formula (7), the multi-indicator unascertained comprehensive measure "" μ " " for the safety and effectiveness evaluation of the bridge navigation safety measures can be calculated.

$$\mu=\omega\times\mu_1 \tag{9}$$

$$=(0.300\ 0.380\ 0.320)\times\begin{bmatrix}0&0&0.0302&0.6759&0.2929\\0&0&0&0.6560&0.3440\\0&0&0&0.6750&0.3250\end{bmatrix}$$

$$=(0\ 0\ 0.0091\ 0.6681\ 0.3226)$$

3.4.3 Bridge safety assessment conclusion

According to the comprehensive measure calculated above, using the formula (7) to identify the multi-indicator comprehensive measure of the safety and effectiveness assessment of the bridge safety guarantee measures [7], the following conclusions can be drawn:

a. $k_0=1$, at this time there is 0.6681 for maximum integration and greater than 0.66. If it is taken $\lambda=0.66$, there is $k_0=1$, $0.6681>\lambda$, that is, the safety effectiveness evaluation level of the bridge navigation safety measure belongs to the " Great effect" level, and the confidence level is not less than 66%.

b. $k_0=2$, at this time there is 0.6681+0.3226=0.9907.

If it is taken $\lambda=0.99$, there is $k_0=1$, $0.9907>\lambda$, that is, the safety effectiveness evaluation level of the bridge navigation safety protection measure belongs to the "Greater effect" level, and the confidence level is not less than 99%.

4 Research conclusions and prospects

This paper takes the various security measures adopted by the Nanjing Yangtze River Bridge as the research object, and uses theoretical analysis methods and questionnaire survey data to evaluate the implementation effectiveness of the bridge navigation safety guarantee measures for the navigation environment of the bridge area of the Nanjing Yangtze River Bridge, and analyze the Nanjing Yangtze River. The safety benefits brought by the bridge in terms of ensuring the safety of the navigation of the ship, the smooth flow of the channel [8], and the construction of the port are finally evaluated to evaluate the effectiveness of the safety measures for the navigation of the bridge; further recommendations for the optimization of relevant safety measures are proposed.

The research results are of great significance for maintaining the safety of other bridges, the navigation order of ships in the bridge area, protecting the navigation resources of the bridge area, and ensuring the navigation safety of the ships. It is of great significance to the development of the Yangtze River trunk shipping and the sustainable development of the port water transport economy. At the same time, it can provide suggestions for maritime affairs and related departments, which has certain practical significance for improving the safety of the bridge.

5 References

1 Qiang.Gao Research on Risk Assessment of Water Traffic Safety in Multi-Bridge Reach. (2010501 Wuhan University of Technology)

2 "Guidelines for the Implementation of Navigation Safety Responsibilities by Bridge Management Units"

3 "Nanjing Yangtze River Bridge Bridge Area Water Navigation Guide". (Nanjing Maritime Safety Bureau issued in 2018)

4 Jiani.Zhao Overview of the Comprehensive Safety Assessment (FSA) method. Navigation Technology, 2005 (2): 77-78

5 Yansheng.Han Research on Yangtze River water traffic safety evaluation index system [D]. Wuhan: Wuhan R&D University, April 2006

6 Jinsong.Zhao Ship-bridge collision and anti-collision problem of Nanjing Yangtze River Bridge [J]. Journal of Dalian Maritime University, 1992(1): 77~81

7 Silveira P A M, Teixeira A P, Guedes S C. Use of AIS Data to Characterise Marine Patterns and Ship Collision Risk offthe Coast of Portugal[J].Journal of Navigation, 2013, 66(6):879-898.

8 Zhong Wei. Current Status and Reform Suggestions for Navigation Safety Maintenance of Local Maritime Underwater Activities[J]. China Water Transport, 2018(11): 42-43.

The Development of A Versatile Movable Barrier Solution

Meng Li[1,a]

[1] *Research Institute of Highway, Ministry of Transport, Xituchenglu8hao, Beijing, China*

Abstract: To reduce the accident risk of work zone on highway, a new movable barrier solution is presented in this paper. The barrier is effective, affordable, environment-friendly and versatile. It can raise the level of highway safety performance in work zone, median opening and tidal lane. The development of the versatile movable barrier took reliability, mobility, economical efficiency, construction convenience into account. Researchers optimized structure by finite element analysis as well as mechanic analysis. The finite element analysis was accomplished by LS-DYNA. This barrier set eventually passed simulated impact tests, which proved that it is able to fulfil the requirement of 160kj-level safety performance.

1 Introductions

Highway barrier is generally used to prevent run-off-the-road crashes, which result from vehicle running off the roadway or into opposite lanes. Run-off-the-road crashes often lead to serious casualty and property damage, and they are an important cause of secondary accidents. Many countries and research agencies have invested resources and man-power on barrier development. However, most of the efforts work on traditional barriers such as concrete barrier, w-beam barriers. Comparing with these traditional barriers, movable barrier did not gain enough attention.

With a boost of highway mileage, the demand of roadway maintenances and highway upgrade construction keeps augmenting. It causes vehicle crashes in work zone and consequently increases the risk of fatal and injury accidents. The accident shown in the Figure 1 is a crash happening on Shenhai freeway in China. There were two lanes in a traveling direction, and one lane is closed due to construction when the crash happened. The driver of a truck at the back of a construction vehicle was too late to divert or stop, The truck rear-ended the construction vehicle and then headed off to the shoulder. After hitting another vehicle, the truck finally stopped by the barrier in the opposite lane. The accident led to one casualty and 3,000,000 Chinese yuan loss.

Figure 1. A work zone crash on Shenhai freeway

To avoid this kind of crashes, a new type of movable barrier, which is able to serve as a set of temporary longitudinal barrier, is required for highway management agencies and contractors. Current countermeasures for work zone risk in China are varisous traffic control device such as signs, plastic drums, lights, cones, barricades, chevron panels. These traffic devices are utilized to regulate, warn and guide drivers. The main purpose of the traffic control devices is not to offer a physical protection but to advise road users how to traverse a work-zone section in a proper manner. The new movable barrier ought to meet the requirement of protection as well as guidance and warning.

[a] Corresponding author: 464374092@QQ.com

Figure 2. Traffic control devices

In addition, the new movable barriers can be widely used as temporary barrier in various locations, such as work zone, highway median opening, tidal lanes and emergency evacuation exits.

2 Structure design

The feasible design scheme of the movable barrier should be free-standing, movable and reliable. It can serve as a set of temporary longitudinal barriers in work zones, tidal lanes and long-term barriers in median opening. The safety performance ought to satisfy the requirement of the 160kj-level barrier, which means the impact energy of the barrier reaches 160kj. As a kind of the reusable facility, the new movable barrier has to guarantee mobility. Developers must limit the weight of the barrier and make them easy to be assembled. The work zone is an ever-changing environment that differs in weather, humidity, location, roadway facilities and traffic flow, therefore the barrier design scheme ought to adapt to various conditions. Moreover, the new movable barrier must be feasible to different construction projects including renewal, upgrading and maintenance of infrastructure and facilities. Last but not least, the cost of the new barrier should be affordable, so it can be adopted in any region and environment. [1]

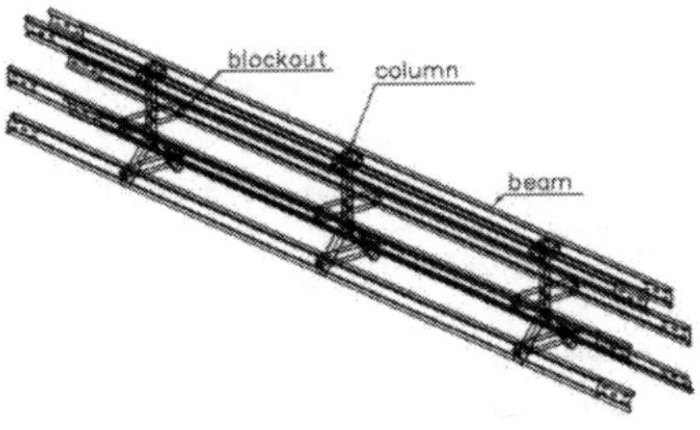

Figure 3. Standard unit

Researchers tried several different design schemes and decided the final structure design. The new barrier system consists of several standard units and two end terminals. As demonstrated in Figure 3, a standard unit is composed of three columns, six curled C beams, nine blockouts and bolts. The length of a standard unit is 6m. By flexible combination, standard units and two end terminals are able to form a set of barriers with any length. The terminal end, shown in Figure 4, is 1.5m in length, and it can be temporarily anchored in ground. The design of terminal end prevents the beam of barriers penetrating into vehicle.

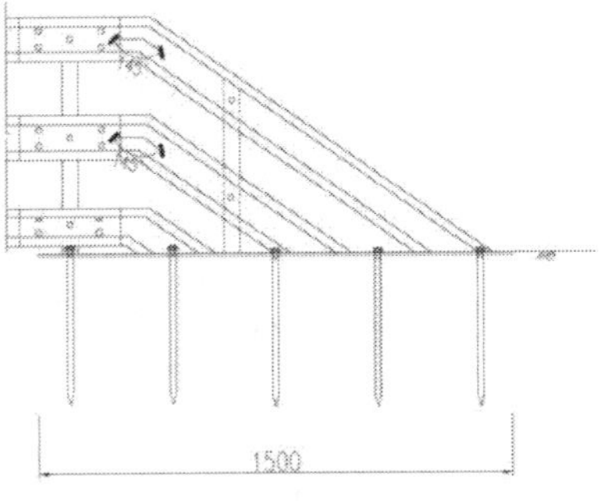

Figure 4. End terminal

The cross section of design is demonstrated in Figure 5. The barrier is 0.75m in height and 0.5m in width, and the unit can fit in a regular truck even a pickup. The small size makes it easy to transport and install. Installing units as a set of longitudinal barriers will only take several minutes with a few workers. Developers adopt a curled beam design scheme to offer better performance of energy absorbing and buffering. Comparing with the traditional w beam, the curled beam of the new movable barrier has not only better mechanical properties but also less weight and cost.

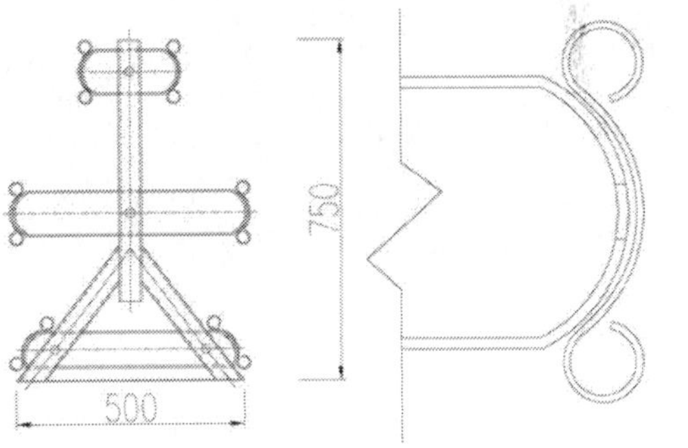

Figure 5. Cross section design

Another important design element to meet potential customer's need is the material. Researchers choose Q235 steel, which is a type of carbon steel. Q235 steel is a common steel material in Chinese market. Its yield strength reaches 235MPa, enough to provide protection performance for barriers. Due to its moderate carbon content, good comprehensive properties, high strength, plasticity and welding properties, it has been widely used in many fields like civil engineering and mechanical engineering. Another advantage of Q235 steel is the lower price because it is widely required and manufactured.

3 Simulation tests

Researchers adopted Finite Element Analysis as a good method to test the safety performance of the new movable barrier. Finite Element Analysis simulates impact tests between vehicle and barriers instead of the full-scale impact test with real vehicles, which costs much more. The FEA method has already been proved reliable and effective. LS-DYNA, a mature finite element analysis program, is selected to accomplish simulation tests. LS-DYNA algorithm is mainly based on Lagrange equation, and the program automatically calculates the time step to satisfy the stability requirement. Test objects are meshed by four-node tetrahedral element, eight-node cubic element, two-node beam element, three-node element and four-node shell element. LS-DYNA contains a lot of material models, contact elements and collision algorithms, which enable researchers to simulate complex impact process. Figure 6 shows the meshed structure model ready for impact simulation of the new movable barrier.

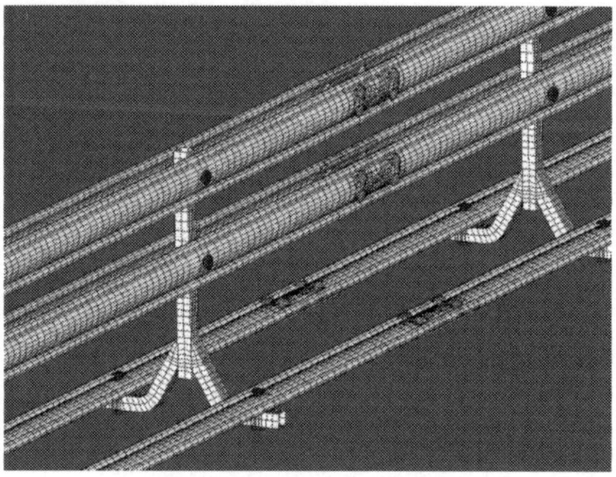

Figure 6. The meshed structure model

The impact conditions of simulation tests are set in accordance with *Standard for Safety Performance Evaluation of Highway Barriers*, which is the current Chinese evaluation standard of highway safety facilities. According to the standard, there are three impact tests required in the barrier evaluation, including car-barrier, truck-barrier and bus-barrier impact tests. Impact conditions are demonstrated as in Table 1. The impact angle for all of three 160kj-level tests is 20° ; the vehicle mass and speed vary basing on different vehicle types. The test is similar but not exactly same with TB11,TB41,TB42 in European evaluation standard. [2][3][4][5]

Table 1. Impact test conditions

Test No	Vehicle	Mass (ton)	Speed (km/h)	impact angle	Energy (kj)
1	car	1.5	100	20°	160
2	midsize truck	10	60	20°	160
3	midsize bus	10	60	20°	160

Figure 7 illustrates the simulation of the car-barrier test. In the FEA simulation, the test car hit the movable barrier and then was stopped by longitudinal barriers. The test vehicle was successfully contained by the movable barrier without structure failure. It did not penetrate, underride, or override the installation. The occupant velocity was 4.78m/s in x axis and 2.48m/s in y axis, less than 12m/s of requirement; the occupant acceleration was 150.32m/s^2 in x axis and 183.62m/s^2 in y axis, less than 200m/s^2 of requirement.

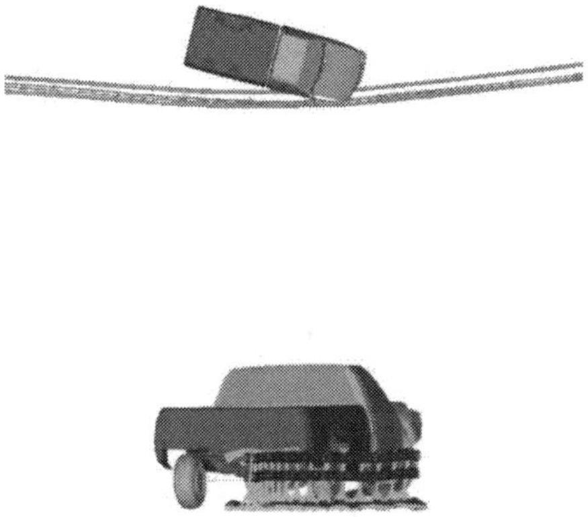

Figure 7. Car-barrier test

Figure 8 illustrates the simulation of the truck-barrier test. In the FEA simulation, the test truck first collided with the movable barrier and then got redirected back to the normal direction. There was no any structure failure such as penetrating, underriding, or overriding the installation. The test truck was successfully contained by the longitudinal barrier.

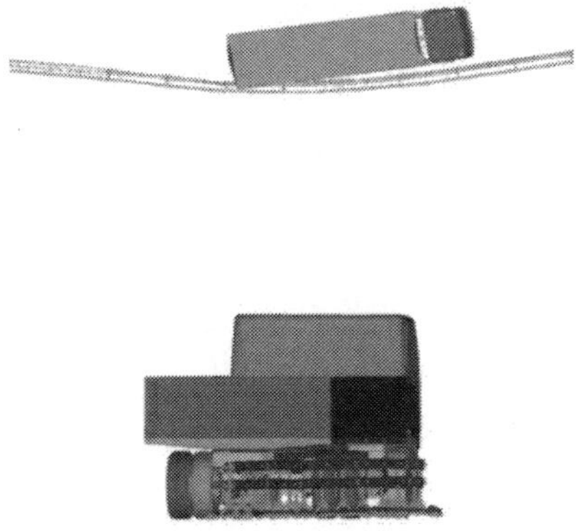

Figure 8. Truck-barrier test

Figure 9 illustrates the simulation of the bus-barrier test. In the FEA simulation, the test bus first hit the movable barrier and then got redirected back to the normal direction. There was no any structure failure such as penetrating, underriding, or overriding the installation. The test bus was contained by the longitudinal barrier, so the test was successful.

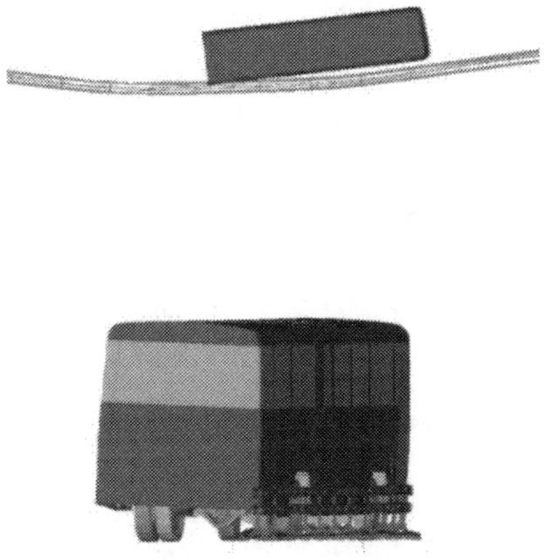

Figure 9. Bus-barrier test

4 Conclusion

Based on the result of impact simulation tests, the development of the new movable barrier is successful. The steel consumption per meter is 52.8 kg, which lowers the cost of the movable barrier to 46$ per meter. The cost-benefit quality is outstanding for a movable barrier solution.

The new barrier solution has many edges in safety performance. First, the protection function is proved by the simulation of impact tests. The barrier can contain vehicles after they run the roadside. It will decrease the severity of run-off-the-road crashes, and thus save road users' life. The application of the movable barrier will greatly raise the safety performance of highway.

Second, the average weight of the barrier is 52.8kg/m, which will allow the barrier to be carried by most of trucks; and therefore the movable barrier has remarkable mobility.

Third, the modular structure makes it convenient to manufacture. The requirement of manufacturing process is simple. Both of these two factors keep the cost of barriers affordable. It means this solution can be implemented in many developing countries. Moreover, the movable barrier is reusable; consequently, it becomes very financially efficient.

Last but not least, the movable barrier is multi-functional. It can be used in various locations other than work zone. For example, a set of movable barriers can be employed in the median opening. It is an effective way to prevent vehicles rushing into opposite traffic flow. It can also be used in the tidal lanes to temporarily separate traveling direction.

The research offers a possible option for those who need barriers that can be quickly installed. There are numerous factors influencing the user's final choice, such as the price of material, the construction capacity, temperature, humidity, traffic volume and so on. The movable barrier discussed in this paper can fulfill most of these demands. Further research will be started to raise the safety performance grade from 160kj-level to 280kj-level.

Acknowledgments

The work was supported by Basic Scientific Research Project of Research Institute of Highway, Ministry of Transport (2017-9018).

Reference

1. Highway Series Editorial Board, *Highway Traffic Engineering and Roadside Facilities* (China Communications Press, Beijing,1999)

2. Beijing shenhuada traffic engineering detection Co Ltd, *JTG B05-01-2013 Standard for Safety Performance Evaluation of Highway Barriers* (Ministry of Transport of the People's Republic of China, Beijing, 2013)
3. The Road of Scientific Research Institute of the Ministry of transport, *JTG D81-2006 Specification for Design of Highway Safety Facilities* (Ministry of Transport of the People's Republic of China, Beijing, 2006)
4. American Association of State Highway and Transportation Officials Executive Committee, *Manual for Assessing Safety Hardware 2009* (AASHTO, Washington DC, 2009)
5. CEN, *EN 1317:1998. Road restraint systems* (CEN, Brussels, 1998)

Assessing a risk-avoidance navigation system based on localized torrential rain data

Sadanori Ito[1,a] and Zettsu Koji[1]

[1]*NICT Big Data Analytics Laboratory, National Institute of Information and Communications Technology, Nukuikitamachi 4-2-1, Koganei, Tokyo 184-8795, Japan*

Abstract. Localized torrential rainfall events and related traffic problems are increasing in Japan, suggesting the need for a navigation-alert system to help drivers avoid such risks. Based on ongoing developments of weather radar systems for early detection of localized torrential rain and a cross-data collaboration platform for traffic optimization, in this study we tested the application of a route-guidance system that can help drivers avoid heavy rainfall. Participants were given equivalent levels of pre-training un the early detection of rainfall and the relationship between rainfall and accidents, then allowed to test a driving simulator set up with four alert methods, three route options, and four levels of possible risk avoidance. Using this system, the heavy rain avoidance rate was 85.63%, suggesting that such a system would be socially acceptable and useful, though further research is needed to refine the specific approach.

1 Introduction

Torrential rain events have recently become more common in Japan [1,2], along with increasing traffic problems caused by unusual weather [3]. Heavy rain raises driving risk [4-6] due to lowered visibility [7,8] and road friction [9]. While localized short-term heavy rain is difficult to predict [10], its narrow coverage makes it possible to avoid through altered travel routes. The Japanese National Institute of Information and Communications Technology (NICT) is developing a weather radar system intended for the early detection and prediction of localized torrential rain [11], while also developing a cross-data collaboration platform [12] to optimize navigation by collecting information related to traffic risks such as various IoT sensor data including those from weather radar (Figure 1). This approach may allow localized torrential rain to be treated as a traffic risk and provide route guidance for avoidance. Although it is widely known that extreme weather affects traffic behaviour [13], it is unclear whether guidance to avoid heavy rain would be used by drivers to avoid accidents and traffic congestion. Thus, in this study we used observed rainfall and actual travel routes to investigate whether heavy rain information would be used by drivers for risk avoidance. In addition, we investigated whether the method of information provision affected driver behaviour.

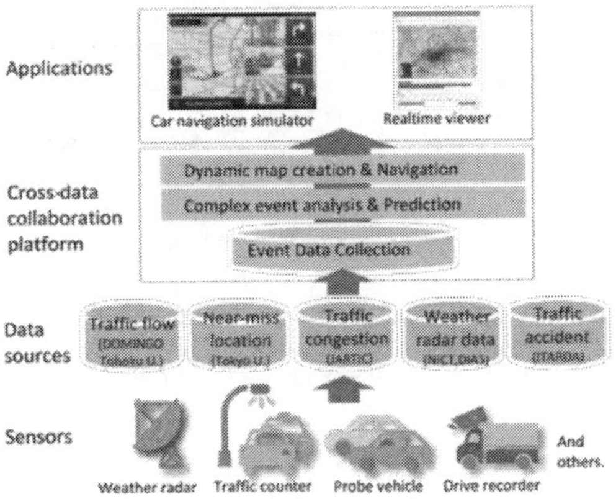

Figure 1. Configuration of cross-data collaboration platform.

2 Methods

2.1 Participants

A total of 30 participants (20 male and 10 female, 20–50 years old) were recruited; each had more than 3 years driving experience.

2.2 Pre-experiment lecture

Decision-making in risk situations relies heavily on personal knowledge and conditions [14]. To ensure an equal experimental baseline, all participants were given a

[a] Corresponding author: sadanori.ito@nict.go.jp

one-hour pre-experimental lecture covering (1) statistical evidence of the relationship between rainfall and accidents, (2) scientific analysis of accidents during rainfall (loss of friction and visual deterioration), (3) reliability of prediction and avoidance of localized heavy rainfall using MP-PAWR weather radar, (4) examples of accidents caused by heavy rain, and (5) example film of a driver's view in heavy rain. Furthermore, as the choice of travel route depends on personal situation (such as commuting or leisure) [15], we instructed all participants to assume that they were driving with family on a holiday, allowing for plenty of time and prioritizing safety. All were instructed on proper use of the experimental apparatus.

2.3 Apparatus

The experimental apparatus (Figure 2) was designed to simulate vehicular operation using a large display and a steering wheel controller (DFKI OpenDS) with map-based navigational information (such as the car's current location) provided by a synchronized display screen (an Android tablet (HUAWEI MadiaPad M3 Lite) with 8" display).

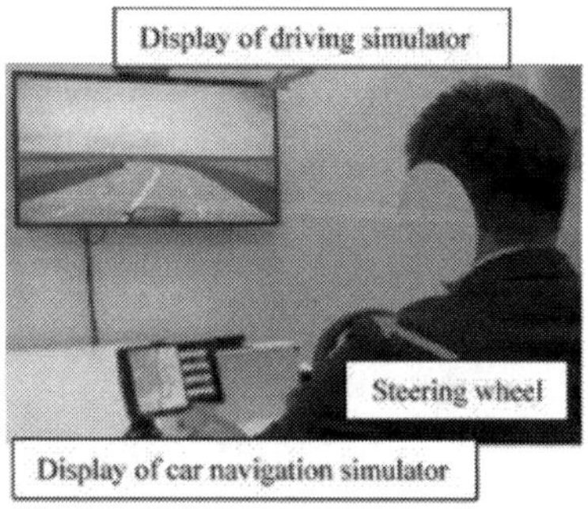

Figure 2. Layout of experimental apparatus.

2.4 Procedure

We designed a procedure based on a dual-task paradigm [16] that allows simultaneous driving and decision-making tasks to best simulate an actual driving situation. The primary driving task required participants to follow the road's centre line while driving a course with gentle curves to left and right that simulated a steady attention load on the driver. The maximum vehicle speed was fixed at 30 km/h by the configuration, but the participant was asked to continue driving at a constant speed by continuing to depress the accelerator. When looking at the route selection display, the participants were instructed to stop as necessary to ensure safety.

The secondary task required the participants to make decisions based on alert messages and route selection (Figure 3). At a pre-set location on the navigation simulator's map displayed the current position of the vehicle on a map. At a pre-set location on the navigation simulator's map, an alert message was displayed, at which point the participant selected whether or not to refer to the route selection screen. In that screen, participants were given the option to select a route that reduced the risk of traffic. We randomly presented 4 types of alert messages and 3 types of route selection screens, performing as many trials as possible within a time limit of about 30 minutes.

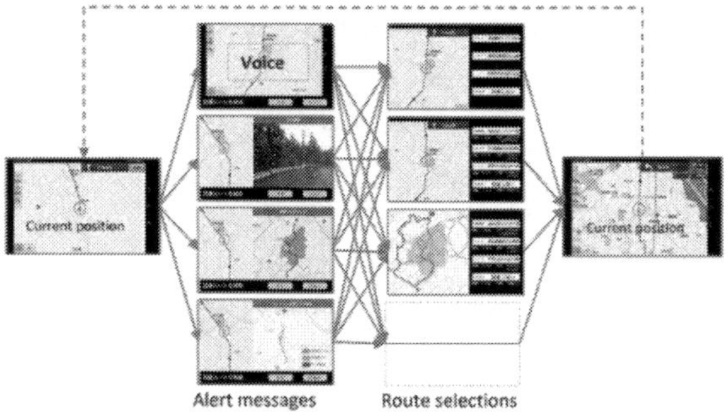

Figure 3. Schematic diagram of the trial procedure, in which drivers are given one of four alerts and one of three route selection options, based on their current simulated location.

2.5 Stimuli

In order to reproduce realistic situations, we used localized torrential rain events recorded by MP-PAWR weather radar [11]. From the radar precipitation maps, we created three traffic risk maps in which the driving environment deterioration was indexed from 0 to 10 (Figure 4) using "Class of rain and wind. [17]" Four possible travel routes were set for each risk map at different levels of risk avoidance (Table 1).

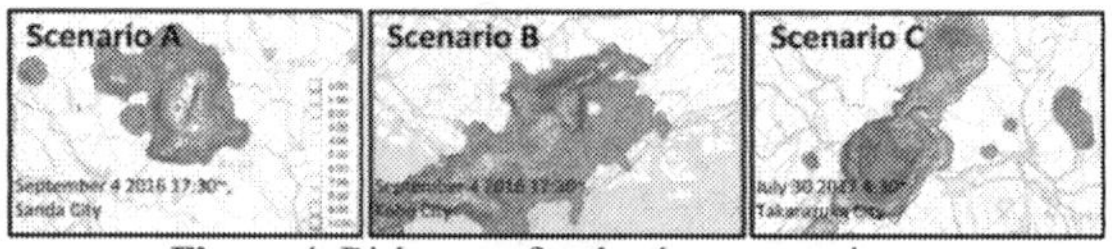

Figure 4. Risk maps for the three scenarios.

Table 1. Travel route specifications.

Scenario	Travel routes			
	Avoid risk Lv. 1 or more	Avoid risk Lv. 3 or more	Avoid risk Lv. 6 or more	Do not avoid risks.
A	41.8 km, 196 min.	49.3 km, 116 min.	41.7 km, 102 min.	27.3 km, 62 min.
B	44.8 km, 118 min.	29.8 km, 70 min.	26.5 km, 62 min.	25.5 km, 54 min.
C	88.5 km, 196 min.	36 km, 84 min.	35.7 km, 83 min.	30.6 km, 41 min.

2.5.1 Alert messages.

In the "Voice only" alert message, a voice stated that "Torrential rain is occurred along your travel route. Do you want to bypass?" In the other three message, this statement was played along with other stimuli. In the "Picture of high-risk spot" message, a simulated picture captured by a recorder on a preceding vehicle was displayed. In the "Risk map" message, the travel route and risk map were displayed. In the "Simplified route with risk" message, color-coded risk values were displayed along the route map [18]. In each case, "See avoidance route (Yes / No)" was displayed at the bottom of the screen. If the button was not pressed after 20 seconds or the "No" button was pressed, the screen was returned to the current position map. If the "Yes" button was pressed, the screen changed to a route selection screen.

2.5.2 Route selection.

The first route selection screen had four buttons with short descriptions of each route: (1) "Avoid areas where accident risk increases due to rainfall (Avoid risk level 1 or more)", (2) "Avoid areas where driving difficulties occur (Avoid risk level 3 or more)", (3) "Avoid dangerous areas (Avoid risk level 6 or more)" and (4) "Continue driving without considering torrential rain." The second type also displayed numerical values for distance and travel time, while the third type included also displayed risk maps and movement forecasts for high-risk areas. The specific situations represented by each evaluation and risk value had been defined in the pre-experimental lectures.

3 Results

The 183 total trials were divided between scenarios A (71), B (55), and C (57) . Each participant conducted 5–8 trials, which varied due to the time limit. The order of trials was considered to avoid bias in order effect.

3.1 Alert messages

The alert was accepted 160 times (87.43%), displaying detailed information on the route selection screen. A binomial test indicated that the proportion of acceptance (.87) was higher than the expected .5, $p < .01$. This may be because heavy rain was the target event of avoidance. A Pearson's chi-square test of independence was performed to examine the relationship between alert message type and acceptance rate, which was not statistically significant (X2 (N = 183) = 2.43, p = .49). The mean reaction time was 9.24 sec (SD=3.23; Table 2 and Figure 5). The Brunner-Munzel test was conducted to compare each alert type; there was no a significant difference between all relationships; $p > .05$. These results suggest that alert type did not affect reaction time.

Table 2. Trial number (Nt), acceptance rate, and reaction time for each stimulus.

Alert type	Nt	Acceptance	Reaction time (sec.)		
			(mean)	(sd)	(mode[1])
Voice	58	48 (82.76%)	9.02	2.93	7.90
Photo	41	37 (90.24%)	8.50	2.40	6.35
Mesh map	44	38 (86.36%)	9.19	2.84	6.10
Simple route	40	37 (92.50%)	10.38	4.31	5.70
Total	183	160 (87.43%)	n/a	n/a	n/a

[1]modal class interval = 100 ms

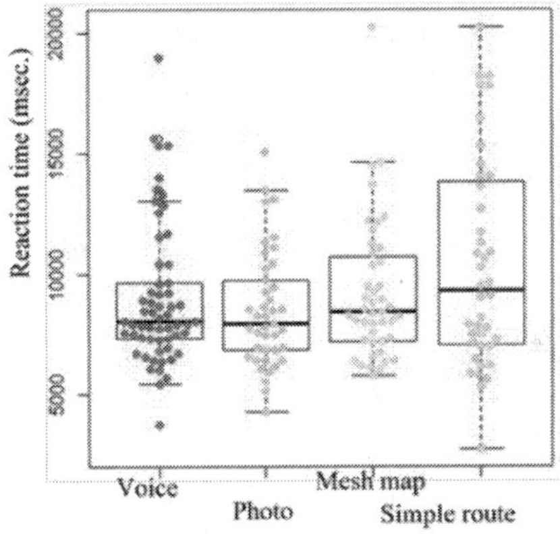

Figure 5. Reaction time to alert messages.

3.2 Route selection

Once the alert was selected, the avoidance rate for all options was 85.63% (Table 3). A binomial test indicated that the proportion of avoidance (.85) was higher than the expected .5, $p < .01$. The Cramer's V was calculated as a measure of association between the type of route selection display and the selection; the results (Cramer's V = .18) showed a moderate relationship.

Table 3. Route selection results.

Selection type	Route selection results				
	Avoid Risk 1 or more	Avoid Risk 3 or more	Avoid Risk 6 or more	Do not avoid risks	Total
Short description	15	29	15	7	66
+ Values of distance and duration	5	18	20	7	50
+ Risk map, routes, and prediction	5	21	9	9	44
Total	25	68	44	23	160

4 Discussion

Our results suggest that localized torrential rain information could be accepted by drivers as a reason for travel route modification. However, differences in information transmission method remain unclear. The difference in the acceptance rate of alert message type was not statistically significant. This may be attributable to the fixation of drivers' situation, the small difference in information reliability due to the pre-experiment lecture, and the low number of trials. Further experiments may be needed to assess driver situation and information reliability.

The alert type did not affect reaction time. The average value of the simplified route type was the longest, but the mode was the shortest (Table 2). F-test results showed that the variances of simplified route type and other types were not equal, which could be affected by inadequate pre-experiment lectures regarding simplified route type representation. In any case, there were no differences worth considering for practical use. Although these results are somewhat unclear, we suggest that the voice option should be adopted, as this is less likely to inhibit driving.

There was a characteristic increase in the avoidance of low-risk situations for the short description type (Figure 6), which may have been caused by the inability to properly compare the risk and cost. This could reduce the reliability of the system in the long term. Also, the "Do not avoid" option increased with a higher level of description, possibly because the risk movement prediction overlaps the route for avoidance. It seems that the risk avoidance benefit was doubted. This acceptability of the movement prediction information requires additional verification.

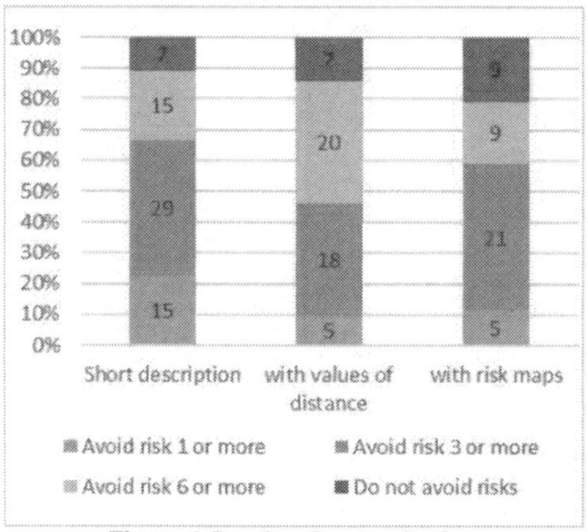

Figure 6. Results of route selection.

We conducted an impression evaluation of participants to examine the relationship between type of route selection display and selection result (Figure 7); the amount of information was highly evaluated. The short description ranked well below the others in sufficiency of information, while the risk map ranked well above the others for clarity of representation, suggesting that cost comparison and graphic representation had a certain effect.

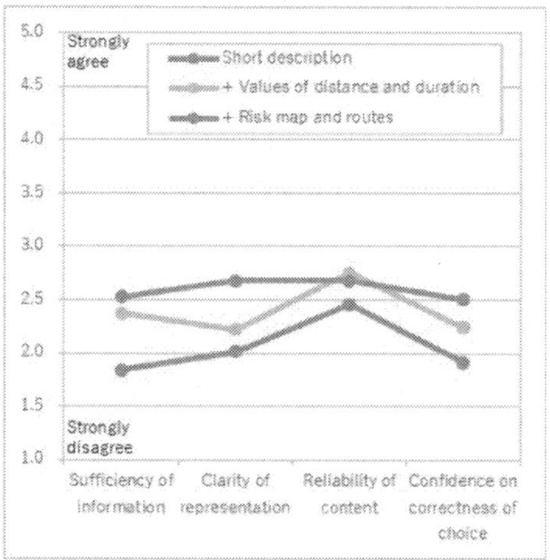

Figure 7. Impression evaluation of representation of route selection display.

5 Conclusion

We tested the acceptability of an application that uses localized torrential rain data as a traffic risk guide driver route selection. Participant behaviour demonstrated the potential value of providing risk information in this way. Based on the results, we suggest that implementing voice alerts along with comparisons of risk and cost data represent the best approach. However, each function's effectiveness should be evaluated separately, and driving distraction should also be considered. In addition, there is a need for sociological research on traffic risks relating to how drivers think about the balance between risk and cost. Events that can tolerate high cost are important risk events. Further research will continue development and experiments aimed at improving the operation of this real-time system.

References

1. Kanae S, Oki T and Kashida A 2004 Changes in hourly heavy precipitation at Tokyo from 1890 to 1999 *J. Meteorol. Soc. Japan* **82** 241–7
2. Japan Meteorological Agency 2019 *Annual occurrence of rain at 50 mm/hour or more* (https://www.data.jma.go.jp/cpdinfo/extreme/extreme_p.html Date accessed: July 24, 2019)
3. Ministry of Land, Infrastructure, Transport and Tourism 201) *White paper 2015* (http://www.mlit.go.jp/hakusyo/mlit/h27/hakusho/h28/data/html/ns009040.html Date accessed: July 24, 2019)
4. Shankar V, Mannering F and Barfield W 1995 Effect of roadway geometrics and environmental factors on rural freeway accident frequencies *Accid. Anal. Prev.* **27** 371–89

5. Edwards J B 1999 The temporal distribution of road accidents in adverse weather *Meteorol. Appl.* **6** 59–68
6. Jaroszweski D and McNamara T 2014 The influence of rainfall on road accidents in urban areas: A weather radar approach *Travel Behav. Soc.* **1** 15–21
7. Ivey D L, Lehtipuu E K and Button J W 1975 Rainfall and visibility: the view from behind the wheel *J. Safety Res.* **7** 156–69
8. Hautière N, Dumont E, Brémond R and Ledoux V 2009 Review of the Mechanisms of Visibility Reduction by Rain and Wet Road
9. Horne W B and Dreher R C 1963 *Phenomena of pneumatic tire hydroplaning* (Washington, D.C.: National Aeronautics and Space Administration)
10. Uesugi T and Tanaka Y 2008 Occurrence structure of short-time heavy Rainfall in the center of Tokyo on 4 July 2000 *Tenki* **55** 23–36 (in Japanese)
11. Takahashi N, Ushio T, Nakagawa K, Mizutani F, Iwanami K, Yamaji A, Kawagoe T, Osada M, Ohta T and Kawasaki M 2019 Development of multi-parameter phased array weather radar (Mp-pawr) and early detection of torrential rainfall and tornado risk *J. Disaster Res.* **14** 235–47
12. Zettsu K 2019 Integrated analysis platform for urban environment big data *J. Environ. Conserv. Eng.* **48** 3 (in Japanese)
13. Kilpeläinen M and Summala H 2007 Effects of weather and weather forecasts on driver behaviour *Transp. Res. Part F Traffic Psychol. Behav.* **10** 288–99
14. Hofinger G, Zinke R and Künzer L 2014 Human factors in evacuation simulation, planning, and guidance *Transp. Res. Proc.* **2** 603–11
15. Kato F, Monden H and Takase N 1992 An analysis of commuters' travel behaviors related to out-of-home, non-work activities under the constraints on a usable time *J. Japan. Soc. Civil Eng.* **449** 97–106 (in Japanese)
16. Kahneman D 1973 *Attention and Effort* (Englewood Cliffs N.J.: Prentice-Hall)
17. Japan Meteorological Agency 2019 *Class of rain and wind* (https://www.jma.go.jp/jma/kishou/books/amekaze/amekaze_index.html Date accessed: July 24, 2019)
18. Ito S and Zettsu K 2019 Report on a hackathon for car navigation using traffic risk data *Proc. 2019 3rd Int. Conf. Intell. Traff. Transport.* (Amsterdam)

Travel decision reversals paradox of competitive metro lines: The Equate-to-differentiate Theory Interpretation

Nian Zhang[1], Yiyuan Zhang[1] and Xia Luo[1,a]

[1] *School of Transportation and Logistics, Southwest Jiaotong University, 999 Xian Road, Chengdu 611756, PRC*

Abstract. Based on the equate-to-differentiate theory, this paper studies the paradox of preference reversal and choice reversal in traffic decision-making in parallel sections of Chengdu Metro Line No.2 and No.4. Travel decision reversals: preference reversals phenomenon (PRP) and choice reversals phenomenon (CRP) are found in the travel investigation for Chengdu Metro Line No.2 and No.4, and the reversals cannot be well explained by random error. Through experimental design of cautious control, the existence of these phenomena is confirmed. This paper establishes the judgment dimension system of option attributes, and designs three groups of psychological experiment in certain, uncertain and risk scenarios. Data analysis shows that PRP and CRP can be explained by the equate-to-differentiate interpretation. The phenomena of reversal are not due to the fact that preference and choice really reverses, but the variable strategic process. The equate-to-differentiate interpretation shows good consistency in the experiments and can predict and explain the preference and choice reversal of travel decision. At last, a comparative analysis which compares the results of the two models of the decision field theory is made and finally the conclusion is obtained.

1 Introduction

Preference reversal is a strange phenomenon of decision behaviours, expressed in the inconformity between the choice and evaluation for the choice game with roughly same expected value, but diverse incomes. Slovic and Lichtenstein (1968) discovered this phenomenon in a risk behaviour research and confirmed its existence through three groups of game experiments (1971) [1]. Preference reversal violated the principle of classical economics theory and decision criterion, thus it was widely oppugned in the economics and psychology fields. For example Greeter and Plott (1979) tried to eliminate preference reversal through the control of economic factors in the experiment [2], and Reilly (1982) tried to improve the communication with subjects, increase their motivation and acquaint them with expected utility theory to eliminate the preference reversal [3]. But the result showed that the preference reversal phenomenon still exists to a considerable extent, which confirmed the universality of its existence. Then the scholars changed to explain the preference reversal. In the field of economics, scholars attributed PRP to the violation of the principles of classical expected utility theory, for example Loomes and Sugden (1982) suggested that PRP violated the principle of transitivity in the regret theory [4]; Holt and Karni(1986) attributed the preference reversal to the violation of the independence principle by analysing the determination of the price pattern (BDM) [5-6]. In recent years, Schmidt and Hey (2004) thought that PRP originated from random error, and an extended error model was proposed to explain PRP [7]. The psychological interpretation of PRP is mainly focused on the development of mental processes. For example, Mellers (1992) proposed the process change theory, which leads to the PRP [8]; Birnbaum (1992) put forward the allocation weight theory by the combination of information decision strategy, and the PRP is well explained, but the model results cannot be predicted [9]. Based on the Equate-to-differentiate approach, Shu Li (2006) explained the PRP [10].

Choice reversals refer to the phenomenon that subsequent choice (preference) changes when decision-makers repeatedly choose the same question in decision-making behaviour. Different from the inconformity between value judgment and choice (preference) in preference reversal, the phenomenon of choice reversal is the inconformity between the previous decision result and the later one, when the decision-maker is facing the essentially same choices [11]. In the field of choice reversals, scholars used the stochastic model to explain the phenomenon of choice reversals and believed that the choice behaviour had random features. For example, Aschenbrenner (1984) proposed that in the either-or problem of multi-dimension options, decision-making is cumulatively formed by the value differences of the different dimension options. Busemeyer (1993) proposed that the individual's preference for choice branches is a probability function of an interval of [0,1]. The formation of choice is the cumulative result of time diffusion [12].

[a] Corresponding author: xia.luo@263.cn

The research on choice reversals in recent years mainly includes that Shu Li (2005-2009) studied the choice reversals of the either-or problem in repetitive experiments under three different scenarios based on the equate-to-differentiate interpretation [13]. It is verified that the equate-to-differentiate interpretation can explain the phenomenon of choice reversals well, and in the framework of equate-to-differentiate theory, the violation of decision axioms in different domain decision problems is also explained. At the same time, in the face of multiple choices, choice reversals were later developed as a study of the inverse problem which is particularly evident in the analytic hierarchy process (3). Shu Li (2015) constructed the "three-member cycle" choice experiment and studied the issue of choice reversals again, further explaining the inconformity nature of the choice [14]. Recent years, Nian Zhang (2018) preliminarily study on preference reversal in traffic under the equate-to-differentiate theory [15].

The economic models and psychological models about PRP and CRP have not yet been widely supported by experiments, and the explanation has not been uniformly agreed. At present, there is less research on the preference and choice reversals of traffic behaviour, but the phenomenon of reversals is not uncommon. This paper is trying to demonstrate the existence of preference and choice reversals phenomenon in the travel decision through "cautious control" and explain the travel decision reversals designs psychological experiment under the equate-to-differentiate theory.

2 PRP and CRP

In our research, the violation phenomenon of preference reversal and choice reversal is first found in the bus lines between the two campuses of Southwest Jiaotong University, and the applicability of equate-to-differentiate theory has been demonstrated partially and preliminarily. This paradox is more prominent in the two choice branches whose expectations are close to each other. In order to validate the universality of the reversals and prove the practicability of equate-to-differentiate theory, the study is extended to the Chengdu Metro Line No.2 and Metro Line No.4 which are almost parallel between Chengdu Sport Institute and Sichuan Provincial Hospital of TCM. As shown in Figure 1.

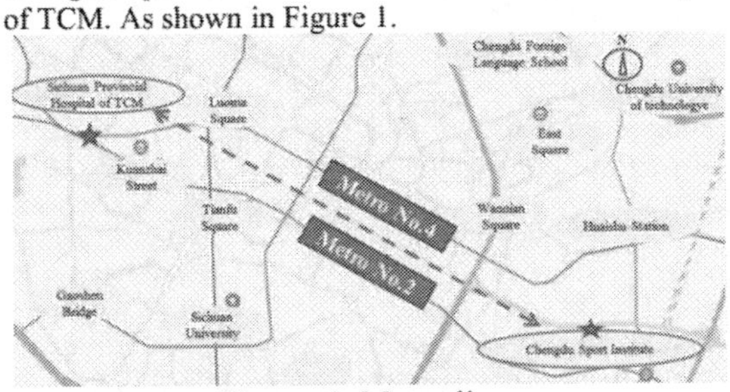

Figure 1. Metro lines

Travelers are asked to evaluate the both metro lines, choose the one they preferred, and the same choice task is taken for the second time 7 days later. The guidance is "Assume that at six o'clock this afternoon you will travel from Chengdu Sport Institute to Sichuan Provincial Hospital of TCM. Metro Line No.2 and Metro Line No.4 are provided, please score the two modes of traffic respectively and choose the one that you preferred according to the attribute information blow" and the result is shown in Table 1-2.

Table 1. Result of preference investigation

		Evaluation		Total
		Better evaluation for Metro Line No.2	Better evaluation for Metro Line No.4	
Preference	Metro Line No.2	29	17	46
	Metro Line No.4	11	13	24
Total		40	30	70

Table 2. Result of repeated choice

	The first round	The second round
Metro Line No.2	43	49
Metro Line No.4	27	21

In the investigation, travellers' preferences and evaluations for the metro lines is inconsistent, 46 travellers choose the Metro Line No.2, while in the evaluation, only 40 travellers rated the Metro Line No.2 better. In choice option, 43 travellers choose Metro Line No.2 and the second increased to 49 in the second time.

3 Equate-to-differentiate theory

As a non-compensatory model, the equate-to-differentiate theory is widely used to explain the framework effect problem and follows the principle of detective dominance, of which the core concept is "Equating to differentiate makes decision, avoiding the complexity does a simple judgment". According to the equate-to-differentiate theory, the travel mode choice model constructed in this paper is based on the following two principles [15].

The choice of transportation mode is not determined by some expected maximum value of the alternative modes, and the traveler is not able to integrate the attributes of the modes of transportation. The difference in one attribute of transportation mode cannot be compensated for by the other attributes.

Travel decision-making allows for variability in selection when repeating choices, and it is not a random model in nature, but a "individual" determinant model. The effect of travelers' instant experience to transportation modes differs due to the different scenes, and the choice of traveler is determined according to the unifying and discrimination of the attribute dimensions of the traffic modes.

The key problem of preference reversal is that if the choice is determined by some desired maximum value. The theory of the Equate-to-differentiate Interpretation does not think that traveller can achieve multi-dimensional integration, and disadvantages of transportation in one dimension cannot be compensated by advantage of another dimension, in face of multidimensional choice, and the decision is more inclined to adopt the non-compensatory decision rules. Meanwhile, the key problem of choice reversal is that if the choice can be obedient to the principle of randomness. Also, the variability of choice decision is allowed in Equate-to-differentiate theory, however, the model is not a probability model essentially, but a decisive model of "individual". The instant experience utility is based on the attribute dimension of different trip scenarios. According to Shu Li's axiomatic analysis of the theory, it is assumed that $A=(A_1, A_2,..., A_n)$ is a limited set of travel options. $A_1, A_2 \in A$ is the two modes of transportation that can be selected and the symbol " > " indicates preference; $B=(B_1, B_2,..., B_m)$ is a subjective space that characterizes the m dimensional attributes of set A and each dimension of option A_i can be any metric level variables (like nominal variable, sequence variable, isometric variable and ratio variable), where A_i can only be represented by the point i of the dimensional space; $C=(C_1, C_2,..., C_m)$ is a set of evaluation for travel modes. The Equate-to-differentiate Interpretation can be described as follows (*20*).

$$E(A_1,A_2)_j=\begin{Bmatrix}(a_1,a_2,...,a_m)^T, when U_{j_0}^{t,p}(O_{1,j})-U_{j_0}^{t,p}(O_{2,j})>0\\(b_1,b_2,...,b_m)^T, when U_{j_0}^{t,p}(O_{1,j})-U_{j_0}^{t,p}(O_{2,j})<0\end{Bmatrix} \quad (1)$$

In the formula: $O_{ij}(j=1,...,m)$ is the objective value level of option A_i in dimension j ; $U_{ij}=U_j^{t,p}(O_{i,j})$ is a monotonic utility function of the subjective value level of A_i in dimension j , and p is the function of the individual formed at a certain time interval t .

$$a_j=\max\left\{0, U_{j_0}^{t,p}(O_{1,j})-U_{j_0}^{t,p}(O_{2,j})\right\} \quad (2)$$

$$b_j=\min\left\{0, U_{j_0}^{t,p}(O_{1,j})-U_{j_0}^{t,p}(O_{2,j})\right\} \quad (3)$$

$$j_0=\arg\max\left\{\left|U_{j_0}^{t,p}(O_{1,j})-U_{j_0}^{t,p}(O_{2,j})\right|, j=1,...,m\right\} \quad (4)$$

In the formula: $a_j b_j$ represent the difference of the option dimensions; j_0 represents the dimension with the greatest difference in subjective utility.

According to the Equate-to-differentiate Interpretation, people can form a preference only if the difference of option dimensions conforms to the principle of dominance, that is

$$\begin{cases}A_1 > A_2, E(A_1,A_2)\geq 0 and E(A_1,A_2)\neq 0\\ A_1 < A_2, E(A_1,A_2)\leq 0 and E(A_1,A_2)\neq 0\end{cases} \quad (5)$$

In order to explore whether preference and choice reversal of travel mode is a random process or an equate-to-differentiate process, we redesign three groups of experiment and analyse the results.

4 Random error interpretation

For this kind of scenario, the result is usually considered as invalid data generated by random errors, but the inconformity ratio in the statistical results is almost close to 10%, the questionnaire design is questioned. Therefore, we try to reduce the errors frequency through the "cautious control" of the investigation. The specific measures are as follows and the problem is set in Table 3.

- Increase the attribute description of the choice branch and strengthen the communication with the subjects(*3*).
- Expand the survey sample to 294 teachers and students who have travelled at least twice in the past month, and use inter-group design, randomly divide them into three groups.
- In order to avoid subjects from rushing to give extreme results as 0 points or 1 point, the evaluation is based on "very bad, bad, moderate, good, very good" and the preference is graded according to "must not choose, will not choose, does not matter, will choose, and must choose".

Table 3. Problems setting of the "Cautious Control"

Guidance: Assume that at six o'clock this afternoon you will travel from Chengdu Sport Institute to Sichuan Provincial Hospital of TCM. Metro Line No.2 and Metro Line No.4 are provided, please score the two modes of traffic respectively and choose the one that you preferred according to the attribute information blow.			
	Ride time (min)	Walking time (min)	Load factor
Metro Line No.2	50	10	60%
Metro Line No.4	40	15	80%

Through the "cautious control", the invalid questionnaire is basically eliminated, but the inconformity does not disappear, and the proportion still reaches to 6%. The result is shown in Table 4-5 (The same score is not considered).

Table 4 Result of preference reversal

		Evaluation for better		**Total**
		Metro Line No.2	Metro Line No.4	
Preference	Metro Line No.2	122	66	188
	Metro Line No.4	50	56	106
Total		172	122	294

The test result ($\chi^2(1)=8.771$, $p<0.01$) does not show significant irrelevance between preference and evaluation. According to the decision theory (axiom), traveller should prefer the traffic mode which they evaluate better, but the statistical analysis results show that this is not the case, which leads us to question the relationship between "preference" and "evaluation", and there is also a problem with the idea of judging "preference" by "evaluation".

Meanwhile, travellers' choice is also inconsistent in twice investigation, when the same investigation is taken in the second time, some travellers change their choices, and the choice variation still exists. The statistical test results show that there is no significant correlation ($phi=0.78$, $sig=0.42$, $p<0.05$) in variation ratio , sample size and interval time. The occurrence of variation cannot be reasonably explained through random errors, and the phenomenon of choice reversal does exist

Table 5 Result of choice reversal

Choice	Group one (N=46, 7 days interval)		Group two (N=99, 10 days interval)		Group three (N=149, 21 days interval)	
	The first investigation	The second investigation	The first investigation	The second investigation	The first investigation	The second investigation
Metro Line No.2	24	28	63	70	111	98
Metro Line No.4	22	18	36	29	38	51
Inconformity ratio	8.7%		7.1%		8.7%	

5 Experiment

In order to explore whether the equate-to-differentiate theory can be used to explain preference reversal of metro travel, we add a psychology matching task experiment that the participants are asked to choose "the option of the greatest difference". Three travel task scenarios under the conditions of certainty, uncertainty and risk are redesigned as follows. In the experiment, A_1 and A_2 respectively mean Metro Line No.2 and Metro Line No.4, B_1 , B_2 and B_3 respectively mean the dimension of attributes difference, C_1 and C_2 mean the travel mode for better evaluation. Before the experiment, each participant was locked by ID number, aiming to track the change of the participant's choice chain.

Scenario I

Students and teachers of Chengdu Sport Institute were informed to go to the Sichuan Provincial Hospital of TCM to participate in a free physical examination. Participants are asked to choose the one they preferred based on the travel information provided, and to circle the matching option which is considered to have the greatest difference, as shown in Table 6.

Table 6. Preference and matching task of scenario I

Preference	Attribute 1	Attribute 2
A_1	arrive in 40 minutes	comfortable inside
A_2	arrive in 30 minutes	crowded inside
Matching	**Differences in attribute dimensions**	
B_1	"arrive in 40 minutes" and "arrive in 30 minutes"	
B_2	"comfortable inside" and "crowded inside"	

46 students participate in the experiment, and the choice option was performed again for the second time after 7 days.

Scenario II

The purpose of setting uncertain scenarios is to reduce the imbalance of dimension difference caused by the "absolutism" attribute in deterministic scenario, and promote the representation of value judgment process. Subjects are informed that a free physical examination will be held after 50 minutes in Sichuan Provincial Hospital of TCM, and the number of participants is limited. Metro Line No.2 and Metro Line No.4 are available to choose, but the load factor is uncertain. In order to explore whether the Equate-to-differentiate model can predict travel choices, we change the sequence of tasks [5]. Participants are asked to perform matching tasks, preference selection, and evaluation tasks in turn. Furthermore, in order to study whether the choice of transportation mode is a multi-dimensional attribute combination process or the equate-to-differentiate process, we add the walking time attribute, to make the overall expected travel time slightly different, as shown in Table 7.

Table 7. Preference and matching task of scenario II

Preference	Attribute 1	Attribute 2	Attribute 3
A_1	arrive in 40 minutes	comfortable	11 minutes on foot
A_2	arrive in 30 minutes	probably crowded	15 minutes on foot
Matching	**Differences in attribute dimensions**		
B_1	"arrive in 40 minutes" and "arrive in 30 minutes"		
B_2	"probably comfortable" and "probably crowded"		
B_3	"11 minutes on foot" and "15 minutes on foot"		

99 travelers participated in the experiment and the same experiment was performed again for the second time after 10 days.

Scenario III

In uncertain conditions, subjects may not have a clear profitability of travel result, in consideration of the uncertain transit time of the bus, a risk task scenario is designed to explore whether the introduction of precise probability can improve the individual's judgment level to reduce the "error" and whether non-compensation principle can explain the preference and choice reversals[6]. 149 graduate students are randomly selected to take the test and informed that an entrance examination will be held in Sichuan Provincial Hospital of TCM after 50 minutes. Now it is accepting register, and the number of register is limited. Late arrivals will be postponed to the next week and the entry may be affected. Two metro lines are available to choose, subjects are required to complete matching, choice, and evaluation tasks in turn. This scenario is shown in Table 8.

Table 8. Preference and matching task of scenario III

Preference	Attribute 1	Attribute 2
A_1	80% possibility of 8 minutes in advance	20% possibility of 2 minutes late
A_2	40% possibility of 10 minutes in advance	10% possibility of 8 minutes late
Matching	**Differences in attribute dimensions**	
B_1	"80% possibility of 8 minutes" or "40% possibility of 10 minutes" in advance	
B_2	"20% possibility of 2 minutes late" or "10% possibility of 8 minutes late"	

In this risk scenario, the expected time is approximately the same. The best dimension is that the Metro Line No.4 arrives 2 minutes ahead of the Metro Line No.2. Therefore, it can help subjects to submit the resume as early as possible and reserve a certain amount of preparation time. The worst result is that the Metro Line No.2 arrives 6 minutes later than the Metro Line No.4, and it is possible to lose the opportunity for register. 149 subjects in the Chengdu Sport Institute take part in the experiment and the same experiment is performed again for the second time after 15 days.

6 Results

In scenario I, Table 9 shows that participants who chose option A_1 pay more attention to the difference of travel time, and the waiting time is "assimilated", and participants who chose option A_2 pay more attention to the difference of waiting time, and the travel time is "assimilated". The result of the chi-squared test ($phi=-0.51$, $\chi^2(1)=15.79$, $p<0.01$) shows that the matching result can explain the choice of traffic mode under uncertain conditions, and it supports the rationality of the equate-to-differentiate interpretation model.

Table 9. Preference and matching result in scenario I

The first experiment		Matching		
		B_1	B_2	Total
Preference	A_1	4	24	28
	A_2	13	5	18
	Total	17	29	46

Table 10. Preference and matching result in scenario II

		Matching			
		B_1	B_2	B_3	Total
Preference	A_1	3	25	28	56
	A_2	36	4	3	43
	Total	39	29	31	99

In scenario II, Table 10 shows that the change of task sequence does not reduce the preference reversal rate. The underlined data in the matching task shows the result consistent with the equate-to-differentiate interpretation, and chi-square test ($phi=-0.72$, $\chi^2(2)=62.67$, $p<0.01$), explicitly explains the correctness of the principle of detective predominance in predicting choice behaviour. The introduction of walking-time attribute does not significantly affect the preference, which proves that individual choice is not a multi-dimensional integration process, but a non-compensatory equate-to-differentiate process.

Table 11. Preference and matching result in scenario III

	Matching	B_1		B_2	
	Evaluation	C_1	C_2	C_1	C_2
Preference	A_1	9	0	78	0
	A_2	0	49	0	13

In scenario III, Table 11 shows that the difference of the best possible results is reduced, that "arriving 8 minutes in advance" and "arriving 10 minutes ahead of time" can be easily eliminated. So, most of the subjects choose the worst results as their basis. The underlined data indicates the result consistent with the equate-to-differentiate interpretation. 58% of the subjects evaluate to the Metro Line No.2 better and choose it at the same time. Participants subjectively believe that the consequences of being late are more severe, and matching task ($phi=-0.69$, $\chi^2(2)=71.84$, $p<0.01$)significantly explains the process of preference. In this scenario, the equate-to-differentiate interpretation can still significantly explain individual preferences, but the data shows that "preference" and "evaluation" are coincident, the so-called phenomenon "preference reversal" disappeared.

Table 12. Preference and matching data of scenario I to III

		Scenario I ($N=46$)		Scenario II ($N=99$)		Scenario III ($N=149$)	
		A_1	A_2	A_1	A_2	A_1	A_2
The first experiment	B_1	4	13	3	36	9	49
	B_2	24	5	25	4	78	13
	B_3	/	/	28	3	/	/
	Correlation φ_1	-0.51		-0.72		-0.69	
The second experiment	B_1	3	15	3	38	5	53
	B_2	21	7	23	5	81	10
	B_3	/	/	27	3	/	/
	Correlation φ_2	-0.50		-0.72		-0.79	

The result of second experiment is shown in Table 12. The underlined data indicates the result which is consistent with the equate-to-differentiate interpretation. In both experiments of the three scenarios, "preference reversal" appear, the test results of matching options significantly explain the equate-to-differentiate process of choice and the equate-to-differentiate interpretation mechanism shows good consistency in the three groups of scenarios. It proves that the choice result can be explained and predicted through the judgment on the difference of travel mode attributes. But in the second experiment, some choices change.

Table 13. Test-retest data of choice consistency

		The first experiment					
		Scenario I ($N=46$)		Scenario II ($N=99$)		Scenario III ($N=149$)	
		A_1	A_2	A_1	A_2	A_1	A_2
The second experiment	A_1	16	8	31	22	52	34
	A_2	12	10	15	21	35	38
Test-retest reliability (r)		0.12 ($sig=0.40$)		0.17 ($sig=0.12$)		0.17 ($sig=0.11$)	

As shown in Table 13, the underlined data indicates the number of subjects who make the same choice in the two experiments. In scenarios I-III, 43%, 37%, and 46% of the subjects change their choice, but the test-retest reliability ($p<0.05$) is not significant. It shows that the judgment of the attribute dimension is a dynamic process in the repeated either-or problem.

Table 14. Test-retest data of consistency between choice and matching

		Choice					
		Scenario I ($N=46$)		Scenario II ($N=99$)		Scenario III ($N=149$)	
		D	S	D	S	D	S
Matching	D	<u>10</u>	7	<u>40</u>	11	<u>47</u>	8
	S	6	<u>23</u>	9	<u>39</u>	18	<u>76</u>
Variation interpretation rate (R^2)		0.15		0.36		0.42	

As shown in Table 14, S and D respectively represent the number of subjects with the same and inconsistent results between choice and matching in the two experiments. The underlined data indicates the number of subjects who change their choice and matching strategies simultaneously. The variability interpretation effect values are respectively 0.15, 0.36, and 0.42 in the three scenarios, and all of them significantly ($p<0.01$) indicate that the choice variation of traffic mode can be explained by the change of the dimension judgment in the matching task under the equate-to-differentiate interpretation. And the summary data of the three groups of experiment is shown in table 15.

Table 15. Summary data of statistical analysis

Scenario	N	D	φ_1	φ_2	r	R^2
I	46	17	-0.51**	-0.50**	0.12	0.15**
II	99	24	-0.72**	-0.72**	0.17	0.36**
III	149	32	-0.69**	-0.79**	0.17	0.42**

7 Comparative analysis

In this paper, the decision field theory is used to simulate above scenarios, making the comparison from the side.

The decision field theory holds that the human decision process is the dynamic accumulation of preference in the activation of the human brain, and the simulation is made by constructing a diffusion process model. Specifically, suppose there are n alternatives, $P(0)$ represents the initial preference strength and define $P(t)=\{p_i(t)\}(i=1,2,\cdots,n)$, where $p_i(t)$ is the preference strength of the alternative i at time t. The evolution of the preference strength is represented by the following equation:

$$P(t)=S\cdot P(t-h)+C\cdot M\cdot W(t)+E(t)\pi \quad (6)$$

In the formula, h is an arbitrarily small unit of time, S is the feedback matrix, M is the attribute matrix, W is the attribute attention transfer matrix, $E(t)\pi$ is the random error term, C is the comparison matrix, which is used to compare the weighted average of each alternative.

Both the decision field theory and the equate-to-differentiate theory recognize the randomness of selection. However, unlike the non-compensatory principle of equate-to-differentiate theory, the decision field theory considers that the formation of preferences is a process of random accumulation of each attribute's valence, and the difference of an attribute can be compensated by the accumulation of another one. Using the traffic attribute data of the initial survey, the simulation subjects are divided into two categories, teachers and students. According to the previous survey of attributes, the main diagonal elements of the feedback matrix are $s_{ii}=0.96$, and the input attention transfer probability matrix is as follows:

$w(t)_1=[0.55,0.22,0.23]$, $w(t)_2=[0.46,0.22,0.32]$.

Table 16 Attributes of two modes

	Ride time (min)	Walking time (min)	Load factor
Metro Line No.2	50	10	60%
Metro Line No.4	40	15	80%

Assuming that the initial preferences are zero, the two groups are simulated 1000 times by MATLAB. The micro-evolutionary map of preference inversion in a repetitive decision-making of the teacher category is shown in Figure. 2 and Figure. 3.

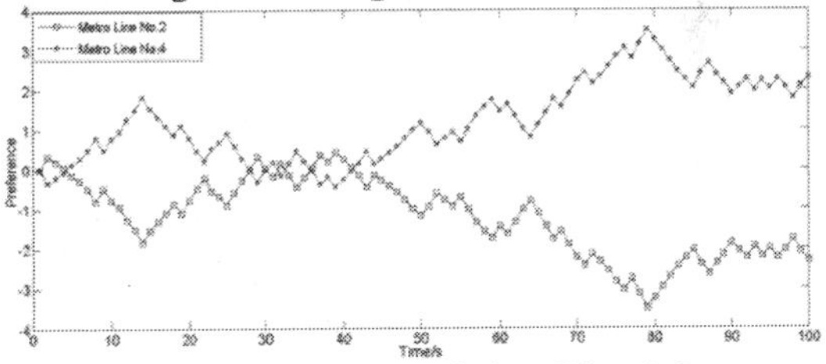

Figure 2. Preference evolution of first choice

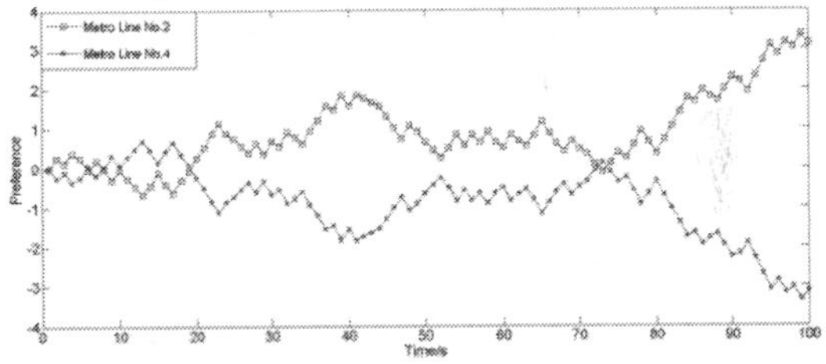

Figure 3. Preference evolution of choice reversal

The decision field theory shows the preference evolution results of two repeated decisions with the same input. For the first decision, Metro line No.4 is chosen, besides, Metro line No.2 is chosen for the second decision. The simulation results are shown in Table 17.

Table 17 The simulation results

Choice	**Group one** ($w(t)_1=[0.55,0.22,0.23]$)		**Group two** ($w(t)_2=[0.46,0.22,0.32]$)	
	Selected number	Percentage	Selected number	Percentage
Metro Line No.2	724	72.4%	388	38.8%
Metro Line No.4	276	27.6%	612	61.2%

Compared with the compensation-based decision field theory, the equate-to-differentiate theory can handle processing of fuzzy non-quantitative attribute indicators more flexibly, while the decision field theory requires strict quantitative data indicators.

The equate-to-differentiate theory attach importance to the statistics of decision-making results and cannot describe the micro-evolution process of preferences. However, the decision field theory can intuitively describe the evolution process of individual preferences.

The randomness in the decision field theory is reflected in the transfer process of attribute attention, and the formation of preference is the result of multi-attribute valence accumulating over time. Moreover, the result of inversion cannot be predicted. However, the equate-to-differentiate theory considers the decision-making process as the value judgment of the single attribute, and the inversion results can be predicted by matching experiments.

7 Conclusion

Some of the reversal phenomena cannot be well explained under the framework of the expected utility theory, and the reversals also cause some confusion and limitations in research and work. The experimental results show that the preference reversal and choice reversal phenomenon actually exist in the travel decision and the variation cannot be reasonably explained by cautious control and random error analysis. Traveler's decision of the transportation mode adopts completely different attitudes, the travel decision is an independent thinking process and the transmissibility principle is not significantly followed.

The analysis shows that, the phenomenon of "reversal" does not really stem from the decision inconformity, but the dimensional differences in the attributes of travel modes. The introduction of precise probability does not necessarily affect the traveler's decision metrics, and the attribute dimension difference is determined by the individual subjective values. If the same attribute dimension is followed in the decision, the reversals do not occur, and if the attribute dimension changes, the reversals do occur.

When faced with the decision that travel modes have the same expectation values, the detective superiority principle under the equate-to-differentiate interpretation can better explain and predict the preference reversal and choice reversal phenomenon. The Equate-to-differentiate Interpretation provides a new perspective for understanding and studying human risk decision-making behavior and provides an alternative explanation for the inconsistence of preferences and choices that shown in travel decision. When confronted with this inconsistence, there is no need to proceed from the "maximization" criterion, travelers can often consider the differences in different dimensions to change their strategies in a timely manner. This result in the behavior is seemingly abnormal but ecologically rational. This article has only carried on the exploration to the travel decision reversal phenomenon of either-or problem, and tried to carry on the theoretical explanation, and we also hope to introduce valuable remarks.

References

1. S. Lichtenstein, P. Slovic, *Journal of Experimental Psychology*, 89(1971)
2. D. M. Grether, C. R. Plott, *American Economic Review*, 69(1979)
3. R. J. Reilly, *American Economic Review*, 72(1982)
4. G. Loomes., R. Sugden., *Economic Journal*,. 92(1982)
5. C. A. Holt, *American Economic Review*, 76(1986)
6. E. Karni, Z. Safra, *The Econometric Society*,. 55(1987)
7. U. Schmidt, J. D. Hey, *Journal of Risk and Uncertainty*, 29(2004)
8. BA. Mellers, LD. Ordóñez, H. M, *Organizational Behavior & Human Decision Processes*, 52(1992),
9. M H. Birnbaum, G. Coffey, B.A. Mellers, et al. *Journal of Experimental Psychology Human Perception & Performance*, 18(1992)
10. S. Li, *Psychological Record*, 56(2006)
11. S. Li, *East China Normal University Press*, (2015)
12. J. R. Busemeyer, J.T. Townsend, *Psychological Review*, 100(1993)
13. L. Shu, *Acta Psychologica Sinica*, 37(2005)
14. S. Li, *International Journal of Creativity & Problem Solving*, 11(2001)
15. N. Zhang, X. Luo, Y.Deng, *Journal of Transportation Systems Engineering and Information Technology*, 18(2018)

The best plan of evacuation in Louvre

Yao Xiao[1,a], Yinqiao Yan[2] and Yiqin Lei[3]

[1]*School of Mathematics and Statistics, Beijing Institute of Technology, China*
[2]*Institute of Statistics and Big Data, Renmin University of China, China*
[3]*Faculty of Science, Leiden University, Netherlands*

Abstract. According to reports, since 2012, there have been 12 terrorist attacks in France. At the same time, the number of visitors of Louvre reached a peak of 8.1 million in 2017. Therefore, making an efficient evacuation plan in Louvre is particularly important. In our paper, we assume that the process of evacuation contains two main parts: the path from the starting point to the exit, and the location where the congestion may occur, such as stairs and exits. For the first part, we use the ant colony optimization algorithm to find the optimal path and get how many flows arrive at each exit. For the second part, as we already know the number of flows, we can use queuing theory to determine whether it is congested or not and calculate the corresponding congestion time. Finally, the optimal evacuation path and the corresponding time are obtained. In addition, associated with the application "Affluences" used in Louvre, some devices are designed to measure the average speed of the crowd movement. If the average speed is below a threshold value, an emergency exit will be opened to improve the evacuation process.

1 Introduction

The increasing number of terrorist attacks in France has led to the need of improving emergency evacuation plans for a variety of popular attractions, allowing visitors to evacuate the building quickly and efficiently in order to avoid casualties. The Louvre is one of the largest and most visited museums in the world, and it receives a huge number of visitors each year. In 2017, it received more than 8.1 million visitors. In addition, the unstable number of visitors in each day and the diversity of visitor types make evacuation in emergencies more challenging. Therefore, it is very important to establish a model and design an effective emergency management policy for the Louvre.

Due to the increasing number of terrorist attacks, the Louvre hopes to get a fast, effective and safe evacuation plan for visitors, so that once the emergency happens, the program can quickly evacuate all the staff and tourists and minimize casualties. We are required to help the supervisor develop an emergency evacuation model that allows staff to evacuate visitors while also allowing emergency personnel to enter the building as fast as they can. For this reason, we carry out two steps as follows: (i) Propose emergency management and procedures to the Louvre for the safety of tourists; (ii) Discuss how to apply the model to other large, crowded structures.

Notations. T_1, T_2, T_3 denote the disaster warning time, evacuation preparation time, and evacuation movement time respectively.

Assumptions. To simplify our model, we give some assumptions as follows without loss of generality. (i) The subway station does not consider the access restrictions in the event of a disaster. (ii) The roads are straight, regardless of the hindrance of the exhibits. (iii) All personnel fully obey the staff command and disabled go to dedicated channel. (iv) After the main export congestion, the crowd orderly queued out of the museum, which means once they start queuing, they no longer leave to other teams. (v) People in lower ground floor go to the Pyramid exit and the Richelieu exit; People in Napoleon hall are equivalent to not entering the museum; People in ground, 1st and 2nd floor are required to evacuation from ground floor. (vi) The elevator is out of service, only the stairs. (vii) Congestion may occur at the entrance and exit, other paths are smooth, therefore, other routes will not cause stamping accidents.

2 Preliminaries

In the case of special disasters (such as natural disasters or terrorist attacks), whether the people in the building can safely evacuate of people depends mainly on two characteristic times: First, the Available Safety Egress Time (*ASET*), which indicates the time required for the disaster to pose a danger to people. The second is the Required Safety Egress Time (*RSET*) [1], which is the time required for evacuating people to a safe area. If it can be ensured that the available safe evacuation time of all personnel in the building is greater than the safe evacuation time, the safety of the personnel in the building is guaranteed in the event of a disaster, which means the evacuation plan of the building is successful.

[a] Corresponding author: xiaoyaoxy1998@sina.com

Assuming the time from the occurrence of the disaster to the detection of the disaster by the indoor personnel is the disaster warning time T_1, the time taken for the staff to prepare for evacuation is the evacuation preparation time T_2, and the time for evacuating the visitors to reach the safety zone outside the building is the evacuation movement time T_3. So the required safety egress time $RSET=T_1+T_2+T_3$ and T_2 are based on the specific situation. Then we will talk about T_3 in the following.

2.1 Emergency evacuation route selection principle

The optimal evacuation route is the best route for emergency evacuation. This route should ensure that the trapped people in the danger zone could be evacuated to safe places in the shortest time. Therefore, in combination with the question above, the best evacuation path should meet the following principles: (i) Try to minimize the time required for evacuation. (ii) Try to avoid the blocking exit. (iii) The evacuation path should have a clear indication. (iv) The building should have an alternate evacuation path.

2.2 Basic theories of evacuation

The basic research on evacuation is actually the study of various parameters in the evacuation process, including effective width, personnel density, personnel movement speed and personnel flow coefficient.

The speed of people flow, according to the literature [2], there is an empirical formula between the flow density and the speed of movement given by

$$v = 1.1\rho^{-0.7954}\,(\mathrm{m/s}),$$

where v refers to the speed of personnel movement and ρ refers to the personnel density.

The personnel flow coefficient refers to the number of people who can pass through the safety passage in unit width and unit time. In practice, it can be represented by the travel speed of the head of the flow. Under normal circumstances, when the flow coefficient of people is $N \leq 1.0\ person/(m \cdot s^{-1})$, there is no delay of people in the safe evacuation passageway. At this time, the flow of people in the building is free [3].

According to the measured statistical data of the personnel flow coefficient, the reference value of the flow coefficient of the entrance and exit is shown in Table 1.

Table 1. Coefficient of personnel flow in channel or passageway.

Flow state	Type of entrance and exit	Personnel flow coefficient ($person \cdot m^{-1} \cdot s^{-1}$)
Emergency evacuation	Corridor	1.5
	Stairwell entrance	1.3
	Refuge entrance	1.5

2.3 The effect of disabled people

The difference of the evacuation ability between disabled individual and healthy individual is the fundamental cause of the particularity of the evacuation of the disabled people. In emergencies, it is less likely that only a single disabled visitor in the Louvre, and disabled people will also hinder the evacuation efficiency in the stairs to some extent. In addition, the actual evacuation ability of disabled persons is affected by many external factors, including the external objective environment and other people in the surrounding area. Therefore, persons with disabilities should take emergency passages with the help of emergency personnel and do not evacuate with normal tourists, which can improve evacuation efficiency.

3 Personnel evacuation optimization model

Nowadays, models for solving emergency evacuation problem of large venues mainly include cellular automata [4], lattice gas model [5] and so on. These models mainly analyze the next evacuation path based on the current situation. However, they are not ideal in providing global optimization of individual space-time path analysis. Therefore, we carried out the following analysis.

In the emergency evacuation of the office building, the evacuation time T_3 can be divided into two parts: The first part is the movement time in the evacuation path t_{31}, including the leveling time and the stair time. Without severe blocking, moving forward at a certain speed can be guaranteed. And there is such a situation in which the queue is waiting long due to the small entrance and exit and the large flow of evacuated people. This refers to as the stagnant time t_{32}. Then we have $T_3=t_{31}+t_{32}$.

For the movement time t_{31}, since it is related to the path selection, the path optimization can shorten the time. Therefore it can be optimized by the adaptive ant colony optimization algorithm. For the stagnant time t_{31}, since it is the waiting time, so it can be optimized by queuing theory.

3.1 Ant colony optimization algorithm

Ant colony optimization algorithm is a bionic optimization algorithm used to find the optimal path in the graph. It was initially proposed by Marco Dorigo [6, 7], inspired by the behavior of ants finding the optimal path by releasing pheromones in searching for food. When solving large-scale complex problems, the effect of ant colony optimization is generally better than that of Floyd algorithm and Dijkstra algorithm due to the randomness, since the former has a certain probability to escape from local optima. Thus, the ant colony optimization algorithm is more robust.

Let m be the number of ants, n be the number of node, and $b_i(t)$ represents the number of ants at node i at the time t. Then we have $m = \sum_{i=1}^{n} b_i(t)$. V donates the set of nodes and E denotes the set of edges between nodes. $\tau_{ij}(t)$ represents the information density at time t on path $\langle i, j\rangle$ and $\Gamma = \{\tau_{ij}(t) \mid v_i, v_j \in V\}$ refers to the set of pheromone concentrations connecting all the nodes remaining on the path at the time t. The pheromone concentration is equal on each path at the initial time; the optimal path selection based on the ant colony optimization is looking for nodes in the directed graph $G=(V,E,\Gamma)$ to achieve the minimum cost of ownership.

During the movement of ant $k \in \{1,2,\cdots m\}$, the direction of the transfer is determined by the concentration of the pheromone on each path. Here, $tabu_k$ is used to record the nodes that the ant k is currently walking. The path set is dynamically adjusted with the evolution process. In the path selection process, the ants calculate the state transition probability according to the pheromone concentration on each path and the heuristic information of the path.

Let $p_{ij}^k(t)$ denote the state-transition probability with which the ant k is transferred from node i to node j at the moment t. In [6], $p_{ij}^k(t)$ is defined by

$$p_{ij}^t = \begin{cases} \dfrac{\left(\tau_{ij}(t)^{\alpha}\right)\left(\eta_{ij}(t)^{\beta}\right)}{\sum\limits_{s \in \text{allowed}_k} \left(\tau_{is}(t)^{\alpha}\right)\left(\eta_{is}(t)^{\beta}\right)}, j \in \text{allowed}_k \\ 0, j \notin \text{allowed}_k \end{cases}. \quad (1)$$

In (1), $allowed_k=\{V\text{-}tabu_k\}$ indicates the node that the ant k is allowed to select in the next step; α represents the information heuristic factor, reflecting the relative importance of the pheromone; β represents the expected heuristic factor, reflecting the relative importance of the pheromone in the ant selection path during the movement; $\eta_{ij}(t)$ represents the local heuristic factor, reflecting the degree of expectation from the transition, usually taking the following expression $\eta_{ij}(t)=1/d_{ij}$, where d_{ij} denotes the Euclidean distance between two adjacent nodes i and j. For ants k, the less d_{ij} is, the larger $\eta_{ij}(t)$ will be, then the larger $p_{ij}^k(t)$ will be. Obviously, the heuristic function represents the degree to which the ant moves from the node i to the node j.

After each ant has completed one step or completed the traversal of all nodes, the residual information is updated. As $\tau_{ij}(t)$ indicates pheromone concentrated on path $\langle i, j\rangle$ at moment t, we have

$$\tau_{ij}(t+1) = (1-\rho)\tau_{ij}(t) + \sum_{k=1}^{m} \Delta\tau_{ij}^k(t), \quad (2)$$

where ρ is the pheromone volatilization coefficient and $\rho \in (0,1]$. $1\text{-}\rho$ represents the pheromone concentration residual factor, indicating the relative importance of the residual pheromone. $\Delta\tau_{ij}(t)$ denotes the increment of pheromone concentration on path $\langle i, j\rangle$ between time t and time $t+1$. In the initial moment $\Delta\tau_{ij}(t)=0$.

In different pheromone update strategies, the Ant-Cycle model is better in global optimization because it utilizes global information. The update strategy is given by

$$\Delta\tau_{ij}^k(t,t+n) = \begin{cases} \dfrac{Q}{L_k}, & (i, j) \text{ on the path of ant } k \\ 0, & \text{otherwise} \end{cases}. \quad (3)$$

In (3), Q represents the total concentration of pheromone released by each ant and L_k represents the total tour length of ant k.

3.2 Queuing theory

A complete queuing system consists of input process, arrival rules, queuing rules, service organization structure, service time and service planning. The queuing system composition is given as follows.

The optimization problem, which is also called the system optimal design problem and the dynamic control problem, is one of the problems solved by the queuing theory. Queuing delay time can be optimized by queuing theory.

Average service time T_f. Average service time refers to the time when evacuation personnel pass the security exit, which is given by

$$T_f = \frac{L}{V},$$

where L represents the length of the evacuation channel (the space connecting the internal exit and the outside world), and V represents the speed at which the personnel evacuate the passage through the exit.

Average service rate μ. The average service rate can be regarded as the average number of people served in a unit of time by the evacuation personnel, which is given by

$$\mu = \frac{1}{T_f},$$

Average arrival time T_d. The average arrival time is the number of evacuations divided by the number of people entering the waiting area by each route per unit time. It is given by

$$T_d = \frac{M}{\sum\limits_{i=1}^{n} N_i B_i} = \frac{\rho L W}{\sum\limits_{i=1}^{n} N_i B_i},$$

where M is the number of people in the evacuation channel; ρ is the population density; L is the length of the evacuation channel for the exit; W is the width of the evacuation channel for the exit; n is the number of branches for the branch; N_i is the flow for the entrance to branch i; and B_i is the width of the entrance to branch i.

Average arrival rate λ . The average arrival rate can be seen as the number of people arriving in a unit of time. It has the following form

$$\lambda = \frac{1}{T_d}.$$

4 Implementation and results

In this section, we will implement our personnel evacuation optimization model and give the numerical results including the optimal evacuation route and the correspondingly time. In the first part, we will demonstrate the results of applying ant colony optimization algorithm. In this case, The nodes represents all the exits and the position of stairs. Since people will not choose another stair if they already arrive at the position of one stair, we do not need to consider the length of stairs and a stair can be simplified as a node when we apply the ant colony optimization algorithm. In the second part, we will give the results of applying queuing theory. The places where queuing would happen in our model are the exits. Then we can obtain the total escaping time. All the results are conducted in MATLAB R2017b in Windows 10 on a Dell Laptop with an Intel(R) Core(TM) i7-5500U CPU at 2.40GHz and 8 GB of RAM.

Implementations. We assume that when a disaster occurs, people on the 1st and 2nd floors will choose the stairs closest to themselves to go downstairs. Therefore, the optimal escape route is such a path starting from the upper of the stairs and ending at the exit. Thus, the problem is transformed into the graph shown in Figure 1, starting from all the nodes at the 2nd layer, and searching for the optimal path to each exit of the ground layer. Selecting the shortest time to use as the optimal escape route for that starting node.

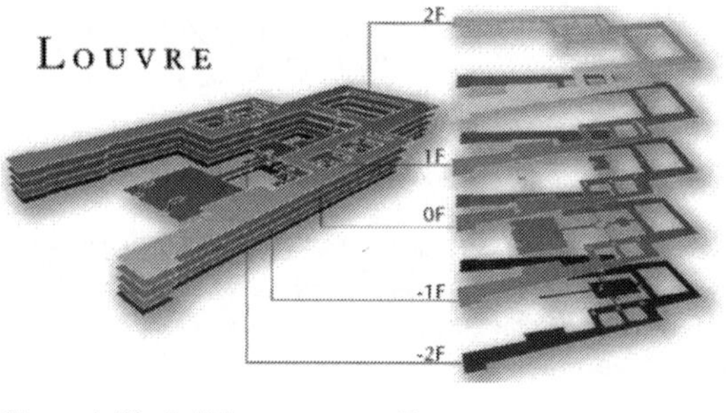

Figure 1. The building structure of Louvre.

4.1 Results of ant colony optimization algorithm

In this part, we will use the ant colony optimization algorithm to derive the optimal evacuation route and the movement time t_{31}. Parameters in this algorithm are set as $m = 1$, $\alpha = 1$, $\beta = 5$, $\rho = 0.1$, $k = 50$, $Q = 100$, and $v_{cor} = 1.0 m/s$. Based on the assumptions in Section 1, we just consider the tourists in the 1st and 2nd floor. We have numbered the stairs sequentially. The results are as follows in Table 2.

Table 2. The optimal evacuation routes and the movement time of the 1st and 2nd floor. St: Stair; Ri: Richelieu; Ar: Arts; Li: Lions.

Floor	Routes	Time
2nd	St18 - St11 - Ri Exit 1	68.83
	St19 - St13 - Ar Exit	143.83
	St20 - St16 - Ar Exit	87.33
1st	St11 - Ri Exit 1	48.17
	St12 - Ri Exit 1	140.67
	St13 - Ar Exit	123.67
	St14 - Ar Exit	113.67
	St15 - Ar Exit	102.67
	St16 - Ar Exit	66.67
	St17 - Li Exit 2	48.67

The visualized map of the evacuation routes are shown in Figure 2.

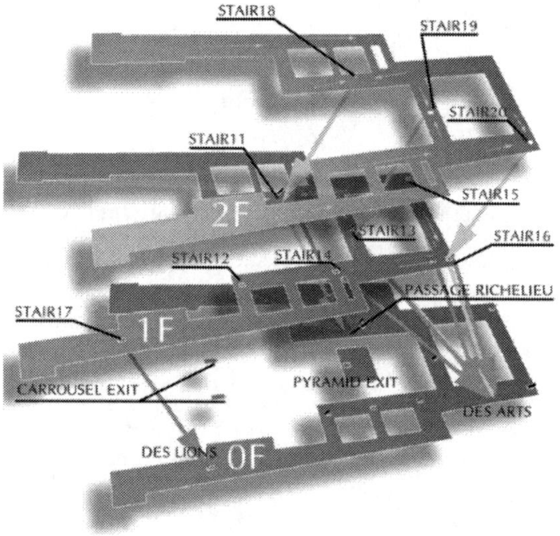

Figure 2. The visualized map of the evacuation routes of 1st and 2nd floor in Louvre.

4.2 Results of queuing theory

In this part, we will use the queuing theory to obtain the waiting time when congestion occurs. For queuing theory, we choose $L = 10$, $V = 0.6$, $W = 1.2$, $N = 1.3$, $B = 0.8$, and $n = 2$. We choose ρ=1.0, 1.5, 2.0 to test our model, representing low, medium and high population densities in practice respectively. The results of stairs and exits are as follows in Table 3, Table 4 and Table 5 respectively.

Table 3. Congestion time at stairs.

	ρ	
Congestion time (s^{-1})	1.0	2.5
	275.33	439.82

Table 4. Congestion situation at exits (ρ=1.0). Br: Branches; Con: Congestion; Ri: Richelieu; Ar: Arts; Li: Lions; Py: Pyramid; Ca: Carrousel.

Exits	μ	λ	Br	Con	Con time	Speed (m/s)
Ri	0.255	0.125	3	No	-	1.80
Ar	0.500	0.333	6	No	-	2.40
Li	0.500	0.055	1	No	-	0.40
Py	0.167	0.074	2	No	-	0.40
Ca	0.500	0.028	1	No	-	0.20

Table 5. Congestion situation at exits (ρ=2.5).

Exits	μ	λ	Br	Con	Con time	Speed (m/s)
Ri	0.080	0.125	3	No	-	0.72
Ar	0.160	0.333	6	Yes	394.83	0.96
Li	0.160	0.055	1	No	-	0.16
Py	0.053	0.074	2	No	-	0.16
Ca	0.160	0.028	1	No	-	0.08

The visualized evacuation situation at 60 seconds and 100 seconds are shown in Figure 3 and Figure 4.

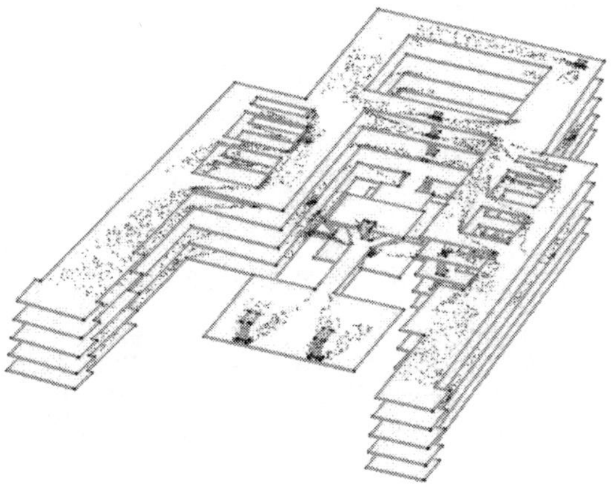

Figure 3. The visualized map of the evacuation routes of 1st and 2nd floor in Louvre (60 seconds).

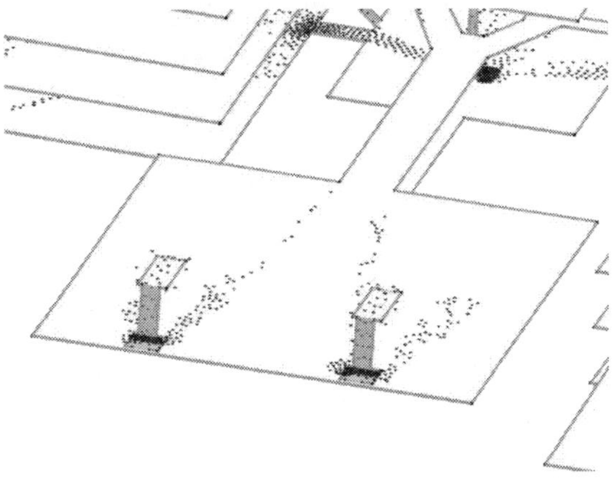

Figure 4. The visualized map of the evacuation routes of 1st and 2nd floor in Louvre (100 seconds).

We have the following observations.

(i) When ρ=1.0, the total escaping time is 148.5s + 143.83s + 275.33s * 2 = 14.05 minutes. When ρ=2.5, the most distant coordinates on the second floor is (457,192), 148.5 meters from the Stair19. After 148.5s, one will pass the route of Stair19-Stair13-Arts Exit. There is a congestion at the art exit and the congestion time is 43.13s. Therefore, the total escaping time is 148.5s + 143.83s + 439.82s * 2+394.83s = 26.11 minutes.

(ii) The total time is then compared to the available evacuation time of the disaster. When the population density is small, it can ensure that everyone can successfully evacuate within the escaping time. When the population density is large, it can guarantee that the number of escapes is as much as possible within the escaping time.

4.3 Emergency exit

According to the assumptions, in the case of high population density, only one emergency door is allowed to open. We implement our model again and see that the evacuation effectiveness is the best when the Emergency Door 1 is opened. The results are as follows in Table 6 and Table 7.

Table 6. The optimal evacuation routes and the movement time of the 1st and 2nd floor when emergency exit 1 is opened. Ri: Richelieu; Ar: Arts; Li: Lions; Em: Emergency.

Floor	Routes	Time
2nd	St18 - St11 - Ri Exit 1	68.83
	St19 - St13 - **Em Exit 1**	62.59
	St20 - St16 - Ar Exit	87.33
1st	St11 - Ri Exit 1	48.17
	St12 - Ri Exit 1	140.67
	St13 - **Em Exit 1**	44.11

 https://doi.org/10.1051/matecconf/202030803008

	St14 - **Em Exit 1**	107.67
	St15 - Ar Exit	102.67
	St16 - Ar Exit	66.67
	St17 - Li Exit 2	48.67

Table 7. Congestion situation at exits (ρ=1.5). Congestion situation at exits. Ri: Richelieu; Ar: Arts; Li: Lions; Py: Pyramid; Ca: Carrousel.

Exits	μ	λ	Br	Con	Con time	Speed (m/s)
Ri	0.12	0062	3	No	-	0.60
Ar	0.24	0.167	6	No	-	0.80
Li	0.24	0.028	1	No	-	0.13
Py	0.08	0.037	2	No	-	0.13
Ca	0.24	0.014	1	No	-	0.07

The farthest point from the lion exist is on the first floor, 148.5m from the exist, speed $1m/s$, 148.5s, and arrive at the lion exist through Stair12 - Lion Exit. Also the exit is not congested.

The final evacuation time: 148.5s + 140.67s + 439.82s * 2 = 22.43 minutes < 26.11 minutes. It can be seen that after opening the Emergency Exist 1, it provides a good diversion effect for the art door, and successfully solved the art door congestion.

4.4 Sensitivity analysis

4.4.1 Characteristics of basic ant colony optimization parameter

Many scholars have discussed the parameters of the basic ant colony optimization algorithm. The parameters Q, C, α, β, and ρ can be determined the optimal combination by experimental methods. There is no theoretical basis at present, and the published results are for specific problems. For the classic TSP, the empirical result $0 \leq \alpha \leq 5$, $1 \leq \beta \leq 5$, Q has little effect and C select relative set based on the fitness of the problem.

4.4.2 Selection of pheromone volatility

In the ant colony optimization, the artificial ant has the function of human memory. As the time pass, the information left before will gradually disappear, and the parameter $1\text{-}\rho$ indicates the disappear degree in the algorithm model. ρ is the signal residual coefficient. Ant colony optimization, like genetic algorithm and other simulation evolution algorithms, also has the disadvantages of slow convergence speed and easy to fall into local optimum. The amount of information volatility $1\text{-}\rho$ is directly related to the global search ability of ant colony optimization and its convergence speed. Due to the existence of information volatility, when the problem is large, it will make the amount of information in those paths that have never been searched reduce to 0, thus reducing the global search ability of the algorithm, and when $1\text{-}\rho$ is too large, the previously searched path is too likely to be selected again, which will also affect the random performance and global search ability of the algorithm. Conversely, by reducing the amount of information volatility $1\text{-}\rho$, although it can improve the random performance and global search ability of the algorithm, the convergence speed of the algorithm is reduced.

5 Conclusion

In this paper, an evacuation model based on the combination of adaptive ant colony optimization algorithm and queuing theory has been proposed as an efficient tool to help the staff in Louvre to solve the difficult escape problem. Our model has the characteristics of information positive feedback and heuristic search, which guarantees that the model can not only solve the evacuation problems under the circumstances of non-specific-location occurrences such as terrorist attacks, but also solve the problems with specific-location occurrences such as fires and stampedes. For the latter case, all we need to do is to cut off the paths where the disaster occurs when applying the ant colony optimization algorithm. Thus, it can design the best escape route of specific disasters.

We also provide the staff with some suggestions:

• Emergency personnel should use emergency exits that are not open during evacuation. When entering the building, their duty is to help the elderly and children to evacuate, and to explain the route for foreign tourists;

• The application "Affluences" can be used as a speed measuring instrument during evacuation. When it shows that a certain exit is blocked, the staff should immediately open the Emergency Exit 1 and guide the visitors to evacuate based on the new evacuation route.

References

1. C. Wentao, X. Dan, Journal of Shenyang Aerospace University, **4**, 76-79 (2011).
2. F. Liu, S. Liao, Chongqing: Chongqing University Press (2007).
3. D. Wu, S. Yang. China Safety Science Journal, **22**, 56-62 (2012).
4. W. Yuan, K.H. Tan, Physica A. Statistical Mechanics and Its Applications, **384**, 549-566 (2007).
5. W. Zhang, L. Zhao, Journal of Safety and Environment, **10**, 169-172 (2010).
6. M. Dorigo, V. Maniezzo, A. Colorni, IEEE Trans. Syst. Man Cybern Part B-Cybern, **26**, 29-41 (1996).
7. M. Dorigo, PhD thesis, Politecnico di Milano, Italy (1992).

An Optimization Model of Railway Empty Container Allocation Based on Customer Satisfaction

Zhiwei Zhang[1], Xiaohong Li [2,a]

[1]*School of Traffic and Transportation, Beijing Jiaotong University, Beijing, China*
[2]*School of Traffic and Transportation, Beijing Jiaotong University, Beijing, China*

Abstract. With the rapid development of modern corporate management, customer satisfaction evaluation has been applied to the management of many companies. Railway empty container allocation not only involves the cost of the railway transportation, but also considers the customer's requirements for time and convenience, etc. Therefore customer satisfaction evaluation also becomes very important for railway container allocation decision. However, it is not easy to properly quantify customer satisfaction, because there are many indicators that cannot be directly quantified. This paper proposes a customer satisfaction evaluation index system based on the RATER index, and adopts a comprehensive evaluation method to achieve quantitative evaluation of customer satisfaction. An empty container allocation optimization model based on customer satisfaction is established. The model can comprehensively consider the interests of both the railway and customers, and provide an effective reference for the railway transportation department to improve the transportation service and optimize empty container allocation plan.

1 Introduction

In the statistics of container transportation expenses of China Railway Container Transport Co., Ltd., the expenses incurred due to empty container allocation accounted for 7.97% of the total payment, while the railway empty container transportation accounted for 21.9% of the total transportation volume [1]. A scientific and reasonable railway empty container allocation plan can not only accelerate the use of railway empty containers, improve the utilization rate of railway empty containers, but also meet the customer's container demand and improve the market service effectively. Therefore, it is necessary to apply customer satisfaction evaluation to optimize the empty container allocation plan.

2 Literature review

Currently, there are many research efforts focused on the customer satisfaction of railway freight transportation. E. Hirata [2] mapped the quality of service dimension to a new set of service characteristics based on the latest priority of shipping companies, and discussed the key service characteristics of the container liner shipping (CLS) industry and its impact on customer satisfaction. J.T. Mentzer et al. [3] constructed the logistics service quality model for evaluating customer logistics service satisfaction based on the characteristics of logistics service quality. Y.Y. Tang et al. [4] established a evaluation index system and combined the fuzzy evaluation with the service quality evaluation model to evaluate the quality of railway freight services.

Scholars also have done a lot on the empty container allocation. J.L. He et al. [5] designed the container empty container allocation route into a transportation network, and established a multi-level container logistics supply chain network (MCSCN) optimization model. E.Y.C. Wong et al. [6] proposed an empty container allocation model based on uncertain factors and made empty container deployment with better routes. T. Luo et al. [7] applied the contract coordination theory to the empty container allocation model to improve the coordination of empty container management and the initiative of each participant. Y. Myung [8] developed a solution algorithm for the two models in literature 7 and proved that the network flow algorithm could also solve the other two models. S.T. Choong et al. [9] established an empty container allocation integer programming model under multimodal transport by analysing the status of empty container allocation in the Great Lakes region of the United States.

3 Customer satisfaction evaluation

One of the foundations of enterprise development is customer satisfaction. In the production and operation, the company should firmly establish the business concept of "customer-centric", and constantly improve product quality to ensure the company's strong market competitiveness.

[a] Corresponding author: xhli1@bjtu.edu.cn

We use the AHP-fuzzy comprehensive evaluation method to calculate the customer satisfaction. Firstly, we establish a customer satisfaction index system for empty container allocation through the RATRA index. Then, we employ the 1-5 scale method to calculate the index weight. Finally, we adopt the evaluation principle with the highest degree of membership in the AHP-fuzzy evaluation method to analyze the evaluation coefficient of customer satisfaction.

3.1 Customer satisfaction evaluation index system

Scientific, reasonable and feasible evaluation indicators are the premise for correctly evaluating customer satisfaction, and also an important basis for establishing an empty container allocation model. The RATER index measures the quality of customer service by the US-based authoritative customer service research organization, Tribune Company, It represents the five major factors that measure the quality of business services and comprehensively demonstrates the partnership between the company and its users. It can fully reflect the customer's satisfaction with the railway transportation department. The scientific application of the RATER index can enhance the competitiveness of the railway market, and be an important method for user management in the railway transportation. This paper embodies the RATER index and obtains an index system that meets the actual situation of railway empty container deployment. The indicator system consists of five first-level indicators and nine second-level indicators. The specific indicator system is shown in Figure 1.

3.2 Customer satisfaction index weight

There are some indicators that cannot be directly quantified in customer satisfaction index system, and the subjectivity of satisfaction evaluation is strong. In order to be more scientific and reasonable, we choose the 1-5 scale method to calculate the weights of the indicators. The weights of the first and second level indicators are determined according to the scores of experts and customers.

Firstly, we construct the index comparison matrix D:

$$D = \begin{bmatrix} d_{11} & \cdots & d_{1s} \\ \vdots & \ddots & \vdots \\ d_{r1} & \cdots & d_{rs} \end{bmatrix} \quad (1)$$

The value of d_{rs} and its meaning are shown in Table 1.

Then, we calculate the maximum eigenvalue of D and its corresponding eigenvector W which is the index weight vector.

Table 1. The value of Comparison matrix D elements and its significance.

Value	Meaning
1	The element *r* is as important as the element *s*
3	The element *r* is slightly more important than the element *s*
5	The element *r* is much more important than the element *s*
2,4	The importance of the element *r* to the element *s* is between the two judgment scales
reciprocal	the comparison scale of the element *r* to *s* is d_{rs}, and the comparison scale of the element *s* to *r* is $1/d_{rs}$

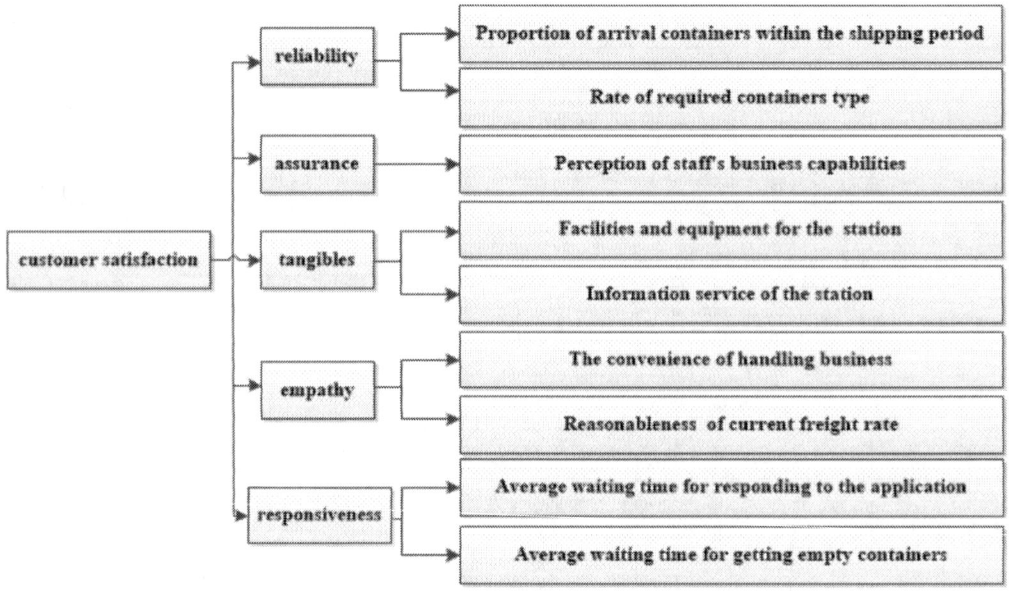

Figure 1. Customer satisfaction evaluation index.

3.3 AHP-fuzzy comprehensive evaluation of customer satisfaction

We assume that the number of the supplying station participating in the evaluation are i (i=1,2,3,4....,m), the number of the demanding station participating in the evaluation are j (j=1,2,3,4,....,n). Customers participating in the survey rate evaluation index for each station according to relevant experience and comment set. We collect the questionnaire and calculate the fuzzy comprehensive evaluation coefficient μ of each station according to the AHP-fuzzy comprehensive evaluation method. We define:

$$\mu_{ij} = \mu_i \cdot \mu_j \tag{2}$$

Eq. (2) means customer satisfaction index from the supplying station i to the demanding station j in the empty container allocation system.

4 Model formulation

4.1 Assumptions

In order to perform the quantitative analysis, the paper give the following assumptions:

(1)The supply-demand relationship between container stations is fixed for a certain period. It is a static system and does not consider dynamic changes caused by uncertain factors.

(2) It is possible to carry out the substitution of the container type and the vehicle type.

(3) There is no empty container convection between the two stations.

4.2 Declaration of parameter variable

List the variables and parameters used in this article as follow.

x_{ij}^{pq} denotes the number of empty p-containers provided by station i to replace the empty q-containers needed by station j. y_{ij}^{gh} denotes the number of g-vehicles to replace h-vehicles from station i to station j. I denotes the set of supply stations. J denotes the set of demand stations. $D_j^q(j = 1,2,3,4,\dots,n)$ denotes the quantity of empty q-containers needed by the station j. $S_i^p(i = 1,2,3,4,\dots,m)$ denotes the quantity of empty p-containers provided by the station i. Ve_i^g denotes the quantity of g-vehicles provided by the station i. c_{ij}^p denotes the cost of transporting one unit of empty p-containers. $c_{ij}^{\prime p}$ denotes the generalized cost of transporting one unit of empty p-containers. c_{ij}^{gh} denotes the additional cost of replacing the h-vehicle by the g-vehicle, When g=h, c_{ij}^{gh}=0. $c_{ij}^{\prime gh}$ denotes the additional generalized cost of replacing the h-vehicle by the g-vehicle, When g=h, $c_{ij}^{\prime gh}$=0. α_j^{pq} denotes whether the empty q-container needed for the station j is replaced by an empty p-container, which can be replaced by α_j^{pq}=1, otherwise, α_j^{pq}=0. φ_i^{gh} denotes whether the h-vehicle is replaced by a g-vehicle provided for the station i, which can be replaced by φ_i^{gh}=1, otherwise, φ_i^{gh}=0. β_j^{pq} denotes the penalty factor when the q-container needed by the station j is replaced by a p-container, when p=q, β_j^{pq}=1. ω denotes the multiple of the number between vehicles and containers. Cap_{ij} denotes the maximum capacity of arc (i, j). u_j denotes the importance coefficient of station j. u_i denotes the importance coefficient of station i. μ_{ij} denotes the customer satisfaction coefficient between station i and station j.

4.3 Objective function

The goal of the optimization model is to minimize the generalized cost required for allocation. The total generalized cost consists of two parts. The first part is the generalized allocation cost generated by empty container transportation, the second part is the generalized substitute cost generated by vehicle substitution.

$$\min Z = \sum_{i=1}^{m}\sum_{j=1}^{n}\sum_{p}\sum_{q} c_{ij}^{\prime p}\cdot x_{ij}^{pq}\cdot \beta_j^{pq} + \sum_{i=1}^{m}\sum_{j=1}^{n}\sum_{g}\sum_{h} c_{ij}^{\prime gh}\cdot y_{ij}^{gh}$$

4.4 Constraints

$$c_{ij}^{\prime p} = \frac{c_{ij}^p}{u_i u_j \mu_{ij}} \quad \forall i,\ j \tag{3}$$

$$c_{ij}^{\prime gh} = \frac{c_{ij}^{gh}}{u_i u_j \mu_{ij}} \quad \forall i,\ j \tag{4}$$

$$\sum_{i=1}^{m}\sum_{p} x_{ij}^{pq}\cdot \alpha_j^{pq} = D_j^q \quad \forall j,\ q \tag{5}$$

$$\sum_{j=1}^{n}\sum_{q} x_{ij}^{pq}\cdot \alpha_j^{pq} \le S_i^p \quad \forall i,\ p \tag{6}$$

$$\sum_{j=1}^{n}\sum_{h} \varphi_i^{gh}\cdot y_{ij}^{gh} \le Ve_i^g \quad \forall i,\ g \tag{7}$$

$$\sum_{g}\sum_{h} y_{ij}^{gh} = \sum_{p}\sum_{q}\frac{x_{ij}^{pq}}{\omega} \quad \forall i,\ j \tag{8}$$

$$\sum_{p}\sum_{q}\frac{x_{ij}^{pq}}{\omega} \le Cap_{ij} \quad \forall i,\ j \tag{9}$$

$$x_{ij}^{pq} \ge 0\ ,\ y_{ij}^{gh} \ge 0 \text{ and integer } \forall i,\ j,\ p,\ q,\ g,\ h \tag{10}$$

Eq. (3)-(4) define the quantitative relationship between empty container allocation cost, vehicle substitute cost and generalized cost; Eq. (5) ensures that

 https://doi.org/10.1051/matecconf/202030804001

the number of empty containers that arrive at station j is the same as the demand for empty containers at that site; Constraint (6) guarantees that the total number of empty containers moved from station i is not more than the empty containers provided by that site; Constraint (7) assures that station i has enough vehicles to transport empty containers; Eq. (8) defines that the sum of vehicles corresponds with the total containers from station i to station j; Constraint (9) denotes the limitation of line capacity; x_{ij}^{pq} and y_{ij}^{gh} are decision variables and non-negative integers.

4.5 Solution

As there is a linear relation between objective function, constraints and decision variables, this is an integer linear programming model. The most common method for solving integer programming problems is the branch and bound method. The branch and bound method, whose basic idea is to enumerate all feasible solutions which means to divide the solution space successively, uses a 'divide and rule' strategy for solving integer programming problems. Therefore, we can solve this integer linear programming model by applying LINGO11.0, which is widely used.

5 Numerical example

In order to verify the validity and rationality of the model, we select eight container-providing stations and seven container-demanding stations, and take vehicle and container-type substitution into account in the study. In this example, we assume that the line capacity meets the transportation need and the importance coefficient of all stations is equal to 1. The container set is {20-foot container, 40-foot container}={T,F}, the container transport vehicle set is { gondola, container flat car}={C,X}, the container substitution judgment parameter matrix is{ $\alpha_j^{TT}, \alpha_j^{TF}; \alpha_j^{FT}, \alpha_j^{FF}$}={1, 0; 1, 1}, j=1, 2, … 7, The vehicle substitute judgment parameter matrix is{ $\varphi_i^{CC}, \varphi_i^{CX}; \varphi_i^{XC}, \varphi_i^{XX}$}={0, 1; 0, 1}, i=1, 2, … 8, The container substitution penalty factor matrix is{ $\beta_j^{TT}, \beta_j^{TF}; \beta_j^{FT}, \beta_j^{FF}$}={1, ∞; 1.21 ,1}, j=1, 2, … 7, The vehicle substitute cost matrix is{ $c_{ij}^{CC}, c_{ij}^{CX}; c_{ij}^{XC}, c_{ij}^{XX}$}={∞, 50; ∞, 0}, i=1, 2, … 8, j=1, 2, … 7. The allocation cost and quantity of supply and demand of empty containers are shown in Table 2. The number of vehicles for the container-providing station is shown in Table 3. Considering the assessment of experts and customers, we calculate a first-level indicator weight set W=(0.3, 0.13, 0.17, 0.2, 0.2), The customer satisfaction evaluation coefficient set of the container-providing stations is ($S_1, S_2, S_3, S_4, S_5, S_6, S_7, S_8$)=(0.42, 0.56, 0.38, 0.41, 0.53, 0.43, 0.53, 0.50), The customer satisfaction evaluation coefficient set of the container-demanding stations is($D_1, D_2, D_3, D_4, D_5, D_6, D_7$) =(0.59, 0.34, 0.51, 0.32, 0.58, 0.43, 0.49).

Table 3. The number of vehicles for the container-providing station.

station	S_1	S_2	S_3	S_4	S_5	S_6	S_7	S_8
gondola	50	15	51	20	40	20	30	10
container flat car	100	70	50	85	99	76	95	50

We solve the empty container allocation optimization model based on customer satisfaction through LINGO11.0 software. The finishing result is shown in Table 4 and Table 5.

Table 2. The allocation cost and quantity of supply and demand.

	Allocation cost（yuan/TEU）								Supply/TEU
	type of container	D_1	D_2	D_3	D_4	D_5	D_6	D_7	
S_1	T	65	108	127	217	266	176	221	100
	F	74	120	139	235	286	191	239	100
S_2	T	62	65	95	165	214	133	169	50
	F	71	75	107	180	231	146	184	60
S_3	T	137	94	174	109	212	51	120	40
	F	150	105	189	120	229	59	132	81
S_4	T	168	125	203	163	180	57	88	56
	F	183	138	220	178	196	66	99	77
S_5	T	286	269	322	186	235	231	186	78
	F	307	289	346	202	254	249	202	100
S_6	T	173	145	210	119	167	77	71	40
	F	189	158	227	131	182	87	81	56
S_7	T	252	219	289	107	155	151	137	66
	F	271	237	310	118	170	165	150	90
S_8	T	121	119	157	108	157	123	121	26
	F	133	132	172	120	171	135	134	35
Demand/TEU	T	80	70	80	110	100	72	40	-
	F	15	45	50	80	70	50	24	-

Table 4. Empty container allocation plan considering customer satisfaction.

	type of container	D_1	D_2	D_3	D_4	D_5	D_6	D_7
S_1	T	12	-	80	-	8	-	-
	F	24	-	18	-	8	-	-
S_2	T	24	50	-	-		-	-
	F	-	36	-	-	-	-	-
S_3	T	-	8	-	32	-	9	-
	F	-	21	-		-	-	-
S_4	T	-	-	-	-	-	56	-
	F	-	-	-	-	-	57	-
S_5	T	-	-	-	78	-	-	-
	F	-	-	-	52	-	-	-
S_6	T	-	-	32	-	-	-	40
	F	-	-	-	-	-	-	24
S_7	T	-	-	-	28	66	-	-
	F	-	-	-	-	62	-	-
S_8	T	35	-	-	-	26	-	-
	F	-	-	-	-	-	-	-

Table 5. Vehicle matching plan considering customer satisfaction.

	type of vehicle	D_1	D_2	D_3	D_4	D_5	D_6	D_7
S_1	C	-	-	-	-	-	-	-
	X	30	-	58	-	12	-	-
S_2	C	15	-	-	-	-	-	-
	X	19	61	-	-	-	-	-
S_3	C	-	-	-	-	-	-	-
	X	-	25	-	16	-	9	-
S_4	C	-	-	-	-	-	-	-
	X	-	-	-	-	-	85	-
S_5	C	-	-	-	-	-	-	-
	X	-	-	-	91	-	-	-
S_6	C	-	-	-	-	-	-	-
	X	-	-	32	-	-	-	44
S_7	C	-	-	-	28	-	-	-
	X	-	-	-	-	95	-	-
S_8	C	-	-	-	-	-	-	-
	X	35	-	-	-	13	-	-

When the customer satisfaction is not considered, we minimize the actual cost as the optimization goal for empty container allocation. The plan obtained by LINGO 11.0 is shown in Table 6 and Table 7.

Overall, the generalized cost of the model considering customer satisfaction is 28,845 yuan less than that of the ordinary model, which is 5.6% lower than the latter. The actual total cost of the allocation model based on customer satisfaction is 4.3% higher than the ordinary model, which is the price of considering the interest of the customer. The costs of the two models are shown in Figure 2. The empty container allocation model based on customer satisfaction can consider the interests of both the railway department and the customer, and strive for a balance between the two by a small increase of actual cost.

Table 6. Empty container allocation plan without considering customer satisfaction.

	type of container	D_1	D_2	D_3	D_4	D_5	D_6	D_7
S_1	T	46	-	54	-	-	-	-
	F	15	-	35	-	-	-	-
S_2	T	-	24	26	-	-	-	-
	F	-	45	15	-	-	-	-
S_3	T	-	46	-	40	-	-	-
	F	-	-	-	35	-	-	-
S_4	T	-	-	-	-	11	56	-
	F	-	-	-	-	-	66	-
S_5	T	-	-	-	44	16	-	-
	F	-	-	-	-	-	-	-
S_6	T	-	-	-	-	32	-	40
	F	-	-	-	-	-	-	24
S_7	T	-	-	-	45	66	-	-
	F	-	-	-	-	45	-	-
S_8	T	34	-	-	26	-	-	-
	F	-	-	-	-	-	-	-

Table 7. Vehicle matching plan without considering customer satisfaction.

	type of vehicle	D_1	D_2	D_3	D_4	D_5	D_6	D_7
S_1	C	-	-	-	-	-	-	-
	X	38	-	62	-	12	-	-
S_2	C	-	15	-	5	-	-	-
	X	-	42	28	50	-	-	-
S_3	C	-	-	-	-	-	-	-
	X	-	46	-	16	-	-	-
S_4	C	-	-	-	-	11	9	-
	X	-	-	-	-	-	85	-
S_5	C	-	-	-	-	-	-	-
	X	-	-	-	22	8	-	-
S_6	C	-	-	-	-	-	-	-
	X	-	-	32	-	32	-	44
S_7	C	-	-	-	28	-	-	-
	X	-	-	-	17	78	-	-
S_8	C	-	-	-	-	-	-	-
	X	34	-	-	13	-	-	-

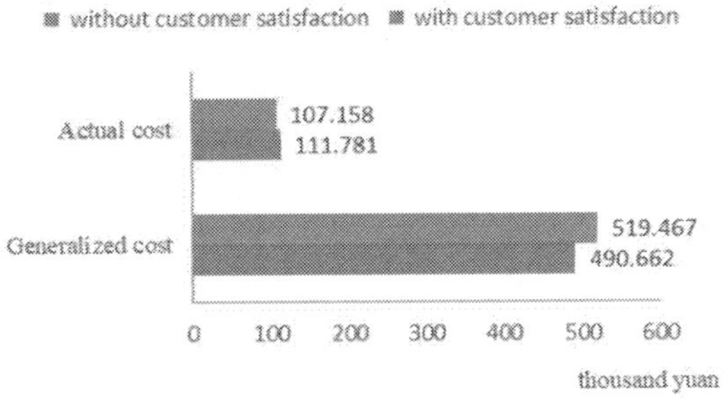

Figure 2. Comparison of costs between two models.

From a part perspective, the customer satisfaction evaluation result of S_2-D_1 is better than that of S_1-D_1. Based on the customer satisfaction, the generalized cost of the empty container for station D_1 decreased by 9.3%, and the actual cost increased by 4.9%. The balance between the empty container allocation cost and customer satisfaction is realized by increasing a small part of the actual cost.

In addition, through the AHP-fuzzy comprehensive evaluation method to quantitatively study customer satisfaction, we can find out the problems existing in the empty container allocation system, and provide the improvement direction and optimization measures for the operation and management of container transportation. For example, the customer satisfaction evaluation results of the station S_1 are (reliability, assurance, tangibles, empathy, responsiveness) = (0.08, 0.13, 0.42, 0.24, 0.13), Therefore, the improvement measures for the empty container deployment service are proposed:

(1) Optimize the empty container allocation transportation organization, ensure that the empty container is transported to the time limit.

(2) Strengthen the concept of service, and improve the transportation service standards.

(3) Further optimize the information service, quickly and conveniently serve customers, and shorten the waiting time to apply for empty containers.

6 Conclusion

This paper proposes an optimization model for the empty container allocation based on customer satisfaction, which provides an effective reference for the railway to improve the transportation service and optimize empty container allocation plan. Firstly, we establish a customer satisfaction index system based on the RATER index. Secondly, we effectively quantify customer satisfaction via AHP-fuzzy comprehensive evaluation method. Finally, the optimization model converts actual costs into generalized costs through customer satisfaction evaluation coefficient, and comprehensively consider the interests of both the railway and the customer to optimize empty container allocation plan. This model gives out the importance coefficient of stations, and the example assumes those are equal. However, the importance coefficients are possibly different in practice. How to quantitatively study the weight of stations is the direction that need further research and discussion in the future.

References

1. C. Zhong, J. Feng, Analysis of Transportation Cost of Railway Empty Container. J. Railway Transport and Economy. **3**, 41-43(2005)
2. E. Hirata, Service characteristics and customer satisfaction in the container liner shipping industry. J. The Asian Journal of Shipping and Logistics. **35(1)**, 24-29(2019)
3. J.T. Mentzer, D.J. Flint, G.T.M. Hult, Logistics Service Quality as a Segment—customized Process. J. The Journal of Marketing. **4**, 82-101(2001)
4. Y.Y. Tang, W.W. Li, Customer Satisfaction Evaluation of Railway Freight Service Quality Based on SERVQUAL. J. Chinese Railways. **8**, 15-19(2016)
5. J.L. He, Y.F. Huagn, D.F. Chang, Simulation-based Heuristic Method for Container Supply Chain Network Optimization. J. Advanced Engineering Informatics. **29(3)**, 339-354 (2015)
6. E.Y.C. Wong, A.H. Tai, M. Raman, A maritime container repositioning yield-based optimization model with uncertain upsurge demand. J. Transportation Research Part E: Logistics and Transportation Review. **82**, 147-161 (2015)
7. T. Luo, D. Chang, Empty container repositioning strategy in intermodal transport with demand switching. J. Advanced Engineering Informatics. **40**, 1-13 (2019)
8. Y. Myung, Efficient solution methods for the integer programming models of relocating empty containers in the hinterland transportation network. J. Transportation Research Part E: Logistics and Transportation Review. **108**, 52-59 (2017)
9. S.T. Choong, H.C. MICHAEL, E. Kutanogul, Empty container management for intermodal transportation networks. J. Transportation Research Part E: Logistics and Transportation Review. **38(11)**, 423-438 (2002)

Freighter Loading Efficiency at Airports for International Air Express

Hua-An Lu[1,a] and Chi-Sheng Chung[2]

[1] *Professor, Department of Shipping and Transportation Management, National Taiwan Ocean University, 2 Pei-Ning Road, Keelung 202, Taiwan, R.O.C.*
[2] *Ramp Agent-lead, Ramp Operations, FedEx Express Co., 8-1 Hang-Chin North Road, Tao Yuan Hsien 337, Taiwan, R.O.C.*

Abstract. Loading efficiency of freighters plays an indispensable role for the service performance of international air express companies. This study examined the priorities of factors to affect freighter loading efficiency at Taoyuan International airport (TPE) with an analytic hierarchy process (AHP). Unit load device (ULD) inventory control, load factor and load plan were ranked top 3 from 15 factors by practical experts. This study used the technique for order preference by similarity to ideal solution (TOPSIS) to evaluate 26 flights of MD11 freighters operated by an international air express company at TPE. The results revealed that the performance ranges between the best and worst flights were obvious.

1 Introduction

International air express provides door-to-door services over the world. Every sector of operation chains is running against time from the beginning of the pickup to the end of final delivery. One of the key factors for a successful service is the process of aircraft loading at airports. Loading performance is not only to expense the least time to load cargo on aircraft but also to spent the least resources to carry as much cargo as available. Aircraft loading efficiency affects the results of express services. There might be many factors to affect the loading efficiency of freighters at airports for international air express. Actually, the freighter capacities are the most important resource supply in the air cargo and logistics community. It has become a crucial topic to keep a well performance of cargo shipping for freighter flights with efficient shipment carries [1-2].

Feng et al. [3] reviewed studies dedicated on air cargo operations from the practical perspectives of airlines, freight forwarders, and terminal service providers. Some mentioned acitivies could affect airlines operation performance at airports, such as improving load factors, allocating cargo space-allotments, resource management of terminal staff, shipments tendered, dangerous goods control, cargo routing and scheduling, packing validation, etc. According to the literature reviews of Brandt and Nickel [4], previous studies focused on the airport operation processes with a variety of operational research techniques. The main topics of air cargo loading can be divided into aircraft configuration problem, build-up scheduling problem, air cargo palletization problem, and weight and balance problem.

Currently most of air cargo are built up into the air unit load devices (ULDs), which mainly include containers and pallets, for loading on the appropriate deck positions in aircraft. ULDs inventory control concerned about equipment provision for loading operations [5]. The assemblies of ULDs in a load plan affect not only available capacities of loaded freights which performance reflects on the load factor of every flight but ULDs repositioning also [6]. Special contours of containers and flexible stacks of pallets affect the sufficient uses of equipment capacity. The selection of ULDs must consider the 3-dimension scales, weight limits, stacking stability, as well as the different characteristics of shipment and mixed build-up with each other of different items [7-8].

While ULDs are loaded on the aircraft, the allotment of ULDs have to be planned in consideration of aircraft interior contour, weight limits on structure, and resulted position of aircraft center of gravity (CG). This problem is referred to the aircraft weight and balance (W&B). These conditions are all set as loading constraints [9-11] based on a decision objective function, such as maximization of cargo transport profits [12], or maximization of shipment loads [13]. The decision of such kind contents will be more complicated if taking into account multiple flight legs [14] or combining with ULDs selection [15].

For the international air express field, most of previous studies explored the flow distribution patterns within a hub-and-spoke network [16-18]. The proposed models in these studies could also decide shipment consolidation and distribution at hub and spoke airports for the container loads and the less than container loads.

Most reviewed papers related to loading operations were focused on all-cargo or combined airlines. However, international air express operates significantly different from both kinds of airlines. This study aims to explore

[a] Corresponding author: halu@ntou.edu.tw

key factors affecting the loading efficiency of freighters at airports for international air express. This ultimate goal can be evaluated with different dimensions of operation processes. Each aspect took into account the attributed considerations that can be evaluated the efficiency in detail. This study applied the analytic hierarchy process (AHP) to investigate experts of an international air express company serving in Taiwan at the Taoyuan International Airport (IATA code: TPE) in constructing an evaluation framework. The results of evaluated weights for factors framework were also applied to compare the loading efficiency of selected MD11F flights using the technique for order preference by similarity to ideal solution (TOPSIS).

2 Methodologies

This study exploited the AHP with a rank pair-wise comparison (RPC) concept [19] to investigate expert opinions and then to calculate the weights of affecting factors. Experts suggested some numerical indexes being the data to represent the factors. The application of TOPSIS is appropriate to evaluate loading efficiency of air express freighters for export operations.

2.1 AHP with RPC

The AHP has been one of the popular multi-criteria decision-making methods to be successfully applied to different fields. The purpose of an AHP is to measure relative comparison among factors and then, if need, to evaluate the possible alternatives [20]. First, an AHP requires to construct a systematic and decisive hierarchical framework that composes of goal, levels for evaluated criteria, and possible selectable alternatives. Then, one can allocate the relative weights, which might be examined through group decision of experts by several academic studies, for evaluated criteria and attributed factors. Furthermore, if there are alternatives required to be ranked, the performance evaluation of all alternatives on all decision criteria need to be further measured.

The pair-wise comparison among entries under a common attribute is the core of executing an AHP. A matrix, called pair-wise comparison matrix, presents the compared results between every two entries in which the values of symmetrical elements in this matrix are reciprocal. Through the linear algebra calculation, one can decide the relative weights for every entry. Normally, the pair-wise comparison results follow the perfect transitivity property, i.e., the value of entry a compared to entry b is equal to the multiple between the value of entry a compared to entry c and the value of entry c compared to entry b. An AHP allows deciders to present a slight variety on this property but to satisfy an acceptably consistent level.

However, there are some experienced drawbacks in executing an AHP to collect the opinions of a large group of respondents. First, too many comparative entries generally make the respondents, particularly those who without any experience on pair-wise comparison, unable to discriminate the relative weights that are necessary to reach the acceptably consistent level. For example, the number of comparisons is $(n^2 - n)/2$ for n entries. Second, inconsistency in responses possibly takes place when respondents encounter entries with closer perceived relationships. The reason for this phenomenon might be that respondents do not rank their priorities in advance. Furthermore, too complicated framework listing too many pair-wise comparison queries make respondents chaos during the process of filling the opinions. The original opinions of respondents might be distorted because of repetitive queries for reaching a minimum consistency.

The RPC [19] is one of approaches to overcome above drawbacks of the traditional pair-wise comparison method. This approach asks respondents to express the priorities of the involved entries and subsequently assess the relative weights for two consecutive ranks. The number of pair-wise comparison is reduced to $n - 1$ for n entries. Using the transitive rule, the scale adjustment rule and the reciprocal principle, one can yield an RPC matrix with complete consistency. This approach is rather appropriate to collect opinions for a large group. The whole execution process is designed as follows.

Step 1: Distribute questionnaires to all respondents and then collect their opinions.

Step 2: Construct all respondents' pair-wise comparison matrixes by the following RCP steps [19].
- (a) According to the transitive rule, calculate the values for other nonconsecutive comparison elements with the consecutive comparison results.
- (b) Adjust the values for these elements according to the scale adjustment rule.
- (c) Fill in the values for other elements according to the reciprocal principle.

Step 3: Combine all respondents' pair-wise comparison matrixes which elements are calculated by the geometric means of all corresponding values in all respondent's pair-wise comparison matrixes.

Step 4: Calculate the relative weights for entries in all tiers with these matrixes, using the average of normalized columns approach as Equation (1).

$$w_i = 1/n \{\Sigma_{j=1 \text{ to } n} (a_{ij}/\Sigma_{i=1 \text{ to } n} a_{ij})\} \quad (1)$$

Step 5: Distribute the overall relative weights for all sub-criteria.

Step 6: Rank the priorities for all sub-criteria with the overall relative weights.

2.2 TOPSIS

The TOPSIS approach appropriates to evaluate a variety of alternatives with multiple quantified factors with positive and negative influence. Its concept is to find an alternative that is closest to the defined positive ideal solution (PIS) and furthest with the defined negative ideal solution (NIS). The defined PIS can be the maximal values of benefit criteria and minimal values of cost criteria and vice versa for the defined NIS.

Following the steps proposed by Hwang and Yoon [21], the implementation of TOPSIS for a given matrix with assigned rates to the criteria and to the alternatives is described as the following:

Step 1: Normalize this evaluation matrix.

Step 2: Construct the weighted decision matrix.

Step 3: Define the PIS and NIS.

Step 4: Calculate the distances to the PIS and NIS for every alternative.

Step 5: Compute the closeness coefficient with the ideal solutions for each alternative.

Step 6: Rank the preference orders of alternatives.

3 Investigation

This research has firstly to construct the evaluation framework for the ultimate goal, i.e. examining the factor priorities influencing the loading efficiency of freighters. Then, this study executed sampling investigation to the studied company. The priorities of sub-criteria can be evaluated with the AHP and RPC.

3.1. Analytic hierarchy

This study firstly examined the practice of aircraft loading for express freighters at the airport, and then to list 5 criteria and 15 attributed sub-criteria for reviews of 6 practical experts in our focus group. The five criteria consist of ULD uses, cargo load planning, aircraft weight and balance, resources utilisation, and operation performance. Except ULD uses with 83.3% and operation performance with 66.7% of agreement, other three criteria were all totally agreed. Nine sub-criteria received entirely agreement, while 5 factors and 1 item obtained 83.3% and 66.7% agreement, respectively. The evaluated hierarchy is shown in Figure 1.

3.2 Sampling

A questionnaire designed for the rank pair-wise comparison with 7-point measure scales from 1 for the same important to 7 for extremely important. Respondents can give the sequence of importance for the evaluated elements, and then express the importance level for each consecutive element pair. This questionnaire evaluates 6 pair-wise comparison items with one matrix for criterion level and 5 matrixes for each sub-criterion levels.

Operation unit at TPE airport of the studied company could provide the required investigation objectives and evaluation data resources at TPE. This availability facilitated the whole studied process. In total, 15 ramp operation technicians of the studied company working at TPE airport responded questionnaires during April 1st to 30th 2019. Twelve experts had the experience of above 20 years and other 3 respondents with less than 10 years. All of questionnaires are effective.

3.3 Factor priorities

The opinions of 15 experts were formulated their own pair-wise comparison matrixes following the calculation mentioned in Section 2.1. Calculating the geometric means for the related elements in the same matrixes to form the final comparison matrixes of group decision. Following Equation (1), the result of factor priorities is shown in Table 1.

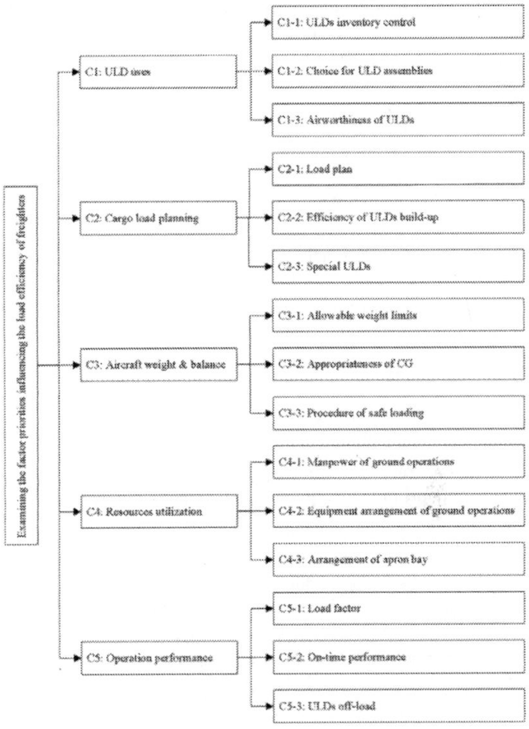

Figure 1. AHP evaluation framework of the study.

Table 1. Weights and ranked priorities.

Criteria	Criterion *weight*	*Sub-criteria*	*Weight* in criterion	*Rank* in criterion	*Distributed weight*	*Rank* in total
C1	0.2026	*C1-1*	0.4270	1	0.0865	1
		C1-2	0.3287	2	0.0666	8
		C1-3	0.2443	3	0.0495	14
C2	0.2145	*C2-1*	0.3758	1	0.0806	3
		C2-2	0.2737	3	0.0587	12
		C2-3	0.3504	2	0.0752	4
C3	0.1984	*C3-1*	0.3284	2	0.0652	9
		C3-2	0.3631	1	0.0720	6
		C3-3	0.3084	3	0.0612	11
C4	0.1949	*C4-1*	0.3810	1	0.0742	5
		C4-2	0.3245	2	0.0632	10
		C4-3	0.2945	3	0.0574	13
C5	0.1897	*C5-1*	0.4392	1	0.0833	2
		C5-2	0.3618	2	0.0686	7
		C5-3	0.1990	3	0.0377	15

Cargo load planning was ranked first in five criteria, following by the criterion of ULD uses. These two criteria share percentages over one fifth. ULDs inventory control, load factor, and load plan were ranked top 3 in fifteen sub-criteria. They all share the weight ratios of more than 0.08. In practice, many flights of the studied company at TPE kept a large amount of ULDs for transferring and positioning. A sufficient amount of ULDs can easily allocate shipment with different sizes and diverse characteristics. In particular, these flights also required more time to build up containers because of operating on early morning. The factor to control ULDs inventory becomes a critical condition. Load factors is the index to reflect the real performance of freighter operations. Each flight might be affected the results of load factor by different reasons, such as weather conditions, flying distances, and of course load control. The whole processes of export operations for a flight are driven by the load plan. Its contents concern about the occurrence of any extraordinary situation on the ramp.

4 Case of MD11F

Some factors influencing freighter loading efficiency might be abstract and negative. Practical experts proposed numerical indexes to express these factors. This study chosen TOPSIS, which can adapt to the sets with positive and negative indexes, to evaluate MD11F flights.

4.1 Indexes of TOPSIS

This study used following numerical data to express sub-criteria in the evaluation hierarchy through suggestion of practical experts.

(1) ULDs inventory control: the number of left empty ULDs after aircraft departure.
(2) Choice for ULD assemblies: the number of loaded ULDs onto aircraft including ULDs built up at TPE, transferred ULDs, reloaded ULDs, etc.
(3) Airworthiness of ULDs: the number of ULDs without airworthiness.
(4) Load plan: the duration between finishing load plan and aircraft pushback.
(5) Efficiency of ULDs build-up: the number of ULDs built up at TPE and really loaded on the aircraft.
(6) Special ULDs: the number of special ULDs loaded onto the aircraft.
(7) Allowable weight limits: maximal aircraft weight allowed to be loaded.
(8) Appropriateness of CG: the distance between actual CG position and the most forward allowance position of CG.
(9) Procedure of safety loading: following the existing procedures of the company, 1 for better application than 2.
(10) Manpower of ground operations: the number of workers for ground operations.
(11) Equipment arrangement of ground operations: the number of ground equipment used.
(12) Arrangement of apron bays: the distance between assigned parking bay with the bay closest with the warehouse.
(13) Load factor: actual cargo load weight divided by the available weight capacity.
(14) On-time performance: delay times of flight departure.
(15) ULD off-load: the number of off-load ULDs.

The positive (benefit) subset of sub-criteria includes factors 1, 2, 4, 5, 6, 7, 8, and 13, other indexes belong to the negative (cost) subset.

4.2 Data Collection

Courtesy of the studied company at TPE airport, this study could collect the data of MD11F flights for a certain period of 2018. These 26 flights, 5 flights per week, departed from TPE and directly landed at the transferring centre of Anchorage airport. They carried transpacific freights to be consolidated with other cargos destined to North America airports.

4.3 Evaluation results

The evaluation results for MD11F flights are shown in Table 2. Flight 16 was ranked first, while flight 12 was ranked worst. The comparison between the best and worst flights is shown in Table 3. Except appropriateness of CG, numerical indexes of flight 16 are all much better than those of flight 12. The gap of their closeness coefficient presents obvious.

Table 2. Evaluation results for MD11F flights.

Flight	Distance to PIS	Distance to NIS	Closeness coefficient	Rank
1	0.01720	0.06656	0.79462	11
2	0.02036	0.06933	0.77301	15
3	0.02020	0.07028	0.77673	14
4	0.01692	0.07179	0.80925	7
5	0.01781	0.07277	0.80336	9
6	0.02621	0.07058	0.72920	22
7	0.02561	0.06688	0.72310	23
8	0.02148	0.07086	0.76741	16
9	0.02691	0.06890	0.71909	24
10	0.02377	0.07311	0.75461	20
11	0.01865	0.07255	0.79552	10
12	**0.06901**	**0.03270**	**0.32152**	**26**
13	0.01933	0.07131	0.78674	13
14	0.01897	0.06208	0.76597	17
15	0.01641	0.07246	0.81534	4
16	**0.01101**	**0.07462**	**0.87141**	**1**
17	0.01430	0.07220	0.83473	2
18	0.02232	0.06962	0.75720	19
19	0.01767	0.07243	0.80384	8
20	0.02458	0.06897	0.73727	21
21	0.01703	0.07290	0.81060	6
22	0.02781	0.06937	0.71386	25
23	0.02188	0.07148	0.76566	18
24	0.01642	0.07262	0.81556	3
25	0.01909	0.07125	0.78865	12
26	0.01691	0.07358	0.81308	5

Table 3. Comparison between the best and worst flights.

Factor priority	Factor	Subset	Flight 16 (best)	Flight 12 (worst)
1	C1-1	P	182	146
2	C5-1	P	1.08653	0.75202
3	C2-1	P	275	248
4	C2-3	P	5	5
5	C4-1	M	15	15
6	C3-2	P	526	581
7	C5-2	M	0	34
8	C1-2	P	50	45
9	C3-1	P	185,510	160,000
10	C4-2	M	8	8
11	C3-3	M	15	15
12	C2-2	P	15	10
13	C4-3	M	0	0
14	C1-3	M	2	0
15	C5-3	M	0.91304	1.00000

5 Conclusion and Suggestion

This study has examined the factors influencing the loading efficiency of freighter operations at airports for international air express. An AHP framework to evaluate factor priorities was constructed. Through an investigation of practical experts with a rank pair-wise comparison questionnaire, factor weights were measured to rank the importance. The dimension of cargo load planning was ranked the most important, while ULDs inventory control, load factor and load plan were top 3 factors.

The AHP framework has been applied to the case of MD11F flights of the studied company at TPE with the TOPSIS. The comparison results between the best and worst flights revealed a large and obvious gap on closeness coefficients. Except appropriateness of CG, numerical indexes of the former were all much better than those of the latter.

This study focused only on the export operation of an international air express company and at a specific airport because of the individual requirement on this topic. However, the AHP and TOPSIS proposed in this study can be applied to practical measures for other freighter flights and similar topics, or even for the operation of other companies and airports, in the future research. It is noted that the weights of factors may be affected and changed because of different flight operation characteristics in air express network. Furthermore, other air cargo business models, such as combination carriers, may require a slight change on the evaluated factors according to their operation properties.

6 Acknowledgement

This study thanks to the financial support of Ministry of Science Technology of Taiwan with the granted number MOST-108-2410-H-019-023-MY2.

7 References

1. R. Merkert, E.V. de Voorde, J. de Wit, J. Air Transp. Manag. **37**, 10-19 (2017)
2. L. Budd, S. Ison, J. Air Transp. Manag. **61**, 34-40 (2017)
3. B. Feng, Y. Li, Z.J.M. Shen, Transp. Res. C **56**, 263-280 (2015)
4. F. Brandt, S. Nickel, Eur. J. Oper. Res., Available online doi.org/10.1016/j.ejor.2018.07.013. (to be published)
5. H.A. Lu, C.Y. Chen, J. Mar. Sci. & Tech. **20(4)**, 431-440 (2012)
6. H.A. Lu, C.Y. Chen, J. Air Transp. Manag. **17**, 94-100 (2011)
7. F.T.S. Chan, R. Bhagwat, N. Kumar, M.K. Tiwari, P. Lam, Expert Syst. Appl. **31**, 472-485 (2006)
8. C. Paquay, M. Schyns, S. Limbourg, Int. Trans. Oper, Res, **23(2)**, 187-213 (2014)
9. O. Larsen, G. Mikkelsen, Eur. J. Oper. Res. **4(6)**, 367-373 (1980)
10. S. Limbourg, M. Schyns, G. Laorte, J. Oper. Res. Soc. **63**, 1271-1283 (2012)
11. W. Vancroonenburg, J. Verstichel, K. Tavernier, G.V. Berghe, Transp. Res. E **65**, 70-83 (2014)
12. I. Brosh, J. Oper. Res. **8(1)**, 40-46 (1981)
13. M. Mongeau, C. Bes, IEEE Trans. on Aero. Ele. Syst. **39(1)**, 140-150 (2003)
14. V. Lurkin, M. Schynes, Eur. J. Oper. Res. **61**, 1-5 (2015)
15. N. Dahmani, S. Krichen, Int. J. Oper. Res. **27(1-2)**, 62-84 (2016)
16. S. Yan, C.-T. Lo, Y.-L. Shih, Transp. Plan. Tech. **29(6)**, 445-470 (2006)
17. S. Yan, Y.-L. Shih, F.-Y. Shiao, Transp. Res. E **44**, 555-575 (2008)
18. C.-H. Tang, Transp. Res. E **47**, 520-531 (2011)
19. H.A. Lu, R.R. Liu, J. Air Transp. Manag. **37**, 10-19 (2014)
20. T.L. Saaty, *The analytic hierarchy process* (McGraw-Hill, New York, 1980)
21. C.L. Hwang, K.P. Yoon, *Multiple attribute decision making: methods and applications, a state-of-the-art survey* (Springer-Verlang, Berline, 1981)

An Intercity Freight Mode Choice Model : A Case Study of High Speed Rail Link Northern Line Thailand (Bangkok – Chiangmai)

Jessada Pochan[1], Wachira Wichitphongsa[1,a]

[1]*Faculty of Industrial Technology, Rajabhat Pibulsongkram University, 156 Moo. 5, Phlai Chumphon, Mueang, Phitsanulok, Thailand*

Abstract. This paper presents a model capturing the intercity freight mode choice behaviour in high-speed rail system from Bangkok to Chiangmai. The model is developed based on the stated preference data collected from 800 freight operators, wholesaler, retailer, and people. The results show that, when the high-speed rail system from Bangkok to Chiangmai are developing in the future, the suitable products for high-speed rail system are types of an express mail service (EMS), air cargo, gold, jewellery, gold accessories, computer circuit boards, high prices agricultural products such as fruit, and flowers. Most of determining the selling price will fluctuate with the speed of transport and damage impairment of the product. With the application of discrete choice models, the results show that, aside from travel cost and time, loading and unloading, delays time, frequency are statistically significant. The application of model indicated that the holder and freight forwarder which in the line of high-speed rail (Bangkok – Chiangmai) tend to use rail-transport such as double-track rail is 27.71%, high-speed rail is 11.18% and the most is trucks 56.51% which is a policy development point loading and unloading, multimodal transportation efficiency and safety of the portion of the freight high-speed rail increased.

1 Introduction

The potential of infrastructure is the scorecard in the economic security and bargaining power of the country. The ranking of efficiency of transport infrastructure system by World Economic Forum (2013-2015) from 144 countries found that Thailand is ranked at 76th (2014-2015) [1] and has a recession trend and lower than other countries in ASEAN. Currently, the government has established a strategic plan of the Ministry of Transport issue 2nd, 2013-2017. That focuses on the development of rail infrastructure which helps reduce transportation costs and increase the country's competitiveness in trade. Since speed is an important factor both of business and lifestyle. So the high speed rails are an alternative way to transport goods, which can connect trade and investment with neighbouring countries and support changes in transportation modes to reduce the cost of logistics and fuel imports of the country.

Thailand also becomes one of the most ambitious countries to develop high speed rail. This transport is under the development of infrastructure and identified as one of the economic development policy. The combination of the project and the development of regional connectivity for social and economic stability provide opportunities for access to economic and social resources. For tourism, trade, education, agriculture and other industries to provide services[1].

From the above data, the researcher has the idea of studying the effect on infrastructure development especially the high speed rail project and the transportation of goods. The purpose is to evaluate and collect the type of goods and develop the behavior modeling of freight vehicle model selection in the future in case of the development of the Bangkok-Chiangmai high speed rail route.

Thus, the objectives of this paper presents a review and explore the behavior of choosing the mode of freight transport under scenario develop a new choice of freight model in the future such as high speed rails Bangkok-Chiangmai route.

2 Literature Review

This research has divided the related literature into 3 topics 1) Review of high speed rail 2) Discrete choice models; and 3) Multinomial logit model (MNL), to guide the development of freight mode choice model, the detail shows as following.

[a] Corresponding author: wachira.tran@psru.ac.th

2.1 High speed rail

High speed trains in Japan and Europe define that high speed trains can reach speed of 200 km/h on the same updated rail and 250 km/h on the new rail, the speed when transporting passengers. High speed trains in various countries mostly used to transport passengers, except Germany and Sweden, which transport people and goods together in the system. Gerhard Troche has determined the type of train according to the speed of service and characteristics of freight train [2].

Table 1. Possible demarcation of different rail freight services

Category	Maximum Speed:	Predominant vehicle concept:
High Speed Rail Freight	>200 km/h	Modified high-speed passenger train
Semi High Speed Rail Freight	140-200 km/h	Both vehicle based on passenger train concepts and further developed freight wagons
Conventional Rail Freight	< 140 km/h	Conventional freight wagons

Source: [3]

Freight experience by high speed train in aboard summarizes from the report "High-Speed Rail Freight" be a part of the study "Efficient train System for Freight Transport" by RTH Railway Group in Sweden in 2015. Which mention Sweden, France, Germany, and international transportation in Europe.

2.2.1 Sweden

Currently the high speed trains in Sweden serviced at 160 km/h because the limit of terrain and technologies. Providing freight services under the slogan "High speed freight, faster than by truck cheaper than by air". There are 2 types of products offered: the postal of Swedish Post Office and express parcel service performed by S.J. company. Mail service transportation uses dedicated mail train but express parcel services load parcels with passenger trains. From these services in 2001-2005 found that they can reduce highway transportation 3.4 million-km-truck equivalent to reducing diesel oil 1.36 million liters and reduce carbon monoxide emissions 3,534 tons per year [2].

2.2.2 France

France, like Sweden, provides postal services with dedicated train (TGV Postal Service). Express parcel services are provided by SNCF (Société nationale des chemins de fer français) as TGV-Fret, which uses the same train as TGV Postal Office. It can contain 10 minicontainers. At present, there is a joint service between countries in Europe by Carex (Cargo Rail Express) company with the goal of continuously delivering and distributing goods (Multimodal transport) from air freight between the main airport and domestic areas [3].

2.2.3 Germany

The interCargo Express (ICGE) freight trains provide general cargo services in 1991 at speed 140-160 km/h by dedicated train, which can load a maximum weight of 500 tons. In 2000, DB cargo company provides postal and domestic parcels transportation services with the goal of attracting transportation from 10,000 container trucks per year equal to the increasing in proportion of the company's rail transportation by 20% [2].

3 Methodology

3.1 Mode choice model

The mode choice model is a disaggregate model, which is created for behavior study in modes selection of travel and goods transportation. Choosing the factors for model analysis consist of speed, travel distance, travel expenses, convenience, and number of vehicle.

3.1.1 Random Utility Theory

Random utility theory is a theory with the assumption that decision maker will consider the options that are most beneficial, which satisfaction can be measured with utility function. The utility function is divided into 2 components: 1) determinant component and 2) random component including errors and survey measurement and individual taste, as shown in equation 1 [4].

$$u_{in} = v_{in} + \varepsilon_{in} \quad (1)$$

3.1.2 Discrete Choice Models

Random utility theory contains assumptions about the distribution of variables that cannot be measured, independent and random variables distributed in normal form or other forms such as probit models (normal distribution) and logit models (Gumbel distribution or type 1 extreme value). In mostly recent research, it is assumed that the variables are Gumbel distribution, which the equation is similar to normal distribution. This study use the logit model, which can show the probability of travelers in choosing the mode of travel, as show in equation 2 [5-8].

$$P_n(i) = \frac{e^{v_{in}}}{\sum_{j \in c_m} e^{v_{jn}}} \quad (2)$$

Logit model consist of travel equation utility model. If the model has 2-way travel options, such as transporting goods by truck and train, it is called binary logit model (BNL). However, the model that has more than 2-way travel options, it is called multinomial logit model (MNL).

3.1.3 Estimating coefficient of variables

Estimating coefficient of variables in utility function has many ways. Maximum likelihood (ML) method is the most used statistical method [9], as shown in equation 3.

$$L(\beta) = \prod_{n=1}^{N} \prod_{i} (P_{ni})^{Y_{ni}} \quad (3)$$

3.1.4 Statistical evaluation

Determining model fit, likelihood ratio index is the ability of the model to explain the variance and concordance (Goodness-of-Fit) of dependent variable. That value was developed to compare with the R^2 in linear regression model. However, logit model is non-linear, so R^2 is not equal to ρ^2, which is calculated by equation 4.

$$\rho^2 = 1 - \frac{L(\hat{\beta})}{L(0)} \quad (4)$$

ρ^2 will have a value between 0 to 1, if value is close to 1, meaning the model that be created can explain the relationship between variables as well, if value is close to 0, meaning that model cannot explain the relationship between variables. ρ^2 should greater than 0.2 [10] and if greater than 0.3, shown that it is appropriate [5,11].

3.1.5 Freight mode choice factor

From the study and survey of shippers in Florida (CUTR, 2004) have ranked the influences of various factors that affect the selection of mode of transportation of goods by entrepreneurs. Found that, main factors are overall cost of logistics and mode of transport. As shown in Table 2 [12].

Table 2. Factors that affect freight mode choice

Category	Factor
Total logistics cost	Transportation charges
	Capital carrying cost in transit
	Service reliability costs
Modal characteristics	Trip time and reliability
Physical attributes of goods	Shipment size
	Package characteristics
	Shipment shelf life
	Shipment value
	Shipment density
Flow & spatial distribution of shipper	Distance of shipment
Firm characteristics	Shippers and receivers situated on rail line
	Shippers near highway
	Firms own small trucks
Flow & spatial distribution of shipper	Shipment frequency
Modal characteristics	Capacity
	Equipment availability
	Order and handling costs
Total logistics cost	Loss and damage costs
	Inventory carrying cost at destination
	Unavailability of equipment costs
	Intangible service costs

Source: [2]

Survey of opinions of government and private freight specialists in the provinces of the Bangkok-Chiangmai high speed rail line found that the affected factors for the modes selection of transportation are 1) Reliability 32.82%; 2) In-vehicle travel time 25.25%; 3) Frequency 18.74%; and 4) Transportation cost 15.33%.

3.2 Sample size

Normally, the sample size 200-500 samples were enough to analyse the disaggregate model but small sample size, 50-70 can give reasonable results also [13]. This study collected 400 samples in the study area, each sample have to answer the question 9 situations. The samples consist of production enterprises, freight forwarders, wholesaler, and retailers in the provinces along the Bangkok-Chiangmai high speed railway. Moreover, the samples have been screened as the authority to decide to choose the type of vehicle to transport products only.

4 Case Study

This research developed the multinomial logit model for reflect the decision, to select a freight transportation vehicle model in Chiangmai and Bangkok according to possibility of goods transportation including 1) high speed rail, 2) Truck, 3) Double track rail, and 4) Plane. Estimation results of the mode selections of freight transport route show details as following.

Asc_tr = Alternative specific constant (Truck)
Asc_pl = Alternative specific constant (Plane)
Asc_hsr = Alternative specific constant (HSR)
Asc_dtr = Alternative specific constant (DTR)
b_cost = Cost coefficient (Baht)
b_time = Time coefficient (Hour)
b_delay = Delay coefficient (Hour)
b_freq = Frequency coefficient (Trip/Day)

Table 3. Estimation results of MNL freight mode choice models.

Variable	Mode	Coefficient	p-value
Asc_tr	Truck	0	-
Asc_pl	Plane	-6.547	0.000
Asc_hsr	HSR	-3.988	0.000
Asc_dtr	DTR	-1.986	0.077
B_Cost	Truck	-0.045	0.000
B_Time	Truck	-0.652	0.000
B_Delay	Truck	-1.379	0.000
B_Freq	Truck	0.104	0.067
B_Cost	Plane	-0.018	0.093
B_Time	Plane	-0.379	0.035
B_Delay	Plane	-1.275	0.000
B_Freq	Plane	0.015	0.884
B_Cost	HSR	-0.040	0.000
B_Time	HSR	-0.762	0.000
B_Delay	HSR	-0.959	0.000
B_Freq	HSR	0.166	0.001
B_Cost	DTR	-0.055	0.000
B_Time	DTR	-0.468	0.000
B_Delay	DTR	-0.869	0.000
B_Freq	DTR	0.095	0.354
Number of observations		7,200	
Adjusted Rho-Squared		0.333	

U_{truck} = - 0.045*Cost - 0.652*Time - 1.379*Delay
U_{plane} = - 6.547 - 0.018*Cost - 0.379*Time - 1.275*Delay
U_{hsr} = -3.988 - 0.040*Cost - 0.762*Time - 0.959*Delay + 0.166*Frequency
U_{dtr} = -1.986 - 0.055*Cost - 0.468*Time -0.869*Delay

Parameters that are used to analyse the influence on the mode of transportation in case of the development of the Bangkok-Chiangmai high speed train system including travel expenses, time travel, belatedness, and round of service shown as Table 4.

Table 4. Application of stated choice experiment

Mode	Cost (Baht)	In-vehicle travel time (hour)	Delay (min)	Frequency (hour)
Truck	50 - 70	6 - 7	0 - 30	2 - 3
Plane	90 - 120	2 - 3	0 - 30	5 - 8
High speed rail	80 - 100	2 - 3	0 - 30	5 - 8
Double track rail	40 - 60	6 - 7	0 - 30	2 - 3

Remark: Freight price per 10 kg
I. Truck calculated from the average shipping rates postal products of Thailand Post Office and private entrepreneurs.
II. Plane calculated from freight rates in warehouses of Chiangmai International Airport.
III. High speed train calculated from 8-10 baht per kg distance from Bangkok to Chiangmai
IV. Double track rail calculated from parcel shipping rate by the State Railway of Thailand

From the proportion of selecting the transportation type found that the most freight forwarders tend to use double track trains (27.71%), truck (56.51%) and high speed train (11.18%), as show in Table 5.

Table 5. Prediction mode share %

Mode	Mode Share %
Truck	45 - 60
Plane	4 - 5
High speed rail	11 - 13
Double track rail	26 - 28

5 Results

The results of the analysis of the potential of goods that have the opportunity to transport by high speed train found that urgent postal products, air freight, jewellery and gold, computer equipment, electrical circuits and agricultural products, such as fruit and high priced flowers. Which mostly needs the transportation speed and some types are easily traumatized. Low priced agricultural products and One Tambon One Product no need the speed to transport and the current shipping cost by truck is much lower than high speed trains. Thus, these goods have low potential for transport by high speed trains. The results of model development by analysing factors such as delivery cost, time travel, loading time, belatedness and service frequency found that these factors has a statistically significant effect on the decision to select a mode of transportation.

The application of the model suggestion a sample group that is more likely to choose to use rail transportation such as double track trains (27.71%), high speed train (11.18%), and most still use freight transportation by truck (56.51%). Because of entrepreneur are still familiar with road transport and mainly used handling time more than transit time. If there is a policy to develop a connection point for handling and modify transportation modes to be efficient and safe. These will increase the proportion of high speed rail transport. Researchers have suggested that the transporting small quantities of goods by high speed train should be made in passenger-designed containers that can be modified to accommodate the transportation of passengers or goods as needed and should choose a period of non-urgent for the best value and efficiency.

6 Conclusions and suggestions

This research focuses on the development and application of the transport and tourism model to select modes of transport. Nowadays, Thailand does not have a high-speed rail system. The researchers used the utility function survey technique by stated preference combined with logit model to develop the model for predicting the decision-making behavior of entrepreneurs in the future, if there are the development of double-track rail and high-speed rail for cargo transportation. The result of model development found that the factors of price, travel time, delay, and frequency of service are factors that freight forwarders are interested. Double-track rail is suitable for capacity. High-speed rail is suitable for travel speed. The modes of transportation are suitable for different products, which gives entrepreneurs a variety of choices. The developed model also helps relevant government agencies to apply for pricing and service policies aim to change mode of low cost transportation and effective, as well as applying to the strategy of the warehouse distribution center and facility to support linking neighboring countries and future needs.

Application of model in this study analyzed initial survey data by multinomial logit models. Some features of the model are limited to simulating the traveler behavior in real situations. The future study will apply advanced discrete choice models both of GEV and mixed logit models to analyzed behavior and taste of travelers caused by taste heterogeneity.

Acknowledgment

This research was supported by Faculty of Industrial Technology, Pibulsongkram Rajabhat University and Excellence Center in Infrastructure Technology and Transportation Engineering (ExCITE), Faculty of Engineering, Chiang Mai University, Thailand.

References

[1] [World Economic Forum, 2015] World Economic Forum, The Global Competitiveness Report, 2015.

[2] CUTR, Analysis of Freight Movement Mode Choice Factors, The Center for Urban Transportation Research, University of South Florida, Award # B-D238, Report for Florida Department of Transportation Rail Planning and Safety, 2004.

[3] Cullinane, K. and Toy, N., “Identifying influential attributes in freight route/mode choice decisions: a content analysis”, Transportation Research Part E: Logistics and Transportation Review, vol. 36, no. 1, pp. 41-53, 2000.

[4] Ortuzar, J. and L, Willumsen., Modelling Transport, 2nd ed: J. Wiley and Sons, New York, 1990.

[5] Murphy, P. R. and Hall, P. K., "The relative importance of cost and service in freight transportation choice before and after: an update", Transportation Journal, vol. 35, no. 1, pp. 30-38, 1995.

[6] Regan, A. C. and Garrido, R. A., Modeling Freight Demand and Shipper Behavior: State of the Art, Future Directions, UCI-ITS-LI-WP-02-2, 2002.

[7] Rose, J. and Hensher, D. A., "Modelling agent interdependency in group decision making", Transportation Research Part E: Logistics and Transportation Review, vol. 40, no. 1, pp. 63-79, 2004.

[8] McFadden D., Conditional Logit Analysis of Qualitative Choice Behavior. In Frontiers in Econometrics. P. Zarembka, ed. Academic Press, New York ,1974.

[9] Gopinath, D. A., Modeling Heterogeneity in Discrete Choice Processes: Application to Travel Demand, Ph.D. Thesis, MIT, Cambridge, MA, 1995.

[10] Train, K., Discrete Choice Methods with Simulation: MIT Press, Cambridge, MA, 2003.

[11] Hensher, D. A., Rose, J. M. and Greene, W. H., "Applied Choice Analysis" Cambridge University Press, Cambridge, 2005.

[12] Lai, K. H., "Service capability and performance of logistics service providers", Transportation Research: Part E, vol. 40, no. 5, pp. 385-399, 2004.

[13] Ben-Akiva, M., Lerman S., Discrete Choice Analysis: Theory and Application to Travel Demand, MIT Press, Cambridge, MA, 1985.

Hijab fashion supply chain: a theoretical framework traversing consumers' knowledge and purchase intention

Eli Sumarliah[1,a], Tieke Li[1] and Bailin Wang[1]
[1]*Donlinks School of Economics and Management, University of Science and Technology Beijing, Xueyuan Road 30, Haidian, Beijing, China*

Abstract. The recent expeditious growth of Muslim population in the world initiates a dramatic upsurge of demand for Hijab fashion. Hijab fashion is a green and sustainable fashion in accordance with Islamic principles on the modesty, safety, cleanliness, and comfort of clothings. Hijab supply chain is an indispensable issue, predominantly for Muslim consumers, manufacturers, and suppliers that oblige global Muslim markets. Nonetheless, research on Hijab fashion is very limited. More specifically, although knowledge of products has a significant consequence on consumers' intention to buy Hijab, no preceding research have attempted to hypothesize the relationship between the knowledge dimensions of Hijab supply chain and consumer purchase intention. This paper attempts to propose an initial framework regarding Hijab purchase behavior. After an overall search of research articles directed in journal reference lists and electronic databases, the conceptual framework was created to hand out a set of propositions highlighting how to strengthen the link between consumers' knowledge dimensions and their intention to purchase Hijab. Recommendations for future research and strategies to boost the advantages for Hijab fashion suppliers, manufacturers, and marketers can be obtained from this model.

1 Introduction

Following numerous headlines about Hijab-wearing women making it into a beauty pageant in America and United Kingdom, New Zealand is the next to succeed. The actuality that an Indonesian-Malaysian bequest wearing Hijab, Nurul Shamsul, has made a history by being a Top 5 in Miss Universe New Zealand 2018 is an inescapable phenomenon indicating that Hijab fashion is not only a leading industry in Muslim populated countries but also an emerging market in Western nations. Islamic fashion has astonishingly achieved $270 billion consumption worldwide in 2017 and been predicted to expand by 5 per cent to US$361 billion by 2023 [1]. Lately the world's renowned fashion brands such as Alberta Ferretti, DKNY, Marks and Spencer, Max Mara and Nike have been operating to tap into the expedient market by producing Hijab fashion for Muslim-focused lines. Other eminent brands such as Dolce and Gabbana, H&M, Carolina Herrera, Massimo Dutti, Macy's and Michael Kors have also created modest collections marketed just for Muslim women [1].

Hijab fashion is a green and sustainable fashion in accordance with Islamic principles on the modesty, safety, cleanliness, and comfort of clothings. Hijab fashion supply chain encompasses everything from the preparation and prosecution of Halal raw materials, e.g. fabrics or textiles which are less animal based, more plant based, and more sustainable to be manufactured, until the delivery of the final products to the customer in line with Sharia (Islamic law). Hijab fashion supply chain is an essential issue, especially for Muslim consumers, manufacturers, and suppliers that accommodate Muslim markets. However, due to very limited studies on Hijab, the principles of the Hijab supply chain are still obscure. Further and more specifically, although consumer is a crucial part of supply chain, no conceptual framework and empirical evidence exist as to translating consumers' knowledge into their purchase intention regarding Hijab supply chain. Meanwhile, knowledge of products has a significant consequence on consumers' intention to buy, especially for new products [2]. Thus, it is presummed that knowledge of the Hijab fashion supply chain, e.g., what and how the products were manufactured, transported and distributed, will impact consumers' purchase intention of Hijab fashion products. Taking the issue into account, this paper ventures to propose an initial framework regarding Hijab fashion purchasing behavior.

2 Methodology

To design the theoretical framework of this study, an overall search of research articles was directed in journal reference lists and electronic databases. Forgoing research with interconnected topics were alienated into two groups. First, prior research that aimed at understanding the relationship between consumers' knowledge and the purchase intention of Islamic (Halal) products were identified. Second, preceding studies that

[a] Corresponding author: eliemulyadi@yahoo.co.id

https://doi.org/10.1051/matecconf/202030804004

aimed at understanding Hijab purchase intention using various independent variables were selected. The result shows that none of the first group of prior studies has explained Hijab fashion (as shown in Table 1) and none of preceding research related to Hijab fashion has used supply chain knowledge as independent variable (as shown in Table 2). Thus, this study sets a foundation for future research by creating an original conceptual framework bridging the relationship between customers' knowledge of Hijab fashion supply chain and their intention to purchase Hijab.

Table 1. Prior research on the links between knowledge and the purchase intention of Islamic products.

Researchers	Year[1]	Product	Findings
Nurhayati and Hendar [3]	2019	Halal foods	This study shows that there is a relationship between Halal product knowledge and Halal purchase intention.
Haro [4]	2018	Halal cosmetics	There are partially positive and significant influence between knowledge towards Halal cosmetics purchase intention.
Briliana and Mursito [5]	2017	Halal cosmetics	Knowledge is significantly related to consumers' attitude toward Halal cosmetics.
Maichum *et al* [6]	2017	Halal foods	Halal knowledge has a significant effect on attitude towards Halal food and had no significant effect on Halal purchase intention. It is due to the young non-Muslim consumers as respondents of the research.
Reni and Ahmad [7]	2016	Halal banking	The result shows that knowledge is statistically significant effect on intention to select Islamic bank in Indonesia.
Yusoff *et al* [8]	2015	Halal foods	This study focused on knowledge supply chains and further identified that knowledge has a direct influence on consumers' purchase intention for Halal food in Selangor, Malaysia.
Yunus *et al* [9]	2014	Halal foods	Halal awareness and product ingredients have significantly influenced Muslim's intention to buy Halal packaged food that are produced by Non-Muslim manufacturers in Malaysia.
Hamdan *et al* [10]	2013	Halal foods	There is a weak relationship between knowledge of Halal foods and purchasing decision.
Aziz and Chok [11]	2013	Halal foods	Knowledge concerning Halal foods has positive influence on attitudes towards Halal foods.
Mukhtar and Butt [12]	2012	Halal foods	Consumers' knowledge related to Halal foods affects their intention to choose Halal foods.

[1] Year of publication

Table 2. Synthesis of the determinants of purchase intentions of Hijab fashion products.

Researchers	Year[1]	Research Society	Research Variables
Ali and Sohail [13]	2018	Pakistani Hijabistas.	Sources of fashion information, fashion motivation , fashion uniqueness , fashion consciousness, Hijab buying behavior.
Rahma and Aprianty [14]	2018	Indonesian customers of Ria Miranda Hijab	Brand image, fashion consciousness, purchase decision.
Wibowo and Masitoh [15]	2018	Indonesian young women.	Religiosity, attitude, Hijab wearing intention
Winahjoe and Sutikno [16]	2018	Indonesian Muslim women not wearing Hijab.	Attitude, subjective norms, perceived personal outcome, perceived behavioral control, purchase intention.
Wibowo [17]	2017	Indonesian Muslim female wearing Hijab.	Religiosity, subjective norm, attitude, Hijab wearing intention.
Hasna and Hati [18]	2017	Muslim women.	Dressing style , fashion consciousness, motivation, uniqueness, Hijab fashion consumption, sources of fashion knowledge, social network sites.
Fadhilla [19]	2017	Customers of two Hijab companies in Indonesia.	Purchase intention, marketing mix (price, products, place, promotion).
Zabeen *et al* [20]	2017	Bangladeshi Hiab-wearing women	Religious duty, beauty, influence of opinion leaders, Hijab's growing popularity.
Hassan and Harun [21]	2016	Malaysian Muslim women.	Dressing style , fashion consciousness, motivation, uniqueness, fashion consumption, sources of fashion knowledge.

Win and Jan [22]	2016	Online shopper in Malaysia.	Online purchase intention, internal influences, external influences, online shopping experience.
Nora and Minarti [23]	2016	Indonesian female students	Religiosity, lifestyle, attitude, purchase intention
Grine and Saeed [24]	2015	Hijab-wearing Muslim women in Malaysia.	Hijab as religious obligation, Hijab as fashion, Hijab acceptance behavior.
Bachleda *et al* [25]	2014	Moroccan Muslim women.	Religiosity, subjective norm, attitude, perceived behavioral control, intention and choice of clothing
Farrag and Hassan [26]	2015	Egyptian Muslim students.	Religiosity, attitude toward fashion.
Nayebzadeh and Jalaly [27]	2014	Iranian Muslim females.	Impulse buying behavior, impulse buying tendency, emotion, self esteem, life satisfaction.

[1]Year of publication

3 Literature Review and Theoretical Framework

3.1 Hijab fashion

Islam regulates every aspect of human's life including the way of clothing. The authentic connotation of "Hijab" is to cover, to screen, or to veil. The English phrase "scarf" and the Arabic expression "Hijab" have been applied jointly, but the Hijab has cultural milieus in many Muslim countries [16]. Generally Hijabs simply allude to the headscarves worn by Muslim women along with loose and modest dresses. However, Islamic dress code requires that both men and women have Hijab; men should cover their bodies from the naval to the knee; women should cover all the bodies except for the face and the hands up to the wirsts. Muslims, both women and men, should wear clothes that are modest and permitted by Islamic law (Sharia) i.e. loose or not tights, not skinny, not transparent, and not resemble that of the opposite gender. Men are not recommended to wear silk and gold because those resemble women's clothings. In addition, Hijab is not only about the clothes but also as a way of life including righteous deeds, behavior, attitude, moral conduct, and intention of the individuals. Hijab also connotes dropping the stare to the differing gender to circumvent the turpitudes of the eye and the heart.

3.2 Hijab fashion supply chain

A supply chain can be outlined as the set of activities that permit the transformation from the original raw materials to the eventual consumption of the final product, reaching out suppliers, manufacturers, distributors and other business players. The conventional supply chain scheme begins with raw material, inbound logistics, repository, manufacturing, depository, outbound logistics, and marketing, and it closes with sales and customer utility [23]. Hijab fashion supply chain starts with raw materials purchasing, Hijab manufacturing, Hijab marketing and ends with Hijab consumers [24].

Previous studies (for example, [8], [26]) have clearly identified the dimensions of Halal supply chain knowledge consisting of Halal animal feed, Halal slaughtering, and proper segregation and it hypothesized the direct relationships between them and consumer purchase intention. This made it easier to develop the framework further. Since this study focuses on Hijab fashion supply chain, it adapted these dimensions to *knowledge* of Halal purchasing, Halal manufacturing, and Halal marketing. Figure 3 indicates the conceptual framework generated in this study.

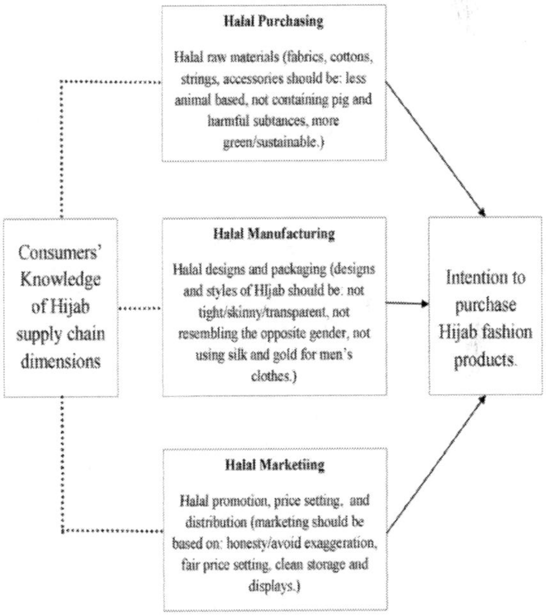

Figure 1. Theoretical framework of this study

3.3 Consumers' knowledge

Knowledge alludes to the truths, experiences or feelings renowned by an individual or a faction of people [27]. Knowledge performs an essential part in consumer behavior studies; it is one major factor that influences consumer's decision to purchase the products [28]. Every Muslim is commanded to wear Hijab clothings, which are the fashion products in accordance with Islamic dress code. Hijab fashion means fashion whose properties and processing techniques include raw materials, handling, the use of various methods from beginning to end, always in reference to Islamic law (Sharia). Prior research (for example, [3], [4], [7-10], [12]) found that there is significant relationship between consumers' knowledge of Halal products and their purchase intention. Hence, knowledge of Hijab fashion supply becomes an imperative part of Muslim consumers's intentions to purchase Hijab fashion.

3.4 Halal purchasing

Purchasing is an indispensable marketing field in prescribing the buyer-supplier connection, and there is capability to estblish a comprehensive ethical brand by unifying Islamic values and features like animal welfare, sustainability, health, less waste, satisfactory working circumstances [29]. In the Hijab fashion context, purchasing is an upstream activity based on suppliers value chain. The purchasing activities include procurement, planning, buying raw materials (e.g., fabrics or cottons, strings, accessories) and inbound logistics i.e. transportation of raw materials delivery to the Hijab company [24]. Halal purchasing for Hijab fashion refers to the raw materials bought by companies that should be permitted by Sharia (e.g., choosing Halal fabrics or textiles instead of non-Halal). Halal materials are green materials which are less animal based, more plant based and more sustainable. It is necesarry to ensure that raw materials used for Hijab are not containing *haram* (forbidden) stuffs such as the skin of pig or of animals that eat filth [30]. Sharia emphasizes that the animals need to be of a permissible species before being used for consumption both as foods and as clothings. Hence, if consumers recognized the specified requirements for Halal raw materials according to Sharia safety, hygienic, health, and sustainability factors, they would be more prompted to purchase Hijab fashion.

3.5 Halal manufacturing

Halal manufacturing refers to at least three aspects of production process. First, the cleanliness of equipments and physical environment are considered for Hijab production. Second, the designs and patterns of Hijab fashion are obliged to follow Sharia i.e. men's clothes should be designed to cover their bodies from the naval to the knee, women's clothes should cover all the bodies except for the face and the hands up to the wirsts, clothes should not resemble that of the opposite gender, clothes should be loose (not tight or skinny) and should not be transparent, and men's wardrobes should not be fashioned by gold and silk. Third, Halal packaging is applied for Hijab fashion products as one source of consumer information about the product quality. The packaging materials for Halal products, i.e., Hijab fashion should be manufactured from materials that are Halal [26]. Taking these procedures into account, it is testified that consumers who know about Halal manufacturing will be more potential to buy Hijab fashion.

3.6 Halal marketing

Marketing is a downstream activity based on consumers value chain in the Hijab fashion industry. Marketing activities includes promotion, price setting, choosing agents or distributors, and outbound logistics [24]. Halal marketing is not only practiced as business orientation but also applied as a way of life, e.g., righteous deeds, no ill intentions to harm, no corruption and fraud [25]. Promotion of Halal products, for example, should be based on honesty thus bombastic and false informations about products is definitely aginst the Sharia principles. Price should not be set too far-exceeding the product quality offered. Outbond logistics e.g. delivering orders should be done on time and in exact quality and quantity promised to the customers. In Islam, keeping the promises and satisfying the customers are parts of righteous deeds. In addition, cleanliness is also a part of faith. Thus, the cleanliness in distribution and retailing of Hijab fashion is also an essential element to be taken into account. Hijab industry should choose partners providing special facilities for Halal logistics to protect Hijab products, such as Halal transportation, Halal shipping, regular cleansing services for storages and displays. It is very admissible for Islamic product sellers to furnish an affirming atmosphere to its customers [31]. Hence, based on the assertions above, consumers amiably choose Hijab fashion if they have knowledge of how Hijab products are managed in the marketing and retailing stage. It is crucial that products that have been guaranteed of 'Halalness' in prior points in the supply chain preserve their value to the end.

4 Conclusion and implication

The fast expansion of the world's Muslim population in recent years results in a dramatic increase of demand for Halal products, especially Hijab fashion products. Halal issues on fashion are a main matter for consumers, manufacturers, and retailers. The upsurge of consumer knowledge regarding Halalness in the supply chain will assist them to purchase products that are Sharia compliant. Based on a complete literature review, this paper, genuinely, designs a conceptual framework. Due to the complex nature of the Hijab supply chain, we focused on the relationships that had the most prior studies support for testifying consumers' knowledge to their Halal purchase intention (for example, [8]). Thus, this conceptual framework brings together key variables in a single model with the purpose of recognizing how to establish a stronger relationship between knowledge dimensions of Hijab supply chain and purchase intention.

The vital contribution of this research is the development of a framework fostered by theory and studies that can be applied by suppliers, manufacturers, and marketers to grasp how to expand purchase intention by analyzing the consumers' knowledge of Hijab fashion supply chain. The implications of this research are that applying this framework can help the players in Hijab supply chain to show them which aspects of product knowledge improve interest in Hijab products, and eventually boost sales. Thus, this study is an invaluable source which should inform marketing plans and blueprints of companies involved in the Hijab business.

In the further investigation the authors will conduct an observed study to test the hypothesized dimensions and variables developed from this research. The authors have planned to carry out a quantitative research study to eventually examine and justify the conceptual framework and the hypothesized variables.

5 Research Guidelines

With this theoretical framework as a foundation, some research guidelines can be provided. The most preeminent recommendion for future studies would be to factually investigate the hypothesized links. Although this model is established from solid theoretical bases, a verifiable investigation with the hypothesized variables in a single model may prove to be an indispensable phase to further enhance this conceptual framework and improve our recognition of how to expand the relationships among knowledge dimensions of Hijab fashion supply chain and purchase intention. As mentioned before, the inquiry for future research can be: does consumers' knowledge of Hijab supply chains influence their purchase intention for Hijab fashion? The future study will, definitely, be the introductory exertion to verify the significance of relationship between consumers' each knowledge dimension as dependent variables and their purchase behavior as independent variable.

Furthermore, verifiable examinations could be conducted through a comparative study (for example, local vs. foreign Hijab brands, luxury vs. non-luxury Hijab products) in order to deliver a more thrifty framework appealing consumers' purchase intention in different conditions. Future studies could also factually investigate these links in different countries, because Hijab fashion has entered global market and adopt Islamic dress code and Halalness as international standards for Muslim consumers across globe. The evolvement of cross-cultural research may provide to easy perceiving the generality of the conceptual framework to different countries. Overall, we expect that this conceptual model accentuates the necessity for an original approach of bridging product knowledge and purchase intention in the Hijab supply chain, and offers a general basis for future empirical research. We anticipate that consecutively, those verifiable investigations of this framework will cultivate more practical advantages for suppliers, manufacturers, and marketers in the Hijab fashion industry.

References

1. T. Reuters, *State of the Global Islamic Economy Report 2018/19* (Dubai: Thomson Reuters, 2019)
2. H.H. Tuu, S.O. Olsen, Asia Pac. J. of Marketing and Logis. **24**, 78–101 (2012)
3. T. Nurhayati, H. Hendar, J. of Islamic Marketing (2019)
4. A. Haro, J. of Entrep.Business and Econ. **6**, 78–91 (2018)
5. V. Briliana, N. Mursito, Asia Pac. Manag. Rev. **22**, 176-184 (2017)
6. K. Maichum, S. Parichatnon, K. Peng, IRA-Int. J. of Manag. and Soc. Science **6**, 354-364 (2017)
7. A. Reni, N.H. Ahmad, J. of Islamic Economics **8**, 137–148 (2016)
8. F.A.M. Yusoff, R.N.R. Yusof, S.R. Hussi, Int. J. of Economics and Manag. **9**, 155-172 (2015)
9. N.S.N.M Yunus, W.E.W Rashid, N.M. Ariffina, N.M. Rashida, Procedia – Soc. and Behav. Sciences, **130**, 145–154 (2014)
10. H. Hamdan, Z.M. Issa, N. Abu, K. Jusoff, J. of Food Products Marketing **19**, 54-61 (2013)
11. Y.A. Aziz, N.V. Chok, *3rd Int. Conf. on Business and Econ. Research Proc.* Bandung (2012)
12. A. Mukhtar, M.M. Butt M M, J. of Islamic Marketing **3**, 108-120 (2012)
13. F. Hasna, S.R.H. Hati, *2nd Int. Conf. on Financial Innovation and Economic Development* **34** (Paris, 2017)
14. F. Fadhilla, Int. J. of Manag. and Appl. Science **3**, 16-20 (2017)
15. A. Zabeen, S. Shams, N. Sultana, Int. J. of Asian Soc. Science **7**, 728-737 (2017)
16. S.H. Hassan, H.Harun, J. of Islamic Marketing **7**, 476-494 (2016)
17. H.T. Win, M.T. Jan, Amity J. of Marketing **1**, 72-81 (2015)
18. L Nora, N.S. Minarti, *Proc. The 2nd Int. Multidisciplinary Conf. November 15th*, 135-148 (Jakarta, 2016)
19. F. Grine, M. Saeed, J. of Islamic Marketing **8**, 430-443 (2017)
20. C. Bachleda, N. Hamelin, O. Benachour, J. of Islamic Marketing **5**, 210-226 (2014)
21. D.A. Farrag, M. Hassan, J. of Islamic Marketing **6**, 95-108 (2015)
22. S. Nayebzadeh, M. Jalaly, J. of Islamic Marketing **5**, 302-320 (2014)
23. M. Cooper, D. Lambert, J. Pagh, Int. J. of Logistics Manag. **8**, 1–14 (1997)
24. M. Prasetyawati, W. Sudarwati, *Proc. SNTI dan SATELIT* (Malang) 4-6 October , 87-95 (2017)
25. J.M. Soon, M. Chandia, J.M. Regenstein, British Food J. **119**, 39-51 (2017)
26. E.N. Omar, H.S. Jaafar, The IEEE Symp. on Business, Engine. and Indust. Appl (Langkawi, 2011)
27. A.A. Rahman, E. Asrarhaghighi, S.A. Rahman. J. of Islamic Marketing **6**, 148-163 (2015)
28. K. Baharuddin, N.A. Kassim, S.K. Nordin, S.Z. Buyong, Int. J. of Academic Research in Business and Soc. Sciences **5**, 170-178 (2015)
29. M. Ismaeel, K. Blaim, J. of Manag.Dev. **31**, 1090-1100 (2012)
30. K. Bonne, W. Verbeke, Meat Science **79**, 113–123 (2008)
31. H. Hashim, S.R. Hussin, N.N. Zainal, Int. J. of Economics and Manag. **8**, 117-136 (2014)

Minimum Required Distance of Strain Gauge from Specimen for Measuring Transmitted Signal in Split Hopkinson Pressure Bar Test

Daesung Kim[1] and Hyunho Shin[1,a]

[1]Mechanics of Materials and Design Laboratory, Department of Materials Engineering, Gangneung-Wonju National University, Gangneung, Gangwon-do 25457, Republic of Korea

Abstract. The minimum required distance of the strain gauge on the transmitted bar of the split Hopkinson bar has been determined from the position of a metallic specimen via an explicit finite element analysis. The minimum required distance was determined when the strain–time profiles at r = 0, 0.5R_o and 1.0R_o, were coincident (r is the radial position and R_o is the radius of the bar.). The determined minimum required distance, $f(x)$, is presented as a function of the relative specimen diameter to that of the bar ($x = D/D_o$): $f(x) = -0.9385x^3 + 0.6624x^2 - 0.7459x + 1.4478$ ($0.1 \leq x \leq 0.9$). This result demonstrates the Saint-Venant's principle of rapid dissipation of localized stress in transient loading. The result will be useful for the design/modification of the pseudo-one-dimensional impact instruments that utilise a stress pulse transmitted through the specimen. The result will also allow one to avoid unnecessarily remote strain gage position from the specimen.

1 Introduction

A strain rate-dependent constitutive equation [1, 2] is indispensable for the modelling and simulation-based design of solids and structures exposed to high strain rate events [3–6] such as crashes in high-speed transportation systems (airplanes, express trains, and automobiles), high-speed machining, blasting of rock and buildings, impact, penetration, and explosion. Stress–strain curves measured under the uniaxial stress condition at a wide range of strain rates are required for the calibration of a strain-rate-dependent constitutive equation. Pseudo-one-dimensional impact systems such as the split Hopkinson bar (SHB) [7–11] and the direct impact bar (DIB) [12–16] are widely used for measuring the uniaxial stress–strain curves of various materials at strain rates of approximately 10^2–10^4 s^{-1} and over 10^4 s^{-1}, respectively.

In the DIB, the striker itself impacts the specimen directly and thus, there is no incident bar (Fig. 1(a)); the strain gauge is attached only to the transmitted bar. In the SHB (Fig. 1(b)), a disk-type specimen is sandwiched between the incident and transmitted bars; the strain gauges are attached to both the incident and transmitted bars (Fig. 1(b)). The strain gauge on the incident bar has to be sufficiently away (a distance of x) from the specimen ($x > L$, where L is the striker length) to avoid the superposition of the incident and reflected pulses. For both the DIB and SHB, the strain gauge in the transmitted bar is essential for measuring the specimen stress.

In these instruments, the diameter of the specimen is usually smaller than that of the bar to avoid the radial expansion of the specimen beyond the bar diameter. According to a recent study [10], the diameter of oxygen-free high thermal conductivity copper specimen needs to be far less than the bar diameter to obtain the stress–strain curve in the SHB test under a constant strain rate.

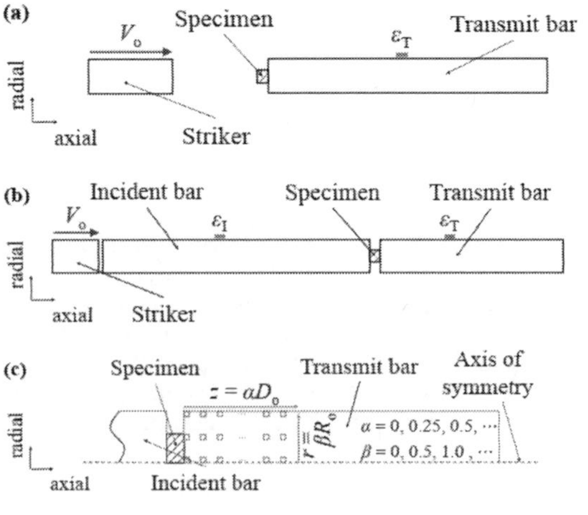

Figure 1. Schematics of (a) direct impact bar and (b) split Hopkinson pressure bar, and (c) considered gauge positions in this study (z and r are the axial and radial positions from the specimen and the axis of symmetry, respectively).

If the specimen diameter is smaller than that of the bar, stresses in the bar are not evenly distributed in the region near the specimen because a part of the bar end region is a free surface which is not in contact with the

[a] Corresponding author: hshin@gwnu.ac.kr

specimen. As a result, the measured stress on the bar surface near the specimen does not represent the bar stress in equilibrium with the specimen. Under such a circumstance, it is desirable to attach the strain gauge in the transmitted bar at a distance sufficiently remote from the specimen, where the stress (strain) is evenly distributed along the radial direction in the bar.

On the other hand, the shape of a stress pulse disperses as it travels the bar [17, 18], resulting in fluctuation in the measured curves of stress–strain and strain rate–strain using the SHB signals. The longer the distance of the strain gauge from the specimen in the transmitted bar, the higher the degree of dispersion, which necessitates dispersion correction to reduce fluctuation in the measured curves.

If the minimum required distance of the strain gage is revealed, it will deepen understanding of the stress wave propagation characteristics in the specimen–bar system, which is essential for the design and modification of the pseudo-one-dimensional impact instruments. Knowledge on minimum required distance will also allow one to avoid unnecessarily remote strain gage position from the specimen because an overly long distance will only increase the necessity of dispersion correction. In these regards, this study reveals the minimum required distance of the strain gauge on the transmitted bar that allows the measurement of the evenly distributed stress (strain) in the bar when the specimen diameter is smaller than that of the bar. In this work, compression is denoted with a positive sign.

2 Numerical analysis

A compression-mode SHB experiment was simulated using an explicit finite element analysis. The diameters (D_o) of the striker, incident bar, and transmitted bar were 20 mm while their lengths were 300, 2,000 and 1,000 mm, respectively. The length of the specimen was 5 mm, while a range of specimen diameters were considered: D/D_o ratios were 0.10, 0.25, 0.5, 0.75 and 0.90 (D and D_o are the initial diameters of the specimen and bar, respectively). The striker was initially 0.1 mm away from the bar, and its initial velocity was set as 15 m/s.

Considering the axial symmetry of the SHB system, half of the two-dimensional axisymmetric geometry was discretised using four-node bilinear axisymmetric quadrilateral elements (CAX4R). The sizes of the elements in the bar and specimen were approximately 1 × 1 and 0.1 × 0.1 mm^2, respectively. The employed elements passed a separate mesh sensitivity test. No friction was considered between the contact surfaces.

Maraging steel (C300) and OFHC copper were considered as the materials for the specimen and bar (striker), respectively. The density, elastic modulus, and Poisson's ratio of maraging steel were considered to be 8,000 kg/m^3, 190 GPa, and 0.29, respectively. The corresponding values for OFHC copper were 8,940 kg/m^3, 115 GPa, and 0.31, respectively. Plastic deformation behaviour of the OFHC copper specimen was described using the Johnson–Cook (JC) constitutive model [1]:

$$\sigma = (A+B\varepsilon^n)(1+C\ln(\dot{\varepsilon}/\dot{\varepsilon}_o))(1-T^{*m}) \quad (1)$$

$$T^* = (T-T_{ref})/(T_m-T_{ref})$$

where ε, $\dot{\varepsilon}$ and T are the true plastic strain, true plastic strain rate, and temperature of the specimen, respectively; A (90 MPa), B (292 MPa), n (0.31), C (0.025), and m (1.09) are material parameters [1]; $\dot{\varepsilon}_o$, T_{ref} are the reference values of the strain rate and temperature for calibration of the model; and T_m (1,356 K) is the melting point of the specimen.

In the post processing stage, the profile of the axial strain in the incident bar (ε_I) was extracted from a surface element at 1,000 mm from the specimen. The profiles of the axial strain in the transmitted bar (ε_T's) were extracted from a series of elements illustrated in Fig. 1(c): the axial distances (z's) of elements varied from zero to 1.5D_o at a step size of 0.25D_o; three radial distances (r's) i.e., r = 0, 0.5R_o and 1.0R_o, were considered (R_o is the radius of the bar).

The extracted signals in the numerical simulation were processed through a one-dimensional analysis [7–11]:

$$s = (A/A_o)e_T(t) \quad (2)$$

$$\dot{e} = -2(C_o/L)e_R(t) \quad (3)$$

$$e = -2(C_o/L)\int_0^t e_R(t)dt \quad (4)$$

where, s, $\dot{e}$, and e are the nominal stress, nominal strain rate, and nominal strain of the specimen, respectively; A and L are the initial cross-sectional area and initial length of the specimen, respectively; e_T and e_R are the nominal strain records of the transmitted and reflected pulses, respectively, and t is time. C_o (=$(E_o/\rho_o)^{1/2}$) and A_o are the velocity of sound and cross-sectional area of the bar, respectively. The true stress–strain and true strain rate–strain curves of the specimen were determined from the corresponding engineering stress–strain and engineering strain rate–strain curves by assuming that the volume of the specimen remains constant during deformation.

To verify whether the strain signals extracted from a given axial position allows the reliable determination of the stress–strain and strain rate–strain curves, the strain rate equation in Ref. [10] was employed:

$$\dot{\varepsilon} = (V_o - 2A\sigma/A_o\rho_oC_o\exp(-\varepsilon)) / L\exp(-\varepsilon) \quad (5)$$

where ε, $\dot{\varepsilon}$ and σ are true strain, true strain rate, and true stress of the specimen, respectively.

3 Results and discussion

3.1 Determination of the gauge position

The stress profiles at two different locations in the transmitted bar are presented in Fig. 2 for the case where the relative specimen diameter (D/D_o) is 0.5. In Fig. 2(a), the magnitudes of stress pulses at z = 0.25D_o differ depending on the radial position. In Fig. 2(b), it can be seen that they coincide at $z = 1.0D_o$

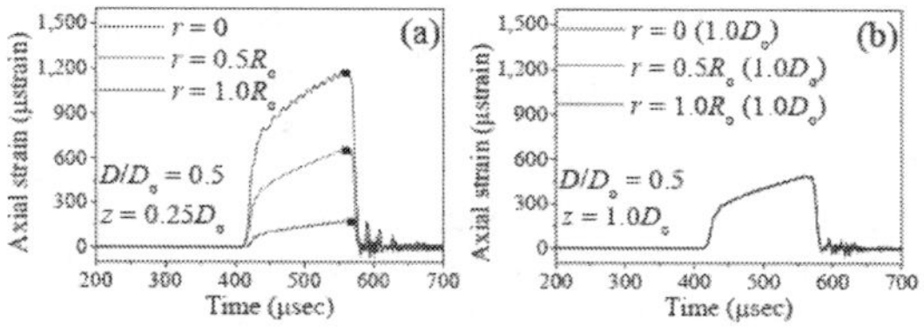

Figure 2. Axial stress profile at (a) $z = 0.25D_o$ and (b) $1.0D_o$ for the case where D/D_o ratio is 0.5.

If the strain signals at $r = 0$, 0.5, and $1.0R_o$ coincide at a given axial distance (z), the stress state in the bar can be regarded as being homogeneously distributed and thus the strain signal extracted at $r \geq R$ may be used to determine the specimen properties such as the dynamic stress–strain curve. To verify this point, the stress–strain and strain rate–strain curves were constructed using strain signals, i.e., ε_R and $\varepsilon_T(z = 1.0D_o;\ r = 1.0R_o)$ for the case where $D/D_o = 0.5$. The result is presented in Fig. 3, where the curves S and R were determined via Eqs. (1)–(3).

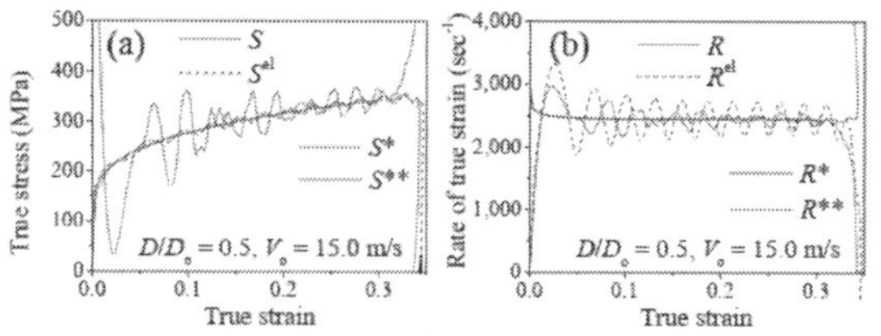

Figure 3. Curves of (a) stress–strain and (b) strain rate–strain determined for the case where $D/D_o = 0.5$.

Included in Fig. 3 are curves S^{el} and R^{el}, which were determined by averaging the values of the axial stress and strain rate of the specimen elements. Curves S^{el} and R^{el} are available only in simulation (not in experiment).

Fig. 3 also includes curves R^* and S^*, which were obtained by simultaneously solving the constitutive equation of the specimen (Eq. (1)) and the strain rate equation (Eq. (5)) using the Excel program available in Ref. [10]. Curves S^* and R^* are the stress–strain and strain rate–strain curves predicted based the theory of specimen strain rate in the SHB[10] and the constitutive equation. In the experiment, the constitutive parameters of the JC model can be reasonably estimated from quasi-static tests via the method described in Ref. [10]. Curves S^* and R^* determined in this way can be referred to before carrying out the SHB experiment.

Finally, Fig. 3 includes curves S^{**} and R^{**}, which were obtained by applying curves S and R, respectively, to Eq. (5). Consistency of curves S^{**} and R^{**} with S and R, respectively, can be used to verify the reliability of the experimental curves of S and R, respectively.

It can be observed from Fig. 3(a) that the curves S, S^{el}, S^* and S^{**} are consistent. In Fig. 3(b), the curves R, R^{el}, R^*, and R^{**} are also consistent. From these observations, it can be confirmed that the strain signals ε_I and especially ε_T at $z = 1.0D_o$ is reliable, where ε_T at r=0, $0.5R_o$ and $1.0R_o$ coincide. Therefore, this study determines the strain gauge position (z_c) on the transmitted bar when $\varepsilon_T(z;\ r = 0)$, $\varepsilon_T(z;\ r = 0.5R_o)$, and $\varepsilon_T(z;\ r = 1.0R_o)$, coincide.

3.2 Required strain gauge distance

In Fig. 2(a), the maximum strain value at a given pulse profile is marked as the closed square. This maximum value of a given profile was normalized to the corresponding value of the profile at $z = z_c$, and the normalized values ($\varepsilon_T(z)/\ \varepsilon_T(z_c)$) at $r = 0$ are presented in Fig. 4(a) for a range of the relative specimen diameters (D/D_o's). The normalized values at $r = 0.5R_o$ and $1.0R_o$ are also presented in Figs. 4(b) and 4(c), respectively.

At $r = 0$ (Fig. 4(a)) and $0.5R_o$ (Fig. 4(b)), interestingly, the normalized value of the axial strain is maximized at the axial position between zero and z_c. This maximization of the axial strain is similar to the phenomenon observed in a quasi-static indentation where the maximum axial stress does not appear at the contact surface but at a distance slightly below the contact surface [19, 20].

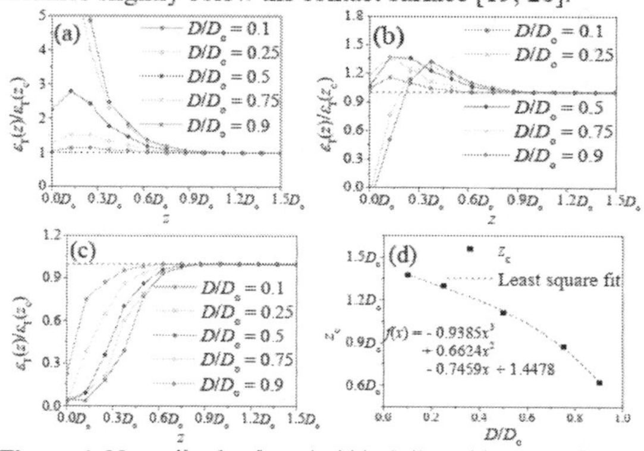

Figure 4. Normalized values ($\varepsilon_T(z)/\varepsilon_T(z_c)$) at (a) $r = 0$, (b) $r = 0.5R_o$, and (c) $r = 1.0R_o$ for a range of D/D_o ratios. (d) Minimum required distance (z_c) for the gauge position in the transmitted bar.

At $r = 1.0R_o$ (Fig. 4(c)), the normalized strain value is generally lower at the axial position (z) near the specimen; the smaller the specimen diameter, the lower the strain value. This observation results from the fact that, when the specimen diameter is smaller than the bar, the stress is not equally distributed in the transmitted bar as a part of the bar end is in contact with the free surface instead of the specimen.

As mentioned, the strain gauge positions (z_c's) in the transmitted bar were determined when $\varepsilon_T(z;\ r = 0)$, $\varepsilon_T(z;\ r = 0.5R_o)$, and $\varepsilon_T(z;\ r = 1.0R_o)$, coincide. The determined result for a range of relative specimen diameters (D/D_o ratios) is shown in Fig. 4(d). As can be seen in Fig. 4(d), the minimum required distance increases as the specimen diameter decreases. Included in Fig 4(d) is the least square fit using a third order polynomial: $f(x) = -0.9385x^3 + 0.6624x^2 - 0.7459x + 1.4478$ $(0.1 \leq x \leq 0.9)$, where $f(x)$ is the minimum required distance and $x = D/D_o$.

https://doi.org/10.1051/matecconf/202030804005

This curve can be referred to for determining the minimum distance for positioning strain gauge on the transmitted bar for specimens with a different diameter from the ones used for numerical analysis in this study.

3.3 Verification for extreme cases

The minimum required distance shown in Fig. 4(d) was obtained for the OFHC specimen with a given plastic property described by the parameters *A*, *B*, *n*, *C*, and *m* in Eq. (1). The parameter *A* determines the yield strength of the material. To verify whether the determined distance indicated in Fig. 4(d) can be used for a higher-strength material, this study additionally considered a hypothetical specimen with an *A* value of 800 MPa (the rest of the JC parameters are the same as the OFHC copper specimen). Example cases of the strain profiles when $D/D_o = 0.5$ at $z = 1.0D_o$ are shown in Fig. 5 (a).

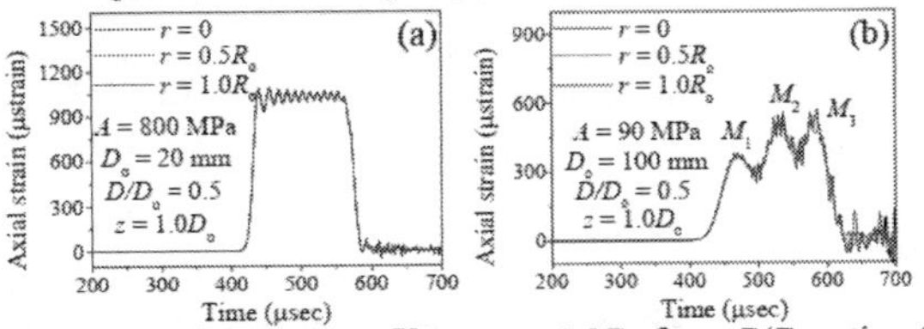

Figure 5. Axial stress profiles at $z = 1.0D_o$ for a D/D_o ratio of 0.5. The yield strength of the specimen and the bar diameter (D_o) were (a) 800 MPa and 20 mm, respectively, and (b) 90 MPa and 100 mm, respectively.

In Fig. 5(a), magnitudes of the ε_T profiles at $r = 0$, 0.5 and $1.0R_o$ reasonably coincide, which is consistent with the result in Fig. 4(d) where $z_c = 1.0D_o$ when $D/D_o = 0.5$ for the OFHC specimen with an *A* value of 90 MPa. Therefore, the quantified minimum distance in Fig. 4(d) can be used for specimens with a yield strength of at least up to 800 MPa.

The minimum required distance shown in Fig. 4(d) was obtained by considering a bar diameter of 20 mm. To verify whether the determined result in Fig. 4(c) can be used for a bar with a larger diameter, this study additionally considered the case where the diameters of the specimen and bar were enlarged while their axial lengths were not varied. Example cases of the strain profiles when $D/D_o = 0.5$ at $z = 1.0D_o$ are shown in Fig. 5 (b) for the case when D_o was 100 mm.

In Fig. 5(b), there are only three peaks in the strain pulse period (marked as M_1, M_2, and M_3). In our separate analysis, number of such peaks decreases with the bar diameter (not shown), which results from the deviation of the impact system from the one-dimensional nature with the increase of the bar diameter. The issue on the allowable maximum bar diameter for the reliable measurement of specimen properties needs further study for clarification. From the observation that the three curves at $r = 0$, $0.5R$, and $1.0R$ in Fig. 5(b) are reasonably coincident, the quantified result in Fig. 4(d) can be used for a bar diameter of up to at least 100 mm. In the literature [21], the necessity of correction for the strain signal measured on the *bar surface* to obtain the signal at the *bar axis* was reported even for the case where the strain gage was sufficiently remote from the specimen. Further detailed numerical verification is necessary to reaveal a more accurate maximum bar diameter, up to which there is no need for the surface signal correction to obtain the bar axis signal.

3.4 Saint-Venant's principle

Saint-Venant's principle [22] states that "*If the forces acting on a small portion of the surface of an elastic body are replaced by another statically equivalent system of forces acting on the same portion of the surface, this redistribution of loading produces substantial changes in the stresses locally, but has a negligible effect on the stresses at distances which are large in comparison with the linear dimensions of the surface on which the forces are changed*". An example of the principle is schematically illustrated in Fig. 6, where the same magnitude of load (100 kgf) is applied to the bar from a small-diameter (high-density) or a large-diameter (low-density) object. Although the local stress fields near the objects are different for the two comparing cases, they are practically the same after a short distance from the bar/object contact surface. In this regard, the principle is conveniently called 'Saint-Venant's principle of rapid dissipation of localized stress' [23]. The principle was generally considered in quasi-static loading. The result presented in this study, Figure 4(d), has a meaning in that it serves as an example of the Saint-Venant's principle in *transient loading*. Figure 4(d) describes the principle on a quantitative base, which supplements the qualitative description of the principle in recent studies [24, 25].

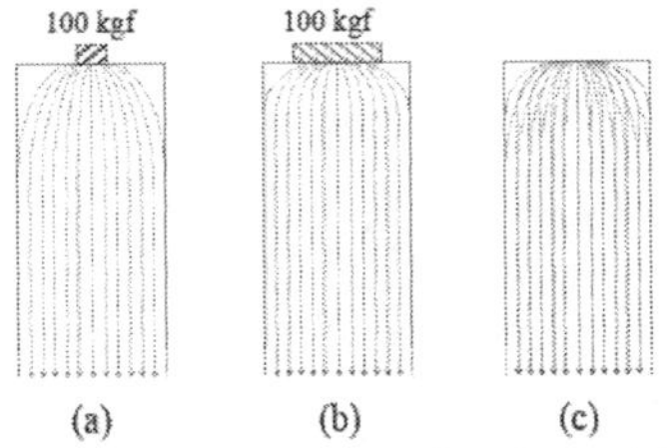

Figure 6. Force lines when a quasi-static load (100 kgf) is applied to the bar from (a) a small-diameter or (b) a large-diameter disk. (c) Comparison of the force lines.

4 Conclusions

Numerical analyses of the split Hopkinson bar experiment were carried out via an explicit finite element analysis for a range of relative specimen diameters to that of the bar (D/D_o). Based on the result of numerical analyses, the minimally required distance of the strain gauge from the specimen surface in the transmitted bar was quantified from the viewpoint of the coincidence of the strain profiles at $r = 0$, $0.5R_o$ and $1.0R_o$. The obtained result is $f(x) = -0.9385x^3 + 0.6624x^2 - 0.7459x + 1.4478$ ($0.1 \leq x \leq 0.9$), where $f(x)$ is the minimum required distance and $x = D/D_o$. This result can be used for specimens with a yield strength of at least up to 800 MPa and for a bar diameter of up to at least 100 mm. The minimum required distance is an example of the Saint-Venant's principle of rapid dissipation of localized stress in transient loading. The result will be useful for the design and modification of the pseudo-one-dimensional impact instruments that utilise a stress pulse transmitted through the specimen. The result will also allow one to avoid unnecessarily remote strain gage position from the specimen because an overly long distance will only increase the necessity of dispersion correction.

Acknowledgement

This study was financially supported by a National Research Foundation of Korea grant under contract No. 2019R1H1A2039686, funded by the Ministry of Science and Technology (Korea).

References

1. G.R. Johnson, W.H. Cook, A Constitutive Model and Data for Metals Subjected to Large Strains, High Strain Rates and High Temperatures. In: *Proc. 7th Int symp ballistics* pp. 541–547 (The Hague, Netherlands, 19–21 April 1983).
2. H. Shin, J.-B. Kim, A Phenomenological Constitutive Equation to Describe Various Flow Stress Behaviors of Materials in Wide Strain Rate and Temperature Regimes, J. Eng. Mater. Technol. **132,** 021009 (2010).
3. Y.-H. Yoo, S.-H. Paik, J.-B. Kim, H. Shin, Performance of a Flying Cross Bar to Incapacitate a Long-Rod Penetrator Based on a Finite Element Model, Eng. Comput., **29**(4), 409–415 (2013).
4. D.-H. Kim, H. Shin, D. Jung, A Numerical Study on Jet Formation and Penetration Characteristics of the Shaped Charge with an Aspect Ratio of 2.73 and a High-Strength Copper Liner (in Korean), Kor. J. Comput. Design Eng., **24**(1), 73–84 (2019).
5. D. Kim, J.-H. Lee, H. Shin et al., A Numerical Study on the Influence of the Flow Stress of Copper Liner on the Penetration Performance of a Small-Caliber High Explosive (in Korean), Kor. J. Comput. Design Eng., **24**(4), 300–310 (2019).
6. H. Shin H, H.-J. Lee, Y.-H. Yoo, W. Lee, A Determination Procedure for Element Elimination Criterion in Finite Element Analysis of High-Strain-Rate Impact/Penetration Phenomena, JSME Int. J. **47**(1), 35–41 (2004).
7. H. Kolsky, An investigation of the Mechanical Properties of Materials at Very High Rates of Loading. Proc. Phys. Soc. Lond. Sect. B, **62**(11), 676–700 (1949).
8. M.A. Meyers, *Dynamic behavior of materials* (John Wiley & Sons, Inc, New York, 1994).
9. M.M. Al-Mousawi, S.R. Reid, W.F. Deans, The Use of the Split Hopkinson Pressure Bar Techniques in High Strain Rate Materials Testing, Proc. IMechE, Part C: J. Mech. Eng. Sci., **211**, 273–292 (1997).
10. H. Shin, J.-B. Kim, Evolution of Specimen Strain Rate in Split Hopkinson Bar Test, Proc. IMechE, Part C: J. Mech. Eng. Sci., **233**(13), 4667–4687 (2019).
11. H. Shin, J.-B. Kim, Understanding the Anomalously Long Duration Time of the Transmitted Pulse from a Soft Specimen in a Kolsky Bar Experiment, Int. J. Precision Eng. Manufact., **17**(2), 203–208 (2016).
12. H. Couque, The Use of the Direct Impact Hopkinson Pressure Bar Technique to Describe Thermally Activated and Viscous Regimes of Metallic Materials, Phil. Trans. R. Soc. A, **372,** 20130218 (2013).
13. L.J. Lea, A.P. Jardine, Application of Photon Doppler Velocimetry to Direct Impact Hopkinson Pressure Bars, Rev. Sci. Instruments **87,** 023101 (2016).
14. R. Othman, S. Aloui, A. Poitou, Identification of Non-Homogeneous Stress Fields in Dynamic Experiments with a Non-Parametric Method, Polym. Testing **29,** 616–623 (2010).
15. J.Z. Malinowski, J.R. Klepaczko, Z.L. Kowalewski, Miniaturized Compression Test at Very High Strain Rates by Direct Impact, Exp. Mech., **47,** 451–463 (2007).
16. F. Kamler, P. Niessen, R.J. Pick, Measurement of the Behaviour of High Purity Copper at Very High Rates of Straining, Canad. J. Phys., **73**(5–6), pp. 295–303 (1995).
17. Gorham DA. A Numerical Method for the Correction of Dispersion in Pressure Bar Signals. J. Phys. E: Sci. Instum., **16,** 477–479 (1983).
18. Z. Li, J. Lambros, Determination of the Dynamic Response of Brittle Composites by the Use of the Split Hopkinson Pressure Bar, Compos. Sci. Tech., **59,** 1097–1107 (1999).
19. Y.-H. Yoo, W. Lee, H. Shin, Spherical Nano-Indentation of a Hard Thin Film/Soft Substrate Layered System: II. Evolution of Stress and Strain Fields, Modelling Simul. Mater. Sci. Eng., **12,** 69–78 (2004).
20. R.A.C. Slater, *Engineering Plasticity, Theory and Application to Metal Forming Processes* (Macmillan, London, 1977).
21. A. Tyas and A.J. Watson, An Investigation of Frequency Domain Dispersion Correction for Pressure Bar Signals, Int. J. Impact Eng., **25,** 87–101 (2001).
22. B. De Saint-Venant, Mémoire sur la Torsion des Prismes (Memory on the Torsion of the Prisms), Mém. des Savants étrangers, **14,** 233-560 (1855).
23. Punmia BC, Jain AK and Jain AK. *Mechanics of Materials.* (Laxmi Publications (P), Ltd., New Delhi 2001).
24. H. Shin, D. Kim, One Dimensional Analyses of Striker Impact on Bar with Different General Impedance, Proc. IMechE, Part C: J. Mech. Eng. Sci., https://doi.org/10.1177/0954406219877210 (2019).
25. H. Shin, J.-H. Lee, J.-B. Kim, S.-I. Sohn, Design Guidelines for the Striker and Transfer Flange of a Split Hopkinson Tension Bar and the Origin of Spurious Waves, Proc. IMechE, Part C: J. Mech. Eng. Sci., https://doi.org/10.1177/0954406219869984 (2019).

MATEC Web of Conferences **308**, 05001 (2020) https://doi.org/10.1051/matecconf/202030805001
ICTTE 2019

Generating relevant scenarios for intelligent transportation service

Ismet Addoui [1,a], Tarek Chouaki [1,2,a] and Ambrogio Delli Colli [1,3,a]

[1] *IRT SystemX, 8 Avenue de la Vauve, 91127 Palaiseau Cedex, France*
[2] *Sorbonne University, 4 Place Jussieu 75005 Paris, France*
[3] *Politecnico di torino, Corso Duca degli Abruzzi, 24 10129 Torino, Italy*

Abstract. This paper addresses risk assessment issues while conceiving complex systems. Indeed, project stakeholders have to share the same problems understanding allowing to undertake rational and optimal decisions. We propose an approach based on Natural Language Processing (NLP) techniques to improve systems quality requirements such as consistency and completeness. We assess the relevancy of our approaches through experimentations and highlighted feedbacks from project stakeholders and players.

1 Introduction

Requirements Engineering aims to define documents and maintain requirements [1]. The number and nature of activities characterizing any process for managing requirements is correlated with the target domain (software, real-time embedded systems, automotive or railway), as well as the importance of the project, its innovative nature and other considerations such as human factors, standards or technologies. Any requirements engineering process includes at least activities of elicitation, specification, validation and change management. These activities are supported by different techniques such as interviews, knowledge engineering, modelling and analysis or simulation.

Having a clear understanding of the customer's needs is not an easy task. The goal of the requirements engineer is to help customers to clarify and to frame the requirements of the system under construction. Building a complete, consistent and feasible requirements base is a challenging issue and a timely research topic. Moreover, the quality of the requirements is of a high importance to avoid snow-ball effects leading to misinterpretation and ineffective realizations.

Without a shared vision, chances to converge quickly to a requirement base that is complete, consistent and feasible are negligible or would require many iterations leading to higher costs, delays and competitiveness. This comes besides with no guarantee on the quality of the requirements. To improve the quality of the requirements, one can consider scenarios with important advantages already addressed in the literature [2, 3]. For instance, scenarios tend to adopt the user's viewpoint, which is a key feature to validate the adequacy between the requirements and the user's needs. Considering the complexity of large systems, the number of scenarios needed to effectively help practitioners to design a system that fits the needs of the customers can be important. The selection of relevant scenarios among possible ones to cover the user's need is not obvious. Today, this task is generally performed in ad-hoc manner by experts that consider the system only through their prism, generally limited to their concerns. We think there is a need to improve this situation and to generate more relevant scenarios considering the full coverage of users' needs and required analysis to be performed to validate their adequacy, their completeness or their feasibility.

2 Motivations and our contribution

Several studies highlighted recurrent problems that occur when large scale projects fail. For instance, the CHAOS Report [4] lists the main reasons IT projects fail. We can cite among them:

Bad vision / understanding of the users' needs

Lack of implication of the various stakeholders early in the development life cycle

Difficulties for the stakeholders to collaborate (silos between experts)

Lack of relevant metrics to take rational decisions

[a] Corresponding author: ismet.addoui@irt-systemx.fr

In practice, most of problems issued from bad decision making are discovered during verification and validation activities. When the first versions of the system have been implemented and deployed. Those late discoveries require most of time re-engineering the system at several levels: operational, functional or physical; or reconsidering the strategy of the enterprise in terms of partnerships or skills to be developed.

From a technical perspective, the design of a complex system driven by scenarios can be helpful if it allows the consideration of any relevant combination of features (events, behaviours, conditions) to really get a deeper understanding of how the system should behave and what it should be made of. The question we shall consider are as follows:

- How many scenarios do we need to consider to embrace all user's needs?
- What are the combinations of events, behaviours or conditions do we need to analyse in priority according to initial hypotheses?

The previous questions are important and necessitate detailed responses. Nevertheless, there are no explicit answers and we propose some experiments with prominent preliminary results providing some hints.

3 Application

We aim to design and implement a mobility service where systems must cooperate to fulfil the expected operation capabilities. We consider five capabilities described in the following:

- Enhance personal mobility: this can be measured through user interviews about comfort or quality of service. Related metrics are the average time of waiting to access the transportation service or the mean time delay to reach the desired destination.
- Ensure the System of System (SoS) integrity: it is about the physical integrity of any human being involved in operational scenarios (user, pedestrian, cyclist, etc.). This latter could be corrupted in many ways: physical attacks on roadside units or vehicles, or cyber-attacks. Meeting this capability can be measured by the reduction of any kind of accidents. This implies also a good crisis management in case of accident, in order to not deteriorate the situation and not create congestions.
- Improve environment impact: this is related to the transportation efficiency and effectiveness of the cooperation between vehicles and roadside units. Meeting this capability can be measured by the reduction of fuel consumption or lower emission of CO_2 fine particles.

Ensure service continuity: this allows to avoid malicious acts that could lead to corruption of human beings, vehicles or roadside units' integrities. It also tries to anticipate damages. Its satisfaction can be measured by the reduction of malicious acts or their associated costs.

Reduce of congestion. It can be measured by the reduction of occupation rates of roads, or by the reduction of annual mean time spent on the roads by users.

Our analysis is driven by usage scenarios leading to ask the following questions:

How systems implied in the scenarios shall behave and cooperate to fulfil each of those operational capabilities?

What are the data they shall exchange? Through which media and what quality of service?

Depending on a described situation, the set of behaviours and interactions that shall be considered can be very large. Hence, this cannot be considered and managed easily at design time. Nevertheless, in this paper, we propose to generate a set of representative scenarios to cover a maximum number of cases, leading to a consistent and complete specification. The quality and the accuracy of simulations' results of the generated scenarios have an important impact on the ability to proceed to relevant architecture choices.

4 Generation of scenarios

Figure 1 depicts and illustrates our approach to generate scenarios. Note that, the definition of scenarios considered is "sequence of events that occurs during a particular execution of a group of objects implied in the system of system". The set of scenarios to drive the analyses at system level, is performed by experts.

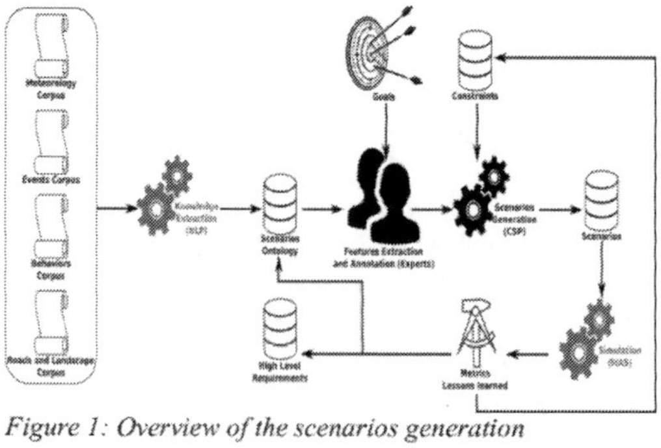

Figure 1: Overview of the scenarios generation

This is a manual activity relying on the knowledge of these experts and their ability to imagine key scenarios to validate users' needs. Pragmatically, the required number as well as the relevancy of scenarios defined by our experts cannot be measured accurately. Knowing that this set of scenarios is produced by humans, we can assume that there are omissions due to the large scope of features. For instance, experts have defined over 50 scenarios ranging from the highest probability of occurrence to the most critical. Are those scenarios enough to cover all the needs? Considering the complexity of the SoS, the answer is obviously no. Then there is a need to assist expert to elicit more relevant scenarios.

4.1 Knowledge Extraction

To improve the quality and the accuracy of scenarios driving key decisions, we have used an approach using knowledge extraction from the dedicated literature in order to capitalize it into an ontology. This ontology classifies all key features characterizing any scenario: users, vehicles, roadside units, road types or traffic conditions. For instance, our ontology describes all kind of:

- Climatic conditions: brightness, rain, snow, wind, fog
- The various roads the vehicles can take; sloppy, curving, muddy or icy
- The taxonomy of all behaviours that can be expected from drivers or pedestrians: careful, distracted or dangerous

This knowledge has been extracted from a set of corpuses using Natural Language Processing with the NLTK framework [5]. The obtained results using NLP are completed with expert recommendations.

As illustrated in Figure 2. We extract key terms from the documents of the mobility corpus after tokenizing, cleaning and lemmatizing content. We rely on the computation of semantic similarities using WordNet [6] to reduce the resulting set of words. Key terms extraction relies on the frequency distribution of remaining words. Finally, we use tags from POS (Part of Speech) Tagging to determine the role of corresponding words in order to generate predicates. These predicates are validated / completed by mobility experts.

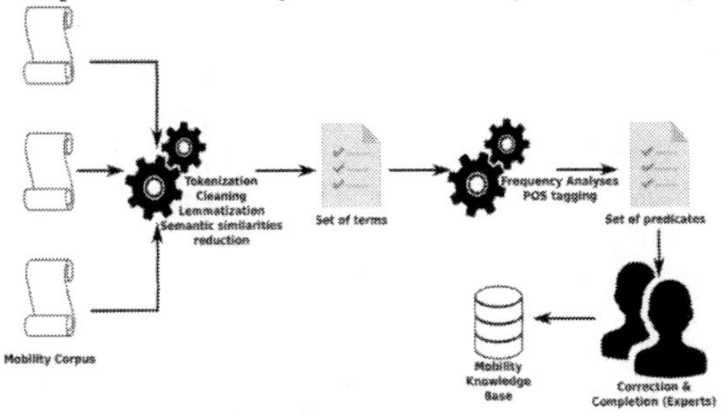

Figure 2: Ontology extraction using NLP

4.2 Generation of scenarios

The extracted ontology is exhaustive and can be used for a huge number of scenarios. Since our experimentation concerns a specific zone in Paris area, we extracted the relevant subset for achieving deeper analyses considering experts' recommendations. For this purpose, we relied on data we collected according to our experimentation: climate, hazards, topology of road, localization of roadside units, etc. We selected 45 features regarding the operational capabilities we wanted to reach. Then, each feature is annotated with a level of probability of occurrence ranging from A (the most probable) to E (the less probable). The features are also annotated with a level of criticality ranging from A (the most critical) to C (the less critical). Table 1 shows the list of selected features along with their respective levels of probability of occurrence and criticality regarding the scope of our experimentation.

Table 1. Scenario Features

Features	Criticality	Probability
Curved Road	B	A
Speed Bump	B	C
Teleoperation Hacking	A	D
CEM	A	E
Cut Out	B	B
Equipements Supervision Hacking	A	C
Impaired Walking Pedestrian	A	C
Overtaking	A	A
Against Rally Driving	A	D
Safety Distance Violation	C	A

In order to generate consistent scenarios, we propose to define some features' constraints. The two following formulas illustrate example of meet constraints.

$$\neg(\text{cross\_road} \land \text{round\_about}) \quad (1)$$

$$\text{Fog} \rightarrow \text{Low\_grip} \quad (2)$$

Formula (1) is used to avoid scenarios with both crossroads and roundabouts. Aware that (in our study) it is impossible to have simultaneously a crossroad and a roundabout.

Formula (2) is used to force the presence of low grip in case of fog.

Adding those constraints allow us to avoid having inconsistent scenarios.

It is important to mention that with 45 selected features and 22 specified constraints; the total number of possible scenarios remains very high (one can consider an exponential number of combinatorial cases scenarios).

https://doi.org/10.1051/matecconf/202030805001

Since we cannot afford to analyse all these scenarios, we have investigated a new approach to prioritize these scenarios using two metrics such as criticality and probability of occurrence. For this purpose, and for each generated scenario, we also propose two scores as indicated by equations (3) and (4):

$$P_g = \prod_{i=1}^{n} P_i \qquad (3)$$

$$C_g = \prod_{i=1}^{n} 2^{C_i} \qquad (4)$$

Equations (3) and (4) provide the global probability (resp. criticality) of a scenario for each of them using the probability of occurrence (resp. criticality) of each feature.

Equation (4) uses an exponential law as it fosters highest levels of criticality. Once a global score for each scenario has been computed in terms of probability of occurrence and criticality, we represent these scenarios in Figure 3.

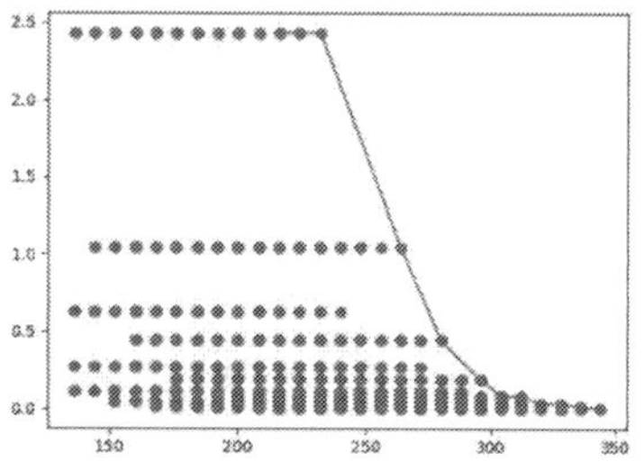

Figure 3: Pareto front of the generated scenarios criticity over probability of occurence

This figure represents the criticality's behaviour according to probability of occurrence variations.

Before going through the details of Figure 3, we propose the following definition [7]

The definition by [7] set of non-dominated solutions, being chosen as optimal, if no objective can be improved without sacrificing at least one other objective. On the other hand, a solution x* is referred to as dominated by another solution x if, and only if, x is equally good or better than x* with respect to all objectives. The Pareto front points (see figure 3) are the non-dominated solutions.

Figure 3 depicts the Pareto front allowing us to extract the most representative scenarios represented by non-dominated points according to the two identified criteria (criticality and probability of occurrence). This Pareto front is an efficient approach to considerably reduce the total number of scenarios and exhibits only the non-dominated points or scenarios that we consider the most important to be investigated. From this 1134 scenarios have been selected from the Pareto front for analysis and study. Items below represent features of a single scenario to be prioritized for analysis:

Impaired Walking Pedestrians
Right Overtaking

Communication Loss
Overtaking
Safety Distance Violation
Direction Change Not Communicated
Wind
Out of Zebra Crossing
Over speed Driving
Glare
Equipment Supervision Hacking
Cut Out
Impaired Driver
Late Obstacle Detection
VRU Obstacle
2 Wheels
Cut In
Emergency Vehicles
Low Grip
Curved Road

Note that items of this example precise only the features that shall appear into the same scenario. They do not reveal information about the causality links between events or behaviours. To cope with this issue, we appealed experts to recommend and select the most relevant scenarios depicted by Pareto front approach.

For instance, we provide in the following an example illustrating a right interpretation of features into a realistic scenario:

"The autonomous vehicle drives on a sloppy slippery road with low visibility because of the fog. Electromagnetic disturbances cause a loss of communication with the control / command center. The road taken is often crossed by wild animals."

Such situation can cause the immediate stop of the transportation service otherwise, it can put the life of users in danger. In this case, we need a set of functions ensuring the good coordination between vehicles and roadside units. This situation requires also a function for the vehicle to park in a safe area and notify the control-command operator as soon as the communication is fully restored. This analysis indeed impacts behaviours and interactions required to ensure safer transportation services.

5 Conclusion

In this paper, we presented some achievements of a novel approach making use of techniques from AI (NLP for instance) to elicit requirements for a mobility service. Compared to traditional approaches, we took the advantage of AI techniques to help practitioner to focus on the most important features when designing a complex system. This approach allows us to select the nature of the scenarios needed. The generation of scenario could focus on the criticity, the probability of occurrence or both. The feedback we had from the experts is considered to improve the quality of the obtained results. Besides, most of techniques presented in this paper have been automated (knowledge extraction, scenario generation, exploitation of results) which is a key point for dynamically managing changes. Obviously, our approach suffers from several limitations in terms of techniques or application that shall be addressed by further works. For instance, we did not take account the strategic aspects of the enterprise into the elicitation of requirements and we do not generate the story of the scenario. Finally, the temporality of the appearance of event (features) is not specified in our method.

6 References

1. I. Sommerville, *Requirements engineering: processes and techniques,* **vol. 10** (plus 0.5em minus 0.4emWiley, 1998).
2. J. C. S. do Prado Leite, G. Rossi, F. Balaguer, V. Maiorana, G. Kaplan, G. Hadad, and A. Oliveros, *Enhancing a requirements baseline with scenarios, Requirements Engineering,* **vol 2**(no 4, pp 184–198, 1997).
3. M. Glinz, *Improving the quality of requirements with scenarios, in Proceedings of the second world congress on software quality*, **vol 9**, (2000, pp 55–60).
4. S. GROUP et al, *The chaos report into project failure, the standish group international inc,* (2014, available on-line at http://www.standishgroup.com/visitor/chaos.html).
5. S. Bird, E. Klein, and E. Loper, *Nltk book*, (2009).
6. C. Fellbaum, *Wordnet, in Theory and applications of ontology: computer applications.* plus 0.5em minus 0.4emSpringer, (2010), pp 231–243.
7. https://www.igi-global.com/dictionary/pareto-front/21878

Research on Traffic Acoustic Event Detection Algorithm Based on Sparse Autoencoder

Xiaodan Zhang[1,a], and Yongsheng Chen[1] and Guichen Tang[2]

[1] *Research Institute of Highway, Ministry of Transport, Beijing 100088, China*
[2] *School of Communication Engineering, Nanjing Institute of Technology, Jiangsu Nanjing, 211167, China*

Abstract. Road traffic monitoring is very important for intelligent transportation. The detection of traffic state based on acoustic information is a new research direction. A vehicles acoustic event classification algorithm based on sparse autoencoder is proposed to analysis the traffic state. Firstly, the multidimensional Mel-cepstrum features and energy features are extracted to form a feature vector of 125 features; Secondly, based on the computed features, the five-layers autoencoder is trained. Finally, vehicle audio samples are collected and the trained autoencoder is tested. The experimental results show that detection rate of the traffic acoustic event reaches 94.9%, which is 12.3% higher than that of the traditional Convolutional Neural Networks (CNN) algorithm.

1 Introduction

In order to provide auxiliary research for intelligent transportation and traffic safety development, an autoencoder based the acoustic feature for traffic event detection is proposed. In order to integrate the dynamic features of the sound, the algorithm extracts the multidimensional Mel-cepstrum features and energy features, and forms a 110 dimension feature vector. Then the 5-layers autoencoder based on the computed features is trained to improve the robustness. The experimental results show that the detection rate of traffic acoustic events reaches 94.9%, and the recognition rate of collision sounds reaches 97. 9%. The proposed audio surveillance system may be used to monitor traffic accidents and save valuable time in rescue mission. In addition, the system can be embedded in the automatic driving system, which is conducive to the timely response of the self-driving car to the traffic state, greatly improving safety.

The detection of traffic state based on acoustic information has been an important research direction for intelligent transportation. Compared to existing monitoring techniques, acoustic signal processing and classification techniques have the advantage of being low cost and unaffected by lighting conditions. Especially in the case of insufficient light or intermediate obstructions, acoustic signals have higher information coverage. Therefore it is an important supplement to existing monitoring methods. However, compared with the laboratory environment, the real traffic environment is complex. For example, the tunnel is a special traffic environment, that is very different from the open road environment. How to effectively process the traffic acoustic data remains a challenge. In 1998, Henryk Maciejewski et al.[1] studied and designed a classification system based on wavelet and neural network. The specific recognition model based the sound signal was constructed for four different vehicles, and the recognition accuracy was 73.68%. Audi Ovox et al. applied sound recognition technology to the field of intelligent transportation[2] and used voice recognition technology in the car phones. Xianglong Luo et al.[3] used empirical mode decomposition (EMD) and support vector machine (SVM) to identify the vehicle state. In recent years, some scholars have tried to apply convolutional neural networks (CNN) to recognize sound event[4]. Compared with traditional classifiers, convolutional neural networks have greatly improved recognition rate and recognition speed. The ConvNet model[5] has improved the accuracy of nearly 20% on Esc-50 database. The LSTM+CNN model proposed by Bae et al. achieved an 84.1% accuracy rate in the DCASE2016 competition[6]. However, there are still some problems when the traditional CNN model is applied to sound event recognition. The CNN adopts a serial stack structure. The convolutional layer in the network transforms the low-level feature map into a high-level feature map layer by layer. Such a network structure will result in the low-level feature information loss in the final extracted features.

For the above mentioned problems, a sparse autoencoder based on acoustic features is proposed to classify the vehicle state. The original acoustic features are multidimensional Mel-cepstrum features and energy features. The autoencoder generates encoded features to classify the vehicle state from the original acoustic features, which improve the robustness of the algorithm. To verify the performance of the proposed algorithm, we collected a total of 829 samples for experiment, including three types of data: engine running (normal driving), brake and crash. Compared with four algorithms, the recognition rate based on the proposed method can reach 94.9%, which is 11.45% higher than that of the traditional

[a] Corresponding author: zhangdaqing_925@163.com

classification algorithm or 12.3% higher than the traditional CNN algorithm.

2 Acoustic Features

For traffic acoustic event detection, three types of vehicles states are important, which are engine running (normal driving) state, crake state and crash state. The waveform and spectrogram of three kinds of signal are shown in Figure 1. From the graph, three states have some obvious difference. For example, the waveform of the running signal is flat and its frequency range is below 800Hz. In addition, the waveform and the spectrogram of the crash signal have the obvious change. After the analysis, the valid features should be selected to reflect these differences.

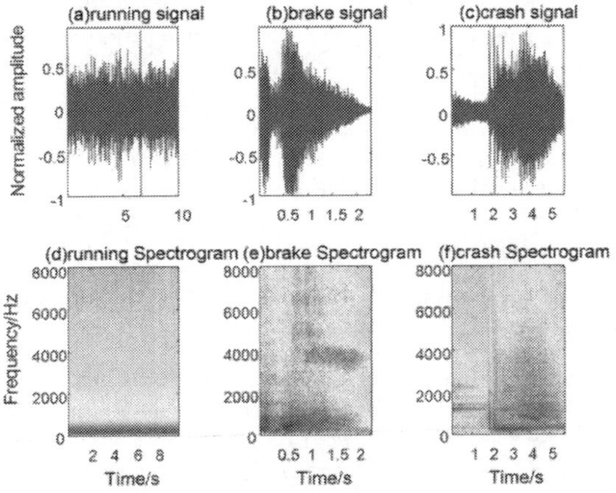

Figure 1. The representation of three types of signals

Among many acoustic features, the Mel-Cepstral feature is widely used in sound event classification and speaker recognition. The Mel-Cepstral is a spectral feature which is calculated base on the non-linear relationship between the human ear's auditory characteristics and signal frequency. However, the standard Mel-Cepstral only reflects the static characteristics of speech parameters. The dynamic characteristics of speech can be described by the differential spectrum of these features. The extracted features are differentiated to further strip the features, and obtain the information such as type and speed changes of the acoustic event. Combining the Mel-Cepstral feature with its difference can improve the recognition performance of the system.

In addition, different types of sounds have different energy values and change trends, so short-time energy and their statistical features are also computed. Table 1 describes the adopted 110 features.

Table 1. List of Acoustic Features.

No.	features
1-95	Mel-Frequency Cepstral Coefficients (MFCC-0 to MFCC-12), averages of their 1st and 2nd order differences, their maximums, minimums, ranges, and standard deviations.
96-110	Short-term energy, averages of its 1st and 2nd order difference, its maximum, minimum, range, and standard deviation.

3 Autoencoder for traffic acoustic event detection

The autoencoder is the neural network composed of several hidden layers and is an unsupervised learning algorithm. The general data representation from the input will be obtained by setting the target value as the input[7]. As shown in Fig.1, the autoencoder includes two parts: encoder and decoder. A number of hidden layers on the left of the graph form an encoder. The right hidden layers form a decoder, which has the same output as the encoder. In the middle of Figure 2, the output of the middle layer is the coded feature. In the process of the input reconfiguration, the data distribution is learned by the encoding and decoding process. At last, the data is compressed and coded, so the more compact features[8] are obtained.

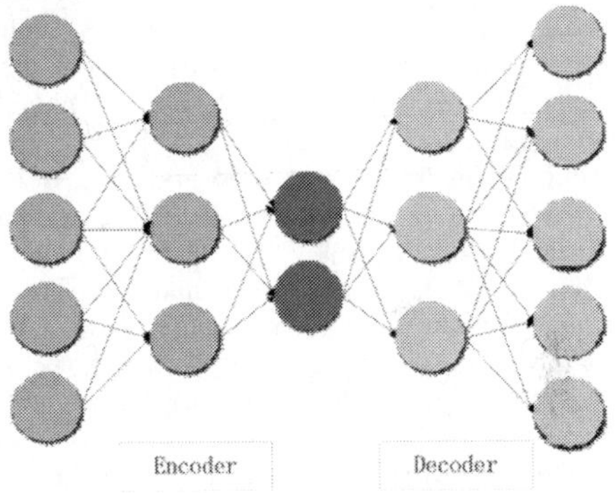

Figure 2. The architecture of autoencoder

For an autoencoder, x_i is the input, y_i is the output and θ represents the parameters. The input x_i represents the 110-dimensional audio feature we extracted, and the output y_i is the optimized feature after the sparse autoencoder, which is also 110-dimensional. During the training, the goal of parameter optimization is:

$$\min_{\theta} \sum_{i=1}^{N} \|x_i - y_i\|^2 \quad (1)$$

By limiting the expected activation of the hidden unit to sparsity, a regularization term[9] is added to punish the deviations between the expected activation degree of the hidden unit and the target sparsity ρ. So equation (1) is:

$$\min_{\theta} \sum_{i=1}^{N} \|x_i - y_i\|^2 + \beta \sum_{j=1}^{m} sp(\rho \,\|\, \hat{\rho}_j) \quad (2)$$

where $sp(\rho \| \hat{\rho}_j)$ is the sparsity penalty term, which is computed as:

$$sp(\rho \| \hat{\rho}_j) = \rho \log \frac{\rho}{\hat{\rho}_j} + (1-\rho)\log \frac{1-\rho}{1-\hat{\rho}_j} \qquad (3)$$

Here, $\hat{\rho}_j$ is the average activation degree for all neurons in the hidden layer, ρ is sparsity, β is the penalty coefficient, and m is the number of all neurons. By adding the sparsity limitation[10], most of the neurons in the autoencoder are suppressed, only a small number of neurons are active, which reduces the redundancy[11] of the network and increases the robustness of the model.

Table2. Parameters Setting.

Component	No. Layer	#Neurons	Activation
Encoder	1 Fully Connected Layer	110	ReLU
	2 Fully Connected Layer	96	ReLU
	3 Fully Connected Layer	64	ReLU
Decoder	4 Fully Connected Layer	96	ReLU
	5 Fully Connected Layer	110	-

The 5-layers autoencoder is constructed for learning the latent feature representation of traffic acoustic event, and its parameter settings are shown in Table 2. The workflow of the proposed autoencoder is shown in Fig.3. The training process is shown in Fig.3(a). At first, the autoencoder is trained to study the latent representation of the original acoustic features from the training set. The learning strategy is to reconstruct input signals to obtain the latent representation, and use gradient descent method to minimize the error between the reconstructed signal and the input signal. When the deviation reaches the set requirements, a trained autoencoder is constructed. Then, a softmax classifier is added to compute the deviation between its output and the real labels. The gradient of the deviation is calculated and the back propagation algorithm is used to tune the parameters of each layer. After the training is finished, its performance of trained autoencoder can be estimated by the test data. The test flow is shown in Fig.3(b).

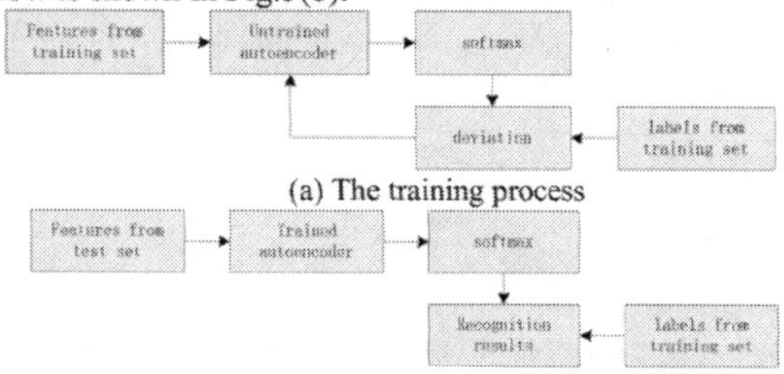

(a) The training process

(b) The test process

Figure 3. The workflow of the proposed model

4 Experimental analysis

4.1. Experimental data and parameter setting

In this experiment, a total of 829 samples were collected for the three types of sound, such as engine running(normal driving), braking and crashing. Among them, there are 442 engine running (normal driving) sounds, 176 brake sounds and 211 crash sounds.

In order to improve the effectiveness of the algorithm, the voice activity detection is firstly done on the sample to remove the silent segment. The sample is then resampled at a frequency of 16000 Hz. Next, the sample is framed and the FFT is calculated, the number of FFT points is 512, and the frame overlap rate is 50%.

The current general deep learning framework Caffe is used to build and train the network. Because the selection of hyperparameters in neural networks has a great impact on the training and the convergence state of the network, the final network hyperparameters are determined through multiple experiments and comparisons.

The comparison classifier and its parameters are: 1) Random forest[12], the maximum depth is 6, and the number of base estimators is 100; 2) CovNet[13], a two-layer convolutional layer with convolution kernels of (110, 6) and (1, 3), with a full layer of 1,000 neurons; 3) k-nearest neighbors (KNN); 4) the support vector machine.

4.2 Comparison of state-of-the-art recognition algorithms

The experimental evaluate the algorithm performance in a five-fold cross-validation manner. For a more valid assessment, Random Forest[14], KNN[15], CNN[16] and ConvNet performed the same experiments and comparisons on the same dataset.

Table 3 shows the experimental results for the five classifiers on the data set. Compared with random forest and KNN, the autoencoder improved the accuracy by 12.2% and 17.4%, respectively. It can be seen that the performance of autoencoder is better than the traditional classifier. The reason is that the data samples are very comprehensive, and there are many recorded data under various environments, while the generalization ability of the traditional classifier is bad. In addition, through data analysis, the traditional classifier has the lowest recognition rate for the brake event category, the highest recognition rate for the driving event category, followed by the collision event category. So, it may also be caused by the fact that the brake data is too small, and the data needs to be supplemented later for further verification.

Compared with traditional CNN and ConvNet, the recognition accuracy is improved by 12.3% and 3.9%, respectively. The possible reason is that the autoencoder uses encoded features to classify the state. These encoded features are more robust than the original features.

Table 3. Recognition rate of five models on the data set.

Classifier	**Overall recognition rate**	**Collision recognition rate**
Random Forests	82. 7%	84. 8%
DNN	77. 5%	85. 6%
CNN	82. 6%	90. 1%
ConvNet	91. 0%	92. 3%
Proposed method	94. 9%	97. 9%

In addition, Table 3 additionally compares the recognition efficiency of various algorithms for collision sounds. The experimental results show that the proposed algorithm's recognition rate of collision sound is 3% more than the overall recognition rate, reaching 97. 9%. This shows that the algorithm can be effectively applied to the traffic incident warning system to realize the timely alarm function of traffic accidents based on traffic state detection.

5 Conclusions

In order to provide auxiliary research for intelligent transportation and traffic safety development, an autoencoder based the acoustic feature for traffic event detection is proposed. In order to integrate the dynamic features of the sound, the algorithm extracts the multidimensional Mel-cepstrum features and energy features, and forms a 110 dimension feature vector. Then the 5-layers autoencoder based on the computed features is trained to improve the robustness. The experimental results show that the detection rate of traffic acoustic events reaches 94.9%, and the recognition rate of collision sounds reaches 97. 9%. The proposed audio surveillance system may be used to monitor traffic accidents and save valuable time in rescue mission. In addition, the system can be embedded in the automatic driving system, which is conducive to the timely response of the self-driving car to the traffic state, greatly improving safety.

Acknowledgments

The work was supported by Science and Technology Innovation Project of Research Institute of Highway, Ministry of Transport (2018-E0021).

References

1. Maciejewski Henryk, Mazurkiewicz Jacek, Skowron Krzysztof, Walkowiak Tomasz, Neural Networks for Vehicle Recognition, in Proc. of the 6th International Conference on Microelectronics for Neural Networks, Evolutionary and Fuzzy Systems, 1998, pp.292–296.
2. Zhang Dian Ye, Jian Prof Jin, Zhi-Zheng Assoc Prof Guo. Exploration into Road Traffic Accident Prevention Research System. China Safety Science Journal, Vol.17, No.7, 2007, pp.132-138.
3. Luo Xiang Long, Niu. Vehicle recognition by acoustic signals based on EMD and SVM. Applied Acoustics, Vol.29, No.3, 2010, pp.178-183.
4. Salamon Justin, Bello Juan. Deep Convolutional Neural Networks and Data Augmentation for Environmental Sound Classification. IEEE Signal Processing Letters, Vol. 99, 2016, pp.1-4.
5. Piczak Karol J. Environmental sound classification with convolutional neural networks. IEEE International Workshop on Machine Learning for Signal Processing, 2015, pp.1-4.
6. H Bae S, I Choi, S Kim N, Acoustic scene classification using parallel combination of LSTM and CNN, Proceedings of the Detection and Classification of Acoustic Scenes and Events, 2016, pp.11-15.
7. Xu J, Xiang L, Liu Q, et al. Stacked Sparse Autoencoder (SSAE) for Nuclei Detection on Breast Cancer Histopathology Images, IEEE Transactions on Medical Imaging, Vol.35, No.1, 2016, pp.119-130.
8. Chandar A P S, Lauly S, Larochelle H, et al. An autoencoder approach to learning bilingual word representations, International Conference on Neural Information Processing Systems. MIT Press, 2014, pp.1853-1861.
9. Goodfellow I J, Le Q V, Saxe A M, et al. Measuring invariances in deep networks, International Conference on Neural Information Processing Systems, 2009, pp.646-654.
10. Mairal J, Bach F, Ponce J, et al. Online Learning for Matrix Factorization and Sparse Coding, Journal of Machine Learning Research, Vol.11, No.1, 2009, pp.19-60.
11. Hinton G E, Salakhutdinov R R. Reducing the Dimensionality of Data with Neural Networks, Science, 313(5786), 2006, pp.504-507.
12. Phan Huy, Maaß Marco, Mazur Radoslaw, Mertins Alfred. Random Regression Forests for Acoustic Event Detection and Classification. IEEE/ACM Transactions on acoustic Speech & Language Processing, Vol.23, No.1, 2015, pp.20-31.
13. Garcia-Pedrajas N, Hervas-Martinez C, Munoz-Perez J. COVNET: a cooperative coevolutionary model for evolving artificial neural networks. IEEE Transactions on Neural Networks, Vol.14, No.3, 2003, pp.575-596.
14. Pal, M. Random forest classifier for remote sensing classification. International Journal of Remote Sensing, 2005, 26(1):217-222.
15. Zhang M L , Zhou Z H . ML-KNN: A lazy learning approach to multi-label learning. Pattern Recognition, 2007, 40(7):2038-2048.
16. Wei Y , Zhao Y , Lu C , et al. Cross-Modal Retrieval With CNN Visual Features: A New Baseline. IEEE Transactions on Cybernetics, 2017, 47(2):449-460.

Impact of intelligent agents on the avoidance of spontaneous traffic jams on two-lane motorways

Jessica Malerczyk[a], Sabine Lerch, Bernd Tibken and Anton Kummert

University of Wuppertal, School of Electrical, Information and Media Engineering, 42119 Wuppertal, Germany

Abstract. This paper approaches the evaluation of intelligent agents for the reduction and avoidance of spontaneous traffic jams, which arise without evident reason. Individual vehicles are regarded as intelligent agents that act autonomously. The basis of this work is the Nagel-Schreckenberg (NaSch) model. Its extensions by the velocity-dependent randomization (VDR) model and multiple lanes allow us to simulate realistic traffic and congestion situations on two-lane motorways. Our concept is applied to the model and analyzed by fundamental diagrams and the average velocity, for example. The results of this paper reveal that traffic congestions are avoided when using swarm intelligence in all vehicles since human behavior, especially misbehavior, is eliminated and the velocities determined by the intelligent vehicle are directly realized. Moreover, an amount of 30% of intelligent vehicles has a significantly positive impact on traffic flow.

1 Introduction

Climate change is a defining keyword of our time. The reduction of carbon dioxide emissions plays a fundamental role here. Compared to pre-industrial times before 1750, CO_2 concentration has increased by 40% due to human activity [1]. The main cause of greenhouse gas is the combustion of fossil fuels from which a significant amount is attributable to road traffic [2]. The reduction of vehicle emissions is, therefore, essential for climate protection. A high proportion of CO_2 emissions are caused by traffic congestions.

In addition to traffic jams caused by accidents or construction sites, so-called phantom traffic jams often occur on motorways. The main reasons for them are overloaded roads and the resulting high traffic density. Then small disturbances, such as overreaction, sudden overtaking or dawdling, are sufficient to form a spontaneous traffic jam [3], [4].

The NaSch model is suitable for macroscopic traffic description by microscopic observation of individual vehicles. It is especially useful to simulate phantom traffic jams. In recent years the model has been widely used and adapted to real measurements, for example in [5], [6]. In addition to its use to investigate traffic flow, the model has also been used to examine different traffic phases and transitions [7]-[9]. The model can also be used to investigate approaches to congestion reduction. In our opinion, there is potential for further investigations, especially when considering phantom jams. For this purpose, on the one hand, the influence of vehicular ad hoc networks has been evaluated (VANETs) [10], on the other hand, the use of reinforcement learning has been analyzed [11] in recent works. Other studies use the open-source traffic simulation package SUMO (Simulation of Urban Mobility) [12], [13], or a swarm-based approach [14].

As far as we know, the avoidance of phantom jams based on the extended NaSch model and the use of intelligent agents in different vehicle types have not yet been sufficiently simulated and investigated.

Here, this paper presents the suitability of intelligent vehicles for avoiding spontaneous traffic jams on motorways, taking into account the obligation to drive on the right. For this purpose, we have implemented the NaSch model and its extensions including human behavior. The application and potential of our concept are analyzed and evaluated using fundamental diagrams and the average velocity, for example.

In chapter 2, the NaSch model and its extensions as well as the functionality of intelligent vehicles we use in our approach is described. In addition, our implementation is presented. In chapter 3, the results are exposed and discussed. The conclusion is revealed in chapter 4.

2 Approach

2.1 Nagel-Schreckenberg model

The NaSch model is a cellular automaton for single-lane traffic. Space and time and also the velocities are discrete. The road is divided into cells that can either be empty or contain exactly one vehicle. The velocity of each vehicle i at time t is described by an inter value $v_i(t) = 0, 1, 2 \dots v_{max,i}$, where $v_{max,i}$ is the maximum velocity. In each time step, the following four rules are applied to all cars in parallel [15]:

[a] Corresponding author: malerczyk@uni-wuppertal.de

(1) Acceleration

$$v_i := \min (v_i + 1, v_{max}) \quad (1)$$

(2) Slowing down due to vehicles driving ahead

$$v_i := \min (v_i, d_i) \quad (2)$$

(3) Randomization parameter for deceleration

$$v_i := v_i - 1 \text{ with probability } p \text{ if } v_i > 0 \quad (3)$$

(4) Vehicle motion

$$x_i(t + 1) := x_i(t) + v_i \quad (4)$$

The first rule clarifies the desire of drivers to move at the maximum possible speed in order to reach their destination as fast as possible. The avoidance of accidents is modeled by the second rule, where d_i is the distance to the next vehicle ahead. By using the probability p in the third rule, the traffic model is adapted to real behavior, since no driver can maintain a constant velocity over a certain distance. The last rule defines the motion. A vehicle at position $x(t)$ at time t is then moved forward by its recalculated velocity v_i during the transition to the next time step $t \rightarrow t + 1$.

2.2 Velocity-dependent randomization model

The NaSch model is not able to reproduce complex phenomena such as hysteresis or metastable states, as they are observed in real traffic situations. Within an extension, the VDR model, this behavior can be simulated [5]. In this model, the randomization parameter $p(v)$ depends on the velocity v, so that delayed moving off is modeled. In an initial step, this parameter is calculated as follows, where $p_0 > p$:

(0) Randomization parameter for deceleration

$$p(v)=\begin{cases} p_0, & \text{if } v=0 \\ p, & \text{if } v \neq 0 \end{cases} \quad (5)$$

After step (0), rules (1)-(4) of the NaSch model are executed.

2.3 Two-lane model

As we aim to investigate the formation of spontaneous traffic jams on motorways, we have to integrate a second traffic lane and introduce lane changing rules [6]. Here, asymmetric rules are used because, in many countries, it is only allowed to overtake on the left lane. In this case, a driver can change from the right to the left lane if the number of cells ahead on the same lane gap_r is too low, compared to a chosen d_r, and the number of cells forward $gap_{l,ahead}$ and backward $gap_{l,back}$ on the other lane are large enough, compared to $d_{l,ahead}$ and $d_{l,back}$, respectively. Furthermore, there is a lane change probability p_{change}, so that lane changes happen stochastically:

$$gap_r < d_r \quad (6)$$

$$gap_{l,ahead} \geq d_{l,ahead} \quad (7)$$

$$gap_{l,back} \geq d_{l,back} \quad (8)$$

$$rand\,() < p_{change} \quad (9)$$

Due to the obligation to drive on the right lane, a vehicle should change to the right, if possible, irrespective of the situation on the left lane. Therefore, the following three conditions are maintained, where the distances on the right lane $gap_{r,ahead}$ and $gap_{r,back}$ are taken into account and compared with $d_{r,ahead}$ and $d_{r,back}$, respectively:

$$gap_{r,ahead} \geq d_{r,ahead} \quad (10)$$

$$gap_{r,back} \geq d_{r,back} \quad (11)$$

$$rand\,() < p_{change} \quad (12)$$

A lane change takes place before the second rule of the NaSch model is applied, which can then be divided into two sub steps: First, a vehicle is moved in parallel but not forwards. Then the safety distance is maintained.

2.4 Autonomous vehicles

This paper investigates the influence of intelligent vehicles on the development of congestion. For this purpose, autonomous vehicles are assumed, which can be interpreted as agents within an intelligent swarm. They drive independently so that human intervention is not necessary. We suppose that the vehicles are equipped with the required sensors to precisely capture the environment, such as ultrasonic and radar sensors, cameras and lidar scanners, for example. That gives them accurate information about the distances to other vehicles and the corresponding velocities within a specific range.

2.5 Implementation

For the implementation in MATLAB, the enhanced NaSch model with a length of 400 cells is used. Each cell has a length of 7.5 m which is equal to the space a car occupies on average in a jam. A total number of 300 time steps are simulated where one time step is equal to 1 s and a closed system with a global density of 12% is used. The average values of 100 executed simulations are considered in the following chapter.

Each vehicle is initialized with a maximum possible velocity $v_{i,max}$ between 3 and 7 cells per time step. It is not possible to accelerate beyond this value. With these different maximum velocities, we are able to model different types of vehicles, such as trucks and sports cars, for example. In this way, we achieve a more precise division of different vehicles than in other works, in which two vehicle types are considered [16], [17].

The velocities are initialized with a normal distribution with the expected value $E = 5$ and the standard deviation $\sigma = 2$. A velocity of $v = 5$ is based on the recommended maximum velocity $v_{rec} = 130$ km/h for German motorways. The corresponding real velocity to the value $v = 0$ is $v_{real} = 0$ km/h and to the value $v = 7$ the real velocity is $v_{real} = 189$ km/h. Inside the integer interval $v = [0,7]$, all further real velocities are distributed linearly.

Besides, each vehicle receives an individual lane change probability p_{change} between 0.6 and 1.0. This value differentiates between drivers who use every possible opportunity to change the lane and drivers who rather stay on their current lane. The randomization parameters for deceleration p and p_0 are selected once between 0.1 and 0.3 and between 0.6 and 0.9, respectively, at the beginning of the simulation. As soon as a vehicle needs more than four time steps to start, it is assigned p, which increases the likelihood to start in the following time step. In contrast to other approaches, we perceive this as more realistic modeling since in real traffic situations, other road users would alert the drivers who are not moving off.

A lane change to the left is considered if the number of free cells to the vehicle in front is less than its velocity value, compare (6). The conditions for a transition to the right lane are checked in each time step. The parameters for the conditions (7), (8) and (10), (11) are set to: $d_{l,ahead} = 3$, $d_{l,back} = 4$, $d_{r,ahead} = v_i$ and $d_{r,back} = 3$, where v_i is the current velocity of the vehicle i. The aim is to simulate more realistic traffic in which vehicles change the lanes even though there is not always enough distance for the rear driver on the other lane to keep its previous velocity. In the applied model no passing zones are not considered.

Our simulated autonomous vehicles operate via two functions: On the one hand, the vehicles only change lanes if they do not force the vehicle approaching from behind to decelerate, neither in the current time step nor in the following time step. On the other hand, the vehicles are able to adapt to the vehicle in front and its driving behavior due to knowing its velocity in each time step.

3 Results and discussion

To analyze the influence of our concept, the model itself is first executed without autonomous vehicles. The resulting congestion situations are evaluated and a reference is obtained. With this reference, the results of the model using autonomous vehicles can be compared. For the purpose of generating the reference, the term congestion is defined. This describes the situation in which at least four vehicles in a row come to a standstill, equivalent to $v = 0$.

3.1 Intelligent vehicle rate of 100%

The use of fully automated vehicles is still a future vision and is considered as a significant research goal. Here, an autonomously driving vehicle replaces the human driver and, thus, eliminates human imperfections such as random behavior. Our results prove these assumptions: Not a single congestion situation is identified, when this concept is applied.

Regarding the average mean velocities in Figure 1, it can be seen that the velocity distribution shifts towards higher speeds. In addition, we can observe that no vehicle drives below a speed of $v = 3$ on average. According to 2.5, this is the minimum initialization velocity. This only occurs due to the fact that no vehicle randomly reduces its speed. As a result, the vehicle behind does not have to decelerate and therefore does not decline beneath the limit of $v = 3$. This implies that no traffic jams are able to occur.

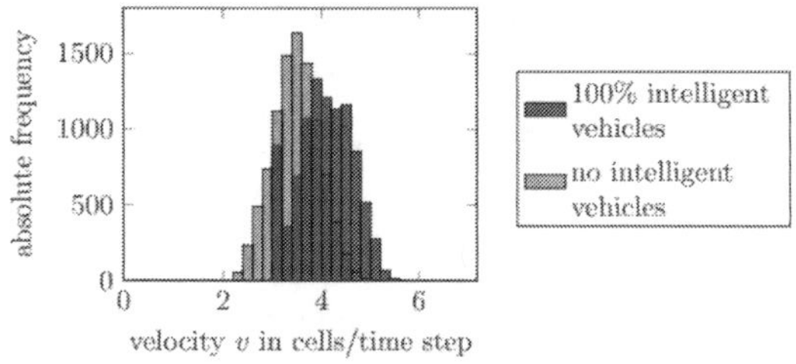

Figure 1. Histogram of average velocities from autonomous and non-autonomous vehicles

For further analysis of the traffic situation, we refer to Kerner's three-phase model, which subdivides traffic into three phases: free traffic [F], synchronized traffic [S] and wide moving jams [J] [18].

Figure 2 shows the fundamental diagrams of the reference on the one hand (top) and the applied concept on the other hand (bottom). Here only the diagrams of the right traffic lane are considered, because those for the left lane are very similar. In the upper diagram, the three phases are plainly visible. Both synchronized traffic and wide moving jam can be identified. In the lower diagram, however, there is only one phase, free traffic. The fact that there are no congested situations is visualized. The difference between the two diagrams, and thus the positive effect of intelligent vehicles is obvious.

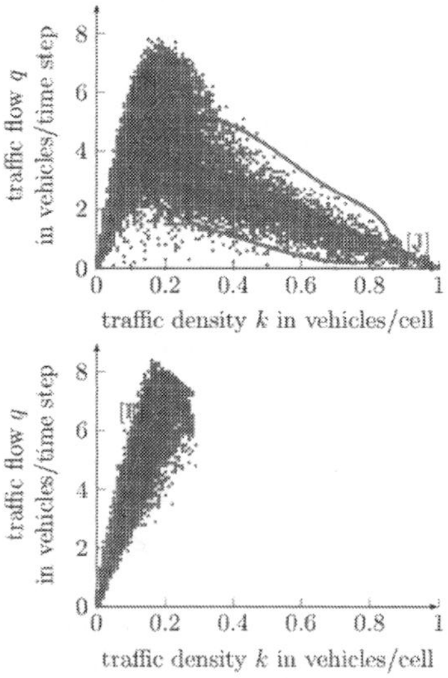

Figure 2. Fundamental diagrams of the reference (top) and of the applied concept, in which all vehicles are equipped with intelligent systems (bottom)

3.2 Intelligent vehicle rate of 5%

Since a complete adoption of autonomous vehicles is very unlikely for the foreseeable future, we consider also the use of just a certain percentage of intelligent vehicles. That allows us to determine to what extent it is worth driving autonomously to avoid traffic jams. So, we combine both intelligent vehicles and human drivers in the simulations.

We take [19] as a basis, which claims that a percentage of 5% of intelligent vehicles is sufficient to harmonize traffic. The analysis of our approach reveals that the number of identified traffic congestions increases by 19.8%, which contradicts the statement in [19].

Over time, the number of traffic jams decreases continuously when no swarm intelligence is used. Considering the use of 5% intelligent vehicles, the number increases initially, instead. Here, most traffic jams occur between the time steps 100 and 150, after which the amount of traffic jams lessens again. This is shown in Figure 3, where the left diagram shows the temporal occurrence of traffic congestion of the reference, and the right diagram illustrates those results when using 5% swarm intelligence. A possible cause for this is the initialization. The vehicles are initialized in a way so they can maintain their initial speed in the first time step in any case. The comparatively few vehicles with integrated intelligent systems then adapt to the vehicles of their environment. They do not push or dawdle. After some time, human behavior predominates, resulting in dawdling or sudden overtaking. The comparatively small number of automated vehicles cannot compensate for this behavior. Traffic jams arise, but at a later point in time.

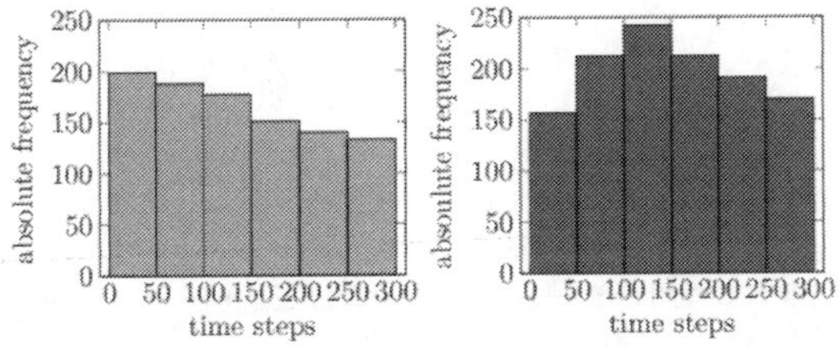

Figure 3. Histograms of time of origin of the identified traffic jams of the reference (left) and when using the concept with 5% intelligent vehicles

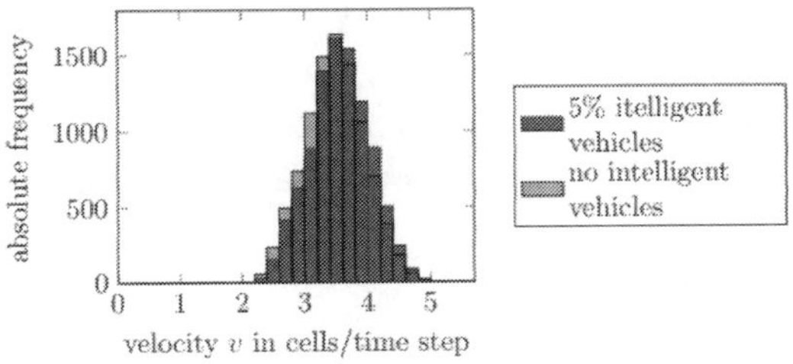

Figure 4. Histogram of average velocities from 5% autonomous and non-autonomous vehicles

The maximum congestion width, as seen in Table 1, which means the number of vehicles involved in a jam, and the length of a congestion are further global characteristics we investigate. Considering these, a comparatively small positive effect caused by the few fully automated vehicles is observed. The comparison of average speeds shows a large similarity between our reference simulations and those with a 5% proportion of intelligent vehicles. This behavior is presented in Figure 4. Although shorter traffic congestions are observed, the increase in the number of traffic jams does not result in a significant difference of the velocity distribution. The vehicles are temporally shorter but more often involved in traffic congestions.

A comparison of the corresponding fundamental diagrams, here again using the example of the right-hand traffic lane, also shows no significant improvements in traffic flow when utilizing intelligent systems in 5% of all vehicles. The diagram in Figure 5 is almost identical with the upper diagram in Figure 2, which illustrates the results of our reference.

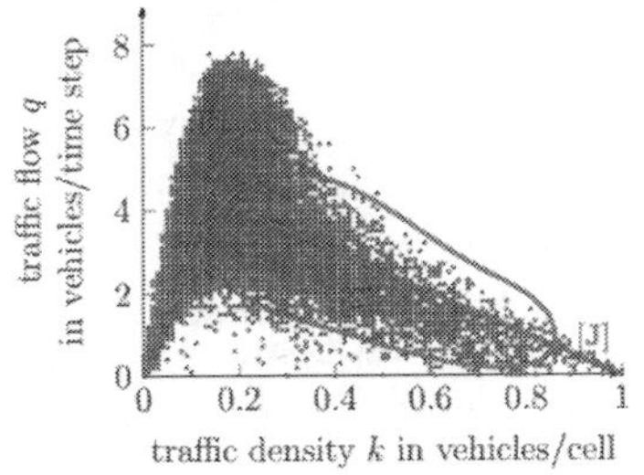

Figure 5. Fundamental diagram of the applied concept, in which 5% of all vehicles are equipped with intelligent systems

3.3 Further intelligent vehicle rates

In this respect, additional intelligent vehicle amounts of 10%, 20%, 30%, 40%, and 50% are investigated. The observed results are presented in Table 1.

Evidently, the number and length of traffic jams decrease as the number of fully automated vehicles increases because they behave predictably. It is noticeable, however, that the maximum congestion width increases and therefore the number of vehicles involved. A possible cause could be convoys, which occur in about one-third of the simulations and are mainly caused by human drivers inside the simulation. Due to the imperfections of such drivers, individual dawdling vehicles may be sufficient to put the whole convoy into congested traffic flow. The tendency for a traffic jam to occur decreases, but if a traffic jam occurs, comparatively more vehicles are involved.

However, the fact that autonomous vehicles can start directly in the next time step means that these traffic jams dissolve more quickly. Furthermore, it can be observed that the average velocity of vehicles increases with a growing percentage of swarm intelligence. The reason for this is the lower number of traffic jams in which the vehicles lose their speed.

Another parameter studied in this paper is the ratio of average velocity to the maximum possible velocity of each vehicle. In addition to an undisturbed and comfortable journey, every driver also wishes to reach his destination as soon as possible. Obviously, this ratio also increases with an enhanced number of autonomous vehicles regarding the comparison.

Taking the results into account, we consider that a number of intelligent vehicles of around 30% has a positive impact on traffic flow. The amount of congestion situations shows a significant decrease. The length of traffic jams is reduced by more than a half and the average speed is increased significantly.

Table 1. Overview of the influence of different ratios of autonomous and non-autonomous vehicles on traffic characteristics

autonomous vehicle ratio	0 %	5 %	10 %	20 %	30 %	40 %	50 %	100 %
number of congestions in %	100	+ 19.8	+ 4.9	- 5.0	- 21.3	- 40.5	- 59.9	-100
temporal congestion length in time steps	56	39	32	22	20	18	16	0
maximum congestion width in vehicles	7.6	7.5	7.5	7.3	7.7	8.0	8.3	0
average velocity in cells/time step	3.4	3.5	3.5	3.6	3.7	3.7	3.8	4.0
ratio of average to maximum velocity in %	69.3	70.6	72.9	74.9	75.6	77.0	78.2	81.1

4 Conclusion

The purpose of this paper is the evaluation of swarm intelligence to reduce and avoid spontaneous traffic jams on two-lane motorways. The basis was the extended NaSch model, which is suitable to model realistic traffic.

By knowing the distances and velocities to other vehicles within a certain range, autonomous vehicles were able to optimize the timing of their overtaking. This prevents the deceleration behavior of the vehicle behind. Besides, a predictive behavior with an orientation based on the velocity of the vehicle ahead has been implemented, resulting in more consistent motion.

At first, the influence of swarm intelligence implemented in all existing vehicles has been evaluated. The results reveal no congested situations. The intelligent vehicles calculate the time to a potential collision in each time step to adjust their speed accordingly. In this way, they prevent abrupt braking of their vehicle and, therefore, of the vehicle behind them, so that traffic congestions cannot occur. The average velocities increase and the traffic flow is undisturbed.

However, the use of completely autonomous driving vehicles for private purposes is currently not permitted due to the non-existing legal situation. The adoption of autonomous vehicles would proceed gradually. For this reason, the influence of swarm intelligence, installed in only 5% of all vehicles, was investigated secondly. However, the results show no positive impact compared to the reference, in which only human drivers were simulated, as human behavior predominates currently.

Only the use of intelligent vehicles in about one-third of all existing vehicles leads to a significant reduction of traffic jams and thus to a harmonization of traffic flow.

To summarize, our results show that future approval would have many advantages in terms of an improved traffic flow. Complete avoidance of congestions would not only lead to a faster arrival time at the destination, but also to a reduction in vehicle emissions.

References

1. P. Ciais, C. Sabine, G. Bala, L. Bopp, V. Brovkin, J. Canadell, A. Chhabra, R. DeFries, J. Galloway, M. Heimann and others, in Climate Change 2013: The Physical Science Basis, 465-570 (2014)
2. R. Sims, R. Schaeffer, F. Creutzig, X. Cruz-Núñez, M. D'Agosto, D. Dimitriu, M. J. F. Meza, L. Fulton, S. Kobayashi, O. Lah, A. McKinnon, P. Newman, M. Ouyang, J. J. Schauer, D. Sperling, G. Tiwari, in Climate Change 2014: Mitigation of Climate Change, 599-670 (2014)
3. G. Orosz, R. E. Wilson, R. Szalai, G. Stépán, Phys. Rev. E, **80**, 046205 (2009)
4. Y. Sugiyama, M. Fukui, M. Kikuchi, K. Hasebe, A. Nakayama, K. Nishinari, S.-i. Tadaki, S. Yukawa, New J. Phys., **10**, 033001 (2008)
5. R. Barlovic, L. Santen, A. Schadschneider, M. Schreckenberg, Eur. Phys. J. B, **5**, 793-800 (1998)
6. M. Rickert, K. Nagel, M. Schreckenberg, A. Latour, Physica A, **231**, 534-550 (1996)
7. R. Barlovic, T. Huisinga, A. Schadschneider und M. Schreckenberg, Phys. Rev E, **66**, 046113 (2002)
8. W. Knospe, L. Santen, A. Schadschneider, M. Schreckenberg, J. Phys. A, **33**, L477 (2000)
9. A. Schadschneider, Physica A, **285**, 101-120 (2000)
10. F. Knorr, D. Baselt, M. Schreckenberg, M. Mauve, IEEE Trans. Veh. Technol., **61**, 3490-3498 (2012)
11. S. Ishikawa, S. Arai, ICA (2016)
12. T. Liebig, M. Sotzny, COSIT (2017)
13. S. Djahel, N. Jabeur, R. Barrett, J. Murphy, ISNCC (2015)
14. C. Stolcis, E. Pfannerstill, IET INTELL TRANSP SYS, **11**, 134-141 (2016)
15. K. Nagel, M. Schreckenberg, J. Phys., **2**, 2221-2229 (1992)
16. D. Chowdhury, D. E. Wolf, M. Schreckenberg, Physica A, **235**, 417-439 (1997)
17. W. Knospe, L. Santen, A. Schadschneider, M. Schreckenberg, Physica A, **265**, 614-633 (1999)
18. B. S. Kerner, Phys. World J., **12**, 25-30 (1999)
19. R. E. Stern, S. Cui, M. L. Delle Monache, R. Bhadani, M. Bunting, M. Churchill, N. Hamilton, H. Pohlmann, F. Wu, B. Piccoli and others, Transp. Res. Part C Emerg. Technol., **89**, 205-221 (2018)

Split allocation model of intersection group oriented to network carrying capacity limit

Jianwei Hu[1,2,a], Ren Zhang[1,2], Ya Li[1,2]

[1]*Traffic Management Research Institute of the Ministry of Public Security, Wuxi 214151, China*

[2]*Wuxi Huatong Intelligent Transportation Technology Development Co., Ltd., Wuxi 214122, China*

Abstract. In view of the network carrying capacity limit under over-saturated condition, a split allocation model of intersection group was developed, which takes total network delay, total output and total queue length as a comprehensive evaluation index. Based on the analysis of the influence of traffic flow turning, road section capacity, signal cycle length and split on traffic control, under the condition of network carrying capacity limit, the global optimization of intersection group splits was realized, taking the network traffic operation state as the basic judgment criterion and the corresponding constraint conditions were constructed. The numerical example analysis and simulation results show that the proposed model can ensure the traffic signal control effect under the premise of different demand states and control objectives, so that the effective prevention and quick dredging of traffic congestion can be realized under the condition of the network carrying capacity limit.

1 Introduction

The traffic congestion at the network level means that the key sections and intersections that play the supporting role of the network skeleton reach the limit of traffic capacity. Under the condition of network carrying capacity limit, the coupling correlation between signalized intersections which can directly intervene and regulate is strengthened, and the adjustment of single intersection signal control scheme is bound to have an impact on its adjacent intersections, while the traditional traffic signal control based on point or line level is limited to the improvement of local traffic conditions, which can not fundamentally solve the traffic congestion problem.

Ernesto [1] and Girianna [2] developed a global optimization model of the split and a regional coordinated control model, respectively. Daganzo [3] proposed an optimal design method of over-saturated coordinated control based on CTM. K. Aboudolas [4-5] proposed a real-time regional signal control strategy based on rolling rule quadratic planning for the congested traffic state of urban large-scale road network. The literature [6] introduced the network analysis method to optimize the traffic system control structure, in view of the lack of the inherent characteristics of the traffic network structure in SCOOT and SCATS control systems.

However, the practical significance of the split optimization for each phase of the intersection under the condition of the network carrying capacity limit is neglected, which is not conducive to reflect the positive impact of network optimization control on alleviating urban traffic congestion. Therefore, how to ensure the smooth operation of traffic under the limit value of network carrying capacity through the reasonable

[a] Corresponding author: hjwtime@163.com

allocation of phase split at each signalized intersection of the network is the key to signal control of traffic network.

2 Basic assumptions

The over-saturated state is a subcategory in the blocking state[7], the performance is that the queue extends to the upstream intersection and has a negative impact on the upstream intersection. Generally speaking, the ultra-long queue spreads to the upstream intersection, which is impossible to dissipate in a single cycle of the intersection, and even causes the deadlock of the intersection. Based on the above research conclusions, the basic assumptions of the model are proposed as follows:

(1) The effective green time of any approach of over-saturated intersection group can be used effectively, and there is no phenomenon of empty green time.

(2) The function matching of intersection approach and exit lane is reasonable, and there is no intertwined phenomenon of traffic flow in and out.

(3) The influence of intersection approach widening and traffic flow reorganization are not taken into account.

(4) The propagation time of the starting wave at the start of green time is not taken into account.

3 Split allocation model

Define an over-saturated intersection group control network *G*=(*P*, *A*), Where *P* is the set of network nodes and *A* is the set of road sections. *I* and *J* are the lane collection of each approach at upstream and downstream intersections, respectively. $S_{\mathbf{u}i}$ is the saturated flow of lane *i* at upstream intersection, $S_{\mathbf{D}j}$ is the saturated flow of lane *j* at downstream intersection. $\lambda_i(k)$ and $\lambda_j(k)$ are the split of the corresponding approach at upstream and downstream intersection in the *k*th signal cycle period, respectively. $q_{\mathbf{D}j}(k)$ and $\widetilde{q}_{\mathbf{D}j}(k)$ are the average arrival and outgoing traffic of lane *j* at downstream intersection in the *k*th signal cycle period, respectively. $U_{\mathbf{D}j}(k)$, $\overline{U}_{\mathbf{D}j}(k)$, and $B_{\mathbf{D}j}(k)$ are the number of vehicles incoming, outgoing and stranded in lane *j* at the downstream intersection in the *k*th signal cycle period, respectively. $C(k)$ is the *k*th signal cycle length.

3.1 Model variable relationship and constraints

1) Incoming and outgoing traffic at the approach

According to the basic assumptions of the model (1), the intersection in the control area is in an over-saturated state at this time, vehicles at each approach is driven out at a saturation flow rate in the release state. Therefore, the average incoming and outgoing traffic of each approach at the upstream and downstream intersections in the control area shall meet the following relationship:

$$\begin{cases} q_{\mathrm{D}j}(k) = \sum_{i=1}^{I} S_{\mathrm{U}i} \cdot \lambda_i(k) \cdot \delta_{(i,j)} \\ \delta_{(i,j)} = \begin{cases} 1, & \text{The traffic flow direction is from the upstream approach } i \text{ to the downstream approach } j; \\ 0, & \text{or else}. \end{cases} \end{cases} \quad (1)$$

$$\widetilde{q}_{\mathbf{D}j}(k) = S_{\mathbf{D}i} \cdot \lambda_j(k) \quad (2)$$

2) Stranded traffic at the approach

As can be seen from the analysis, the stranded traffic at the end of a certain cycle of the signalized intersection is equal to the sum of the stranded traffic in the previous cycle and the difference of incoming and outgoing traffic in the current cycle, the lower limit of the stranded traffic is 0. Thus, the relationship of the stranded traffic at the approach can be obtained as follows:

$$B_{\mathrm{D}j}(k) = \max\{0, B_{\mathrm{D}j}(k-1) + [q_{\mathrm{D}j}(k) - \widetilde{q}_{\mathrm{D}j}(k)] \cdot C(k)\} \quad (3)$$

3) Road section capacity constraints

In order to maintain the dynamic balance of incoming and outgoing traffic between any node in the network for a certain period of time, the capacity constraint relationship of lane in the network is as follows:

$$B_{\mathrm{D}j}(k) \cdot l_{veh} \cdot \varphi_j \le l_{\mathrm{D}j} \ , \quad \forall l_{\mathrm{D}j} \in A \quad (4)$$

where l_{veh} is the average space headway of the queuing traffic; φ_j is the corresponding road section capacity reduction factor $(0 < \alpha_j < 1)$, the value depends on the importance of the road section; $l_{\mathbf{D}j}$ is the distance between corresponding intersections.

4) Cycle constraints

In order to ensure the overall control effect of over-saturated intersection group and take into account the control benefit of each network node, according to the actual operation condition of intersection, the signal cycle length shall meet a certain range of values, that is:

$$C_{p\min}(k) \le C_p(k) \le C_{p\max}(k) \ , \quad \forall p \in P \quad (5)$$

In this formula, $C_{p\max}(k)$ and $C_{p\min}(k)$ are the upper and lower limits of the kth signal cycle length of node p, respectively.

5) Split constraints

In order to ensure the safety of vehicles and pedestrians at intersections, it is necessary to set the minimum green time for each intersection, that is the split of each approach of the intersection should be greater than or equal to its minimum split, and the sum of the split of each phase of the intersection should meet certain upper limit value requirements, and the constraint relationship is as follows:

$$\begin{cases} \sum_{j=1}^{J^*} \lambda_j(k) = 1 - \lambda_{pl}(k) \quad (J^* \in p(j)) \\ \lambda_{j\min}(k) \le \lambda_j(k) \le \lambda_{j\max}(k) \end{cases} \tag{6}$$

In this formula, J^* is the set of key traffic lane groups of node p; $\lambda_{pl}(k)$ is the loss of split of node p; $\lambda_{j\max}(k)$ and $\lambda_{j\min}(k)$ are the upper and lower limits of the kth split of node p, respectively.

3.2 Criteria for judging the state of network traffic

Under the condition of the network carrying capacity limit, the traffic operation states of each node and road section in the network are mainly divided into three types: dynamic diffusion state, dynamic equilibrium state and dynamic dissipation state, the traffic flow contained in the network is also related to these three states. When the network state changes from the dynamic diffusion state to the dynamic dissipation state, the total input of the network traffic flow is less than the total output, the total network queuing traffic is decreased gradually. Therefore, the dynamic variation of network carrying traffic and total queue vehicle length can be selected as the evaluation index to judge the operation state of the network traffic.

Define $Q(K)$ and $\overline{Q}(K)$ are the total input and output traffic flow accumulated for K signal cycles of the network, respectively. E and $\overline{E}$ are the collection of traffic lane groups for traffic flow input and output at the boundary of the network, respectively. $L(K)$ and $D(K)$ are the total queue length and total vehicle delay accumulated for K signal cycles in the network, respectively. According to the basic relation (2), the number of incoming and outgoing vehicles of lane j at the downstream intersection in the kth signal cycle period can be calculated as follows:

$$\begin{cases} U_{Dj}(k) = q_{Dj}(k)\cdot C(k) = C(k)\cdot \sum_{i=1}^{I} S_{Ui}\cdot \lambda_i(k)\cdot \delta_{(i,j)} \\ \delta_{(i,j)} = \begin{cases} 1, & \text{The traffic flow direction is from the upstream approach } i \text{ to the downstream approach } j; \\ 0, & \text{or else.} \end{cases} \end{cases} \tag{7}$$

$$\overline{U}_{Dj}(k) = \tilde{q}_{Dj}(k)\cdot C(k) = S_{Dj}\cdot \lambda_j(k)\cdot C(k) \tag{8}$$

The total network input traffic:

$$Q(K) = \sum_{p\in G}\sum_{k=1}^{K}\sum_{j\in E} U_{Dj}(k) \tag{9}$$

The total network output traffic:

$$\overline{Q}(K) = \sum_{p\in G}\sum_{k=1}^{K}\sum_{j\in \overline{E}} \overline{U}_{Dj}(k) \tag{10}$$

The total network queue length:

$$L(K) = l_{veh}\cdot \sum_{p\in G}\sum_{k=1}^{K}\sum_{j=1}^{J} B_{Dj}(k) \tag{11}$$

1) Criteria for judging the dynamic diffusion state:

$$\begin{cases} L(K)-L(K-1)-[L(K-1)-L(K-2)]>0, \quad K\ge 2 \\ Q(K)-Q(K-1)-[\overline{Q}(K)-\overline{Q}(K-1)]>0, \quad K\ge 1 \end{cases} \tag{12}$$

where $L(0)$, $Q(0)$ and $\overline{Q}(0)$ are the initial state of the network.

2) Criteria for judging the dynamic equilibrium state:

$$\begin{cases} L(K)-L(K-1)-[L(K-1)-L(K-2)]=0, \quad K\ge 2 \\ Q(K)-Q(K-1)-[\overline{Q}(K)-\overline{Q}(K-1)]=0, \quad K\ge 1 \end{cases} \tag{13}$$

where $L(0)$, $Q(0)$ and $\overline{Q}(0)$ are the initial state of the network.

3) Criteria for judging the dynamic dissipation state:

$$\begin{cases} L(K)-L(K-1)-[L(K-1)-L(K-2)]<0, \quad K\ge 2 \\ Q(K)-Q(K-1)-[\overline{Q}(K)-\overline{Q}(K-1)]<0, \quad K\ge 1 \end{cases} \tag{14}$$

where $L(0)$, $Q(0)$ and $\overline{Q}(0)$ are the initial state of the network.

3.3 Objective function

To ensure that the split allocation of intersections in the network is reasonable, and maximize the efficiency of the network, according to the different control objectives, the total network delay, the total network output traffic and the total network queue length are selected as the comprehensive evaluation index to construct the objective function. At the same time, different weight values are given to express their importance. The expression of the objective function is as follows:

$$\begin{aligned}\min Z &= \alpha \cdot D(K) + \beta \cdot L(K) - \gamma \cdot \overline{Q}(K) \\ &= \alpha \cdot \sum_{p\in G}\sum_{k=1}^{K}\sum_{j=1}^{J}[B_{\mathrm{D}j}(k-1)+B_{\mathrm{D}j}(k)]\cdot C(k)/2 + \beta \cdot l_{veh} \cdot \sum_{p\in G}\sum_{k=1}^{K}\sum_{j=1}^{J} B_{\mathrm{D}j}(k) - \gamma \cdot \sum_{p\in G}\sum_{k=1}^{K}\sum_{j\in E}\overline{U}_{\mathrm{D}j}(k)\end{aligned} \tag{15}$$

In this formula, α is the weight coefficient for the total network delay ($0 \le \alpha \le 1$); β is the weight coefficient of the total network queue length ($0 \le \beta \le 1$); γ is the weight coefficient of the total network output traffic ($0 \le \gamma \le 1, \alpha + \beta + \gamma = 1$).

4 Case study

To illustrate the applicability of the proposed model, a simplified 3 × 3 network model has been outlined for experimental tests, as shown in Figure 1. It is assumed that the turning traffic flow of left (*L*), through (*T*) and right (*R*) at each intersection satisfies the relationship: $L:T:R = 5:9:6$; the total lanes capacity of each section is *S*=4 × 1800 pcu/h. The minimum split allowed for each phase of each intersection is set to $\lambda_{\min} = 0.15$, to ensure the normal operation of traffic flow; the loss split of each node in the network is $\lambda_{pl} = 0.05$.

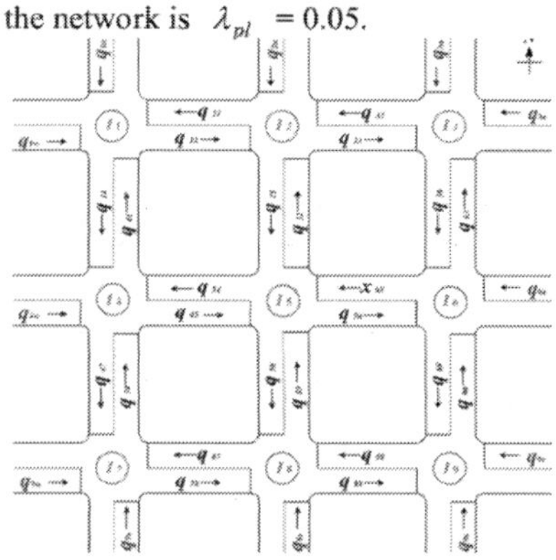

Figure 1. Network traffic flow diagram

According to the criterion of network dynamic equilibrium state and the relationship of the model, the maximum total network output traffic per unit hour is taken as the objective function, and the optimized solution is programmed with LINGO software. The maximum total network output traffic per unit hour of the network is calculated as $\overline{Q}_{\mathrm{T}}$ = 20719 pcu/h, the split allocation of each intersection and traffic flow allocation of each directed section in the network satisfying with the maximum output traffic $\overline{Q}_{\mathrm{T}}$ are shown in Table 1 and Table 2, respectively.

Table 1. Split allocation of each approach direction in the network

Intersection	Eastbound	Southbound	Westbound	Northbound
I_1	0.25	0.23	0.15	0.32
I_2	0.25	0.24	0.22	0.24
I_3	0.33	0.24	0.23	0.15
I_4	0.24	0.22	0.24	0.25
I_5	0.24	0.24	0.23	0.24
I_6	0.27	0.24	0.22	0.22
I_7	0.24	0.15	0.31	0.25
I_8	0.26	0.20	0.25	0.24
I_9	0.33	0.15	0.23	0.24

Table 2. Traffic flow allocation of each road section in the network

Road section	Traffic flow(pcu/h)	Road section	Traffic flow(pcu/h)
q_{12} / q_{21}	1566 / 1770	q_{23} / q_{32}	1652 / 1824
q_{36} / q_{63}	1573 / 1738	q_{69} / q_{96}	1705 / 1619
q_{98} / q_{89}	1850 / 1679	q_{87} / q_{78}	1719 / 1789
q_{74} / q_{47}	1569 / 1768	q_{41} / q_{14}	1656 / 1818
q_{25} / q_{52}	1711 / 1704	q_{58} / q_{85}	1720 / 1668
q_{45} / q_{54}	1705 / 1720	q_{56} / q_{65}	1696 / 1757

Under the condition of dynamic equilibrium, the relationship between the actual average traffic flow of the network optimized by the model and the capacity of each approach at the intersection is compared, as shown in Figure 2.

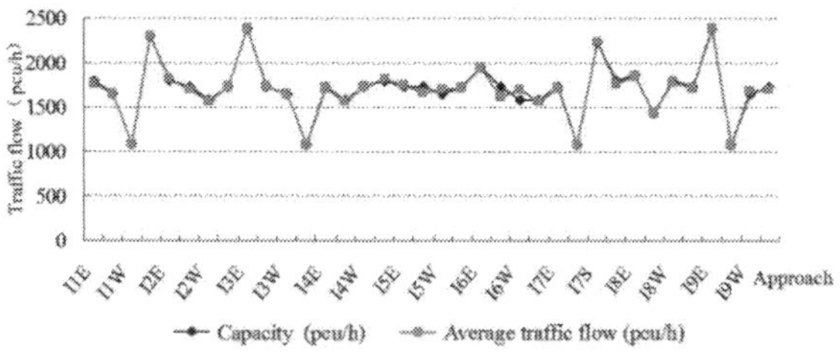

Figure 2. Relationship between actual average traffic flow and capacity after optimization

It can be seen from Figure 2 that under the condition of judging the dynamic equilibrium state of the network, the actual average traffic flow of the network optimized by the model is similar to that of each road section, and the network traffic is in a dynamic equilibrium state. At this time, the split allocation of the intersection group in the network is reasonable, which can not only ensure that the network traffic is in a stable and orderly operation state, but also realize the maximization of the total network traffic output.

5 Simulation experiment

Based on the given intersection geometric conditions and traffic flow input parameters, the simulation road network is established in VISSIM 4.3 environment, and the total network output vehicle number per unit hour is selected as the final simulation evaluation output index. The output traffic volume per hour and the deviation rate relative to the theoretical value are simulated in any random phase sequence group, as shown in Table 3

Table 3. Comparison experiment of VISSIM simulation

Random signal phase sequence	Phase sequence 1	Phase sequence 2	Phase sequence 3	Phase sequence 4	Phase sequence 5
Experiment value（pcu/h）	20813	20742	20339	20284	20327
Deviation rate（relative to	+0.45%	+0.11%	-1.83%	-2.10%	-1.89%

It can be seen from Table 3 that the calculation results of split allocation model of intersection group oriented to network carrying capacity limit proposed in this paper are similar to those of simulation experiments, and the deviation rates are all less than 3%. Thus, the accuracy and reliability of the model are verified, and the output traffic maximization of the network during the peak period of the over-saturated intersection group is effectively guaranteed.

6 Conclusions

In this paper, a split allocation model of intersection group oriented to the network carrying capacity limit is constructed, and the signal control of traffic network under the premise of different traffic demand states and control objectives is optimized. Compared with the existing models and algorithms, the model proposed in this paper can select the corresponding optimal control criteria according to the difference of the control objectives and the network traffic demand. Through optimizing the signal cycle length and split of over-saturated intersection group, the effective prevention and quick dredging of network traffic congestion can be realized, and finally the utilization benefit of network time-space resources can be maximized.

7 Acknowledgments

This research is supported by National Engineering Laboratory Funded Project “Road network bearing capacity and operation reliability evaluation methods based on critical path analysis” (Project No. 2019SJGC01).

References

1. Daganzo C F. The cell transmission model, part II: Network traffic. Transportation Research Part B, 29(2):79-93, (1995).
2. Girianna M, Benekohal R F. Using genetic alorithms to design signal coordination for over-saturated networks. Journal of Intelligent Transporation Systems: Technology Planning and Operations, 8(2): 117-129, (2004).
3. K. Aboudolas, M. Papageorgiou, E. Kosmatopoulos. Store-and-forward based methods for the signal control problem in large-scale congested urban road networks. Transportation Research Part C, 163-174, (2009).
4. K. Aboudolas, M. Papageorgiou, A. Kouvelas, et al. A rolling-horizon quadratic-programming approach to the signal control problem in large-scale congested urban road networks. Transportation Research Part C, 680-694, (2010).
5. Ernesto C, Gaetano F. Combined signal setting design and traffic assignment problem. European Journal of Operation Research, 155(3): 569-583, (2004).
6. Wang L, Liu X M, Wang J Z, et al. Urban traffic signal system control structure optimization based on network analysis. Mathematical Problems in Engineering, 2013: 1-9, (2013).
7. Pignataro L J, Mcshane W R, Crowley K W, et al. Traffic control in over saturated street networks. Transportation Research Board, (1978)

Application of Intelligent Transportation Systems (ITS) for Regional Development: Case Studies in Integrated Functional Approach

Koorosh Gharehbaghi[1], Neville Hurst[1,a] and Kathryn Robson[1]
[1]*RMIT University, Melbourne, Australia*

Abstract. This paper evaluates the application of Intelligent Transportation Systems (ITS) specifically as a part of regional development. Although ITS has been extensively studied in the context of city-wide developments, research on such developments for the regional enhancement is scarce. Importantly, transportation development for regional areas differs to that of traditional urban provinces. Whilst the transportation developments in urban areas need to concentrate on a range of diverse community requirements, for regional areas, such population are less diverse and more alike. Consequently, developed ITS in the regional areas needs to concentrate on more homogeneous communities and their needs. There are many factors influencing regional transportation developments, however functionality is one of the most critical aspects. Although functionality is an important aspect of all developments, it is considered to be the most crucial factor for regional areas. Fittingly, this paper will examine some of the most important transportation functionality issues for regional areas. To further explore this theme, the surrounding areas of Fukushima and Tokyo were studied to highlight their regional transportation requirements. These two case studies were then compared using factor analysis method. The comparison generally found that, due to its geographical location Fukushima aligned its regional functionalities more broadly. Further, this paper proposes a streamlined approach to analyze the successful application of transportation development in the Fukushima and Tokyo regional areas. To do so, an integrated functional approach was carefully undertaken. The proposed approach specifically focused on incorporating mobility and ITS as the key ingredients for the regional developments. It was found that, such integrated functional approach would therefore increase innovation and productivity as the result of successful application of ITS for regional development.

1. Introduction

As highlighted by Andres et al., (2016) [1] the expansion of transportation infrastructure is an essential component for the successful development of rural and regional communities. Bullock et al., (2011) [2] together with Wasson et al., (2011) [3] highlighted that an important attribute of Intelligent Transportation Systems (ITS) is the ability of such technological advancements for both urban and regional areas. Further, key aspects of transportation infrastructure for the regional areas encompass:

- Ensuring effective interconnection of remote areas [4]. This includes embracing and interconnecting the rural communities ensure that all regional transportation issues are vigilantly considered.
- Proficiently utilizing the appropriate ITS as the basis of regional development [5]. Although, the inclusion of ITS is typically aligned with urban transport, as a part of the regional transportation considerations, ITS could also provide important advantages.
- Successfully linking regional communities to urban areas for improved accessibility. Perhaps the most important aspect of transportation infrastructure for regional areas is the inclusion of effective accessibility [6]. Generally speaking, transportation accessibility requires both short and long-term considerations [7]. While short-term considerations include the convenience of transportation accessibility, long-term deliberation includes encompassing foreseeable regional growth and developments.
- Warranting long-term planning and appropriate enduring investment. This comprises implementation of various planning initiatives for rural communities. Here, transportation planning includes the consideration of rural environment and subsequent surrounding as the basis of the regional transportation development.

Appropriately, developed ITS in regional areas needs to concentrate in a more homogeneous manner towards community and its needs. Although other authors including Shu and Yao-Jan, (2013) [8] along Bhaskar and Chung (2013) [9] among many, emphasized the additional ITS abilities for the regional developments, more study is required. Accordingly, this paper aims to

[a] Corresponding author: neville.hurst@rmit.edu.au

https://doi.org/10.1051/matecconf/202030805005

review the application of ITS for regional development and in doing so examines a number of case studies.

2. Literature review

As already conversed, transportation development for regional areas differs to that of traditional urban areas. Whilst transportation developments in urban areas need to concentrate on a range of diverse community requirements, for regional areas, such communities are less diverse [10, 11]. As Cao and Orru (2014) [12] highlighted, the most important transportation accessibility issues in regional areas include functionality together with performance. However, to further validate such hypothesis, a closer examination of the regional transportation functionality is required.

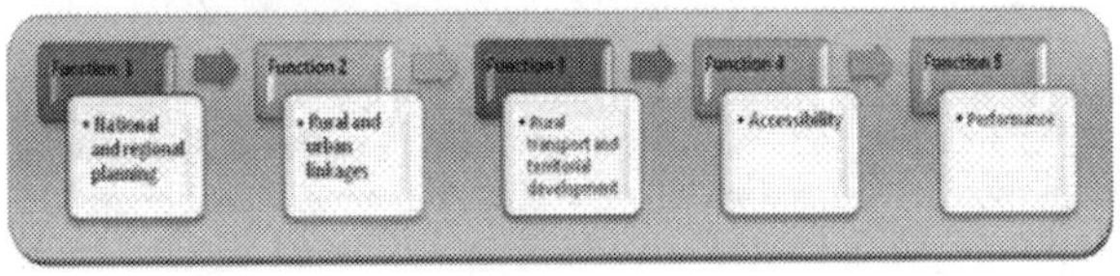

Figure 1: the regional transportation functionality.

It can be observed from Figure 1; regional transportation functionality consists of five main components:

1. National and regional planning, which includes a cluster rather than individual development [13, 14]. The jurisdictional concerns may be difficult; however, this approach is necessary to bring about absolute regional amplification.
2. Rural and urban linkages. The alignment and connecting of both the rural and urban transportation systems is an important strategy [15]. Such strategy ensures that the regional transportation systems are not neglected, but rather cared for.
3. Rural transport and territorial development. The development of regional areas is highly dependent upon adequate transportation [16]. Insufficient rural transport not only obscures any regional development, but also limits territorial growth.
4. Accessibility. Of all the main functionalities, accessibility is perhaps the most challenging component [17, 18]. This is due to the remoteness of some locations. The regional transportation accessibility maybe somewhat inadequate and/or not insufficient.
5. Performance. Of all the main functionalities, performance is the most immensely deliberated [19, 20]. For the regional areas in particular, performance of the transportation services could undeniably make or break any future proposal and spending.

Accordingly, as the basis of ITS for regional development, this paper considers these five functions as the foundation of an integrated approach. To fully understand such integration, the transportation development for regional areas need to be carefully reviewed. Figure 2 articulates the asymmetry of regional transportation developments with the inclusion of ITS.

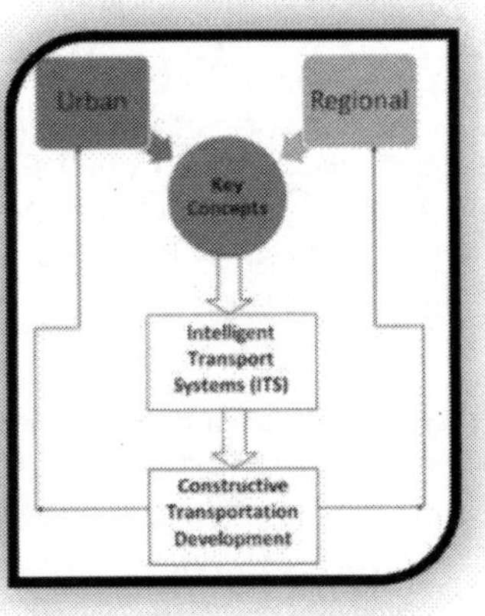

Figure 2: asymmetrical of regional transportation developments.

The proposed asymmetry is skewed towards the urban rather than rural development. Such orientation is due to a greater focus upon ITS in urban areas [21, 22]. Nevertheless, Gudmundsson et al., (2016) [23] highlighted that the importance of transportation for regional development be categorized into the following areas: a) Freight management, b) Road safety and security, c) Traffic management, d) Environment protection, e) Public transport, f) Automotive telematics, g) Parking management and h) Road user charging. Appropriately, these categories are represented in Figure 3.

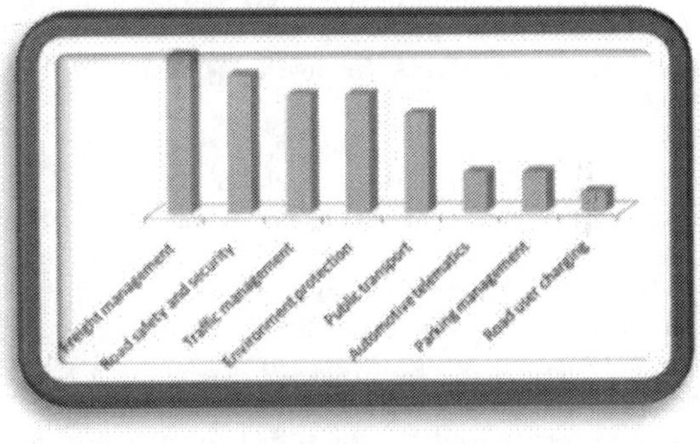

Figure 3: Transportation development for regional areas.

Figure 3 symbolizes the main areas of the regional development. The analyses provided here, is based on reviewed literature and a number of interpreted case studies. Hardisty (2010) [24] highlighted freight management as of absolute importance for the regional development, whilst Thekdi and Lambert (2012) [25] claimed that the road user charging has the least significance. Such analysis could be due to a small number of road users within the regional areas.

On one hand, road safety and security will be the second most important transportation factor for any regional developments [26]. On the other, Kahraman and Cevik Onar (2015) [27], reasoned that the effective traffic management along with environmental protections possess lesser effects than the previous two transportation factors. While it can be worked out that, due to lesser demand than the urban areas, public transport, automotive telematics and parking managements are the succeeding regional transportation issues (in that order). Subsequently, to fully demonstrate the importance of the ITS as a part of transportation development for the regional areas, a integrated functional approach is indispensable.

2.1 ITS for the regional development: Integrated functional approach

Conventionally, ITS can be defined as an amalgam of various methods, such as computational technologies, used to further enhance the overall functionality of all types of transport [27]. Such computational technologies integrate various traffic and transit management techniques as a part of overall functional enhancement approach and includes superior accessibility [28]. Nevertheless, to further ensure the successful application of transportation development for the regional areas, an integrated functional approach need to be carefully executed [26]. Such alignment will specifically focus on incorporating mobility and ITS as key ingredients for regional developments.

ITS also utilizes various methods such as computational technologies to further enhance the overall functionality of all types of transport [29]. Such computational technologies integrate various traffic and transit management techniques as a part of overall ITS functional enhancement approach including superior accessibility [30].

Although ITS has been extensively studied in the context of city-wide developments, research on ITS as part of regional enhancement is scarce. Importantly, transportation development for regional areas differs to that of traditional urban areas [31]. Whilst transportation developments in urban areas need to concentrate on a range of diverse community requirements, for regional areas, such communities are less diverse and more alike. Accordingly, developed ITS in regional areas needs to concentrate on a more homogeneous community and its needs [25]. On one hand, the concept of urbanization, which brings increased demand for the regional areas, needs to be carefully considered [32]. On the other, such regional development needs to prudently position, the competing drivers of engineering, environmental, economic and social. Subsequently, these factors then need to be integrated holistically, that is based on functional approach. This is presented in Figure 4.

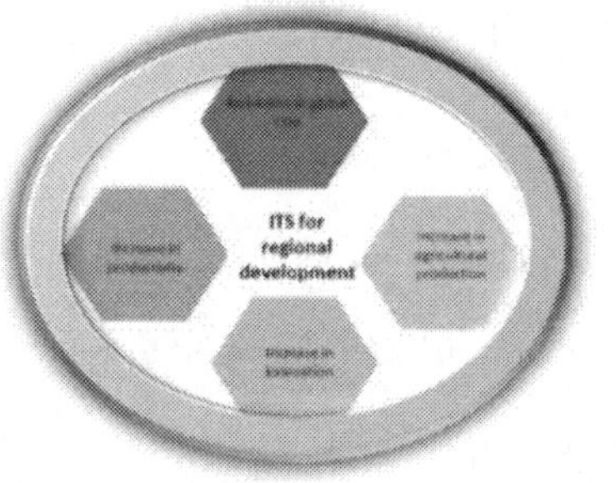

Figure 4: Integrated functional approach for regional development.

As identified, at the centre of this the integrated functional approach, lies the ITS. In other words, ITS is the essential element of this approach. Nonetheless, there are also four other important elements which complete this process. Using advanced transportation systems, could ultimately reduce the CO_2 production through innovative emissions control management. Subsequently, the inclusion of ITS specifically for regional development, would also benefit from such advancement.

On one hand, increase in innovation can dramatically reduce losses and thus improve production and productivity rate. While, this will ultimately lead to better long-term revenue and thus increase economical benefits [33]. In addition, a successful application of ITS for regional development, requires a cautious alignment of transportation functionality and integrated functional approach. To investigate such theory, the following section (case studies) has been developed.

3. Research methodology

To further explore the alignment of 'transportation functionality' and 'integrated functional approach', the regional areas of Fukushima and Tokyo were studied. Such investigation would also carefully highlight the regional transportation requirements. These case studies evaluate the adaptation of ITS as a part of successful regional development and in doing so assess their regional transportation functionalities.

To effectively compare the transportation development for regional areas, the five main components of regional transportation functionality (see section 2.0) are compared on the Fukushima and Tokyo rural zones. It is important to highlight that such analysis was assisted by reports produced by *2012 Citylab* together with *2017 Ministry of LITT* (Land, Infrastructure, Transport and Tourism). These concentrated and inclusive reports were carefully studied and interpreted to determine the key ITS variances for both cities' regional development. Moreover, factor analysis was then used as the basis of comparing these two case studies and their subsequent findings carefully discussed.

4. Case Studies

4.1 Fukushima

Fukushima is the capital city of Fukushima prefecture in Japan. This city has a transportation service containing rail and bus systems. The city uses the Iizaka rail line and an extensive bus network, to serve its regions. Figure 5 represents the general topography of Fukushima and its regional rail transportation.

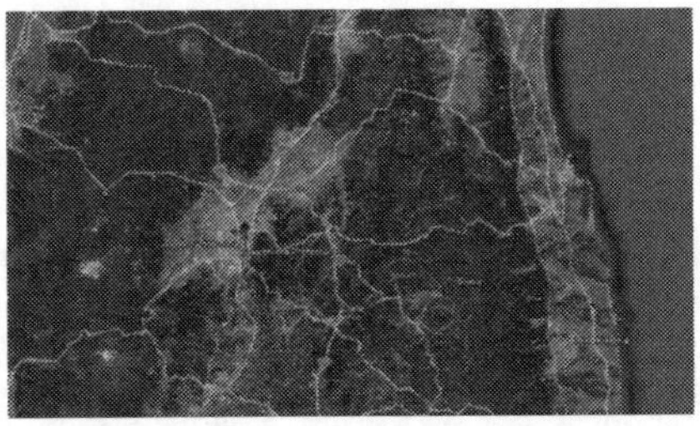

Figure 5: Topography of Fukushima and its regional rail transportation.

As it can be observed, Fukushima and its regional rail transportation is well spread. Such distribution, however, will also create ITS difficulties due to its scale. Not only does such ITS require significant funding but complication exists within such system integration.

4.2 Tokyo

Tokyo is the capital city of Japan and its greater area is the most populated metropolitan area in the world [34]. The transport network in Greater Tokyo includes public and private rail and highway networks; airports for international, domestic, and general aviation; buses; motorcycle delivery services, walking, bicycling, and commercial shipping. Moreover, the public transport systems within Greater Tokyo is serviced by 158 lines, 48 operators, 4,714.5 km of operational track and 2,210 stations [34]. Approximately, 40 million passengers use Tokyo's rail system daily (Citylab, 2017). Figure 6 represents the general topography of Tokyo and its regional rail transportation.

Figure 6, Topography of Tokyo and its regional rail transportation.

As it can noticed, Tokyo and its regional rail transportation is extremely spread. Such distribution will also significantly complicate the ITS. This is possibly due to many different system requirements, particularly dealing with functionality and overall performance.

4.3 Case studies comparison

In comparing the Fukushima and Tokyo case studies, specific regional transportation consideration needs to be measured. These considerations include:

- Foremost, the rural strategic alignment needs to be carefully deliberated. This consideration needs to cautiously contemplate not only land issues, but also alignment of all jurisdictional governance and any future national frameworks.
- The inclusion of innovative transport tools such as ITS, to thoroughly advance and integrate future regional developments.

Nevertheless, as already noted, the regional transportation functionality consists of five main components which are, National and regional planning; Rural and urban linkages; Rural transport and territorial development; Accessibility; and Performance.

Accordingly, to effectively compare the transportation development for regional areas, the five main components are contrasted on the Fukushima and Tokyo peri-urban zones. It is important to highlight that such analysis where assisted by reports produced by Citylab and Japan Ministry of Land, Infrastructure, Transport and Tourism [34]. Before commencing the functionalities analysis of the both cities, their trajectories of change were also explored. Figure 7 provides the projected transportation change for both cities via spatial presentation.

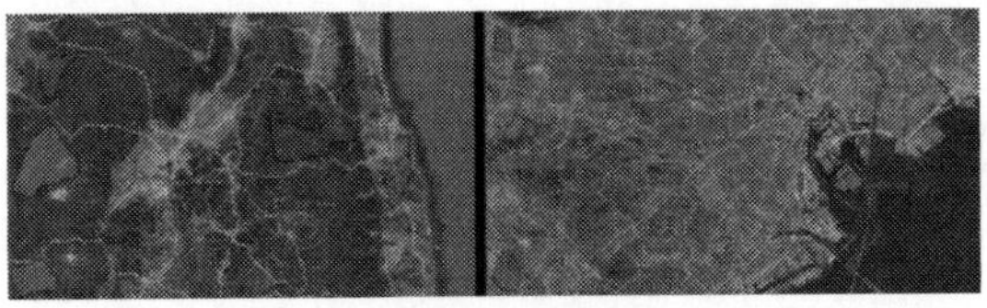

Figure 7: spatial representation of the projected transportation change for both cities.

As it can be summarized, for Fukushima, the projected transportation change is well spread with eastern suburbs having the most extreme (red) trajectories. This is most likely due to the 2011 nuclear disaster which mostly effected the eastern areas. Conversely, for Tokyo, the projected transportation change is generally low (blue) with extreme being planned for a small area. This is due to Tokyo's outstanding transportation infrastructure.

4.3.1 Factor analysis

Once the projected transportation of the both cities been briefly discussed, the actual functionalities analysis for the both metropolitans was then carried out. This comparison was based on the available data from reports produced by Citylab [35] and Japan Ministry of Land, Infrastructure, Transport and Tourism [34]. In addition, a factor analysis was used to characterize the correlations between the variables X_a (for 5 main components and 4 functional approaches) of which the X_{ai} (specific instances). The following conventional factor analysis model was thus used:

$$F_{x_{..}} = \sum_{p} \ell_{ap} F_{pi} + \epsilon_{ai_{..}} \qquad [1]$$

To streamline the modeling process, a maximum value of 1.0 was used as the basis of total output. This output value was thus the basis of the regional functionalities and integrated functional approach factors. This analysis is presented in Table 1.

Table 1, case studies comparison.

Regional functionalities	Fukushima	Tokyo
Accessibility	0.213	0.236
National and regional planning	0.219	0.111
Performance	0.332	0.532
Rural and urban linkages	0.127	0.045
Rural transport and territorial development	0.109	0.076
Integrated functional approach	Fukushima	Tokyo
Increased in agricultural production	0.136	0.09
Increased in innovation	0.565	0.601
Increased in productivity	0.167	0.203
Reduction of global Co_2	0.132	0.106

Table 1 represents the summary of the factor analysis of the regional functionalities and integrated functional approach for both cities. A *Mplus* software was used to calculate the presented analysis. To further demonstrate the findings Figures 8a and 8b are also presented.

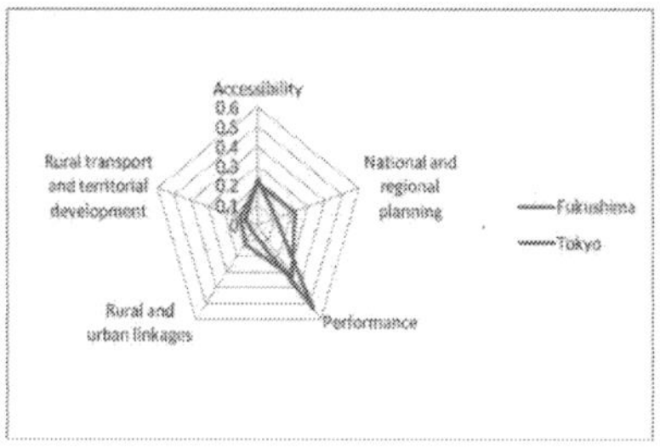

Figure 8a: 'regional functionalities' comparison.

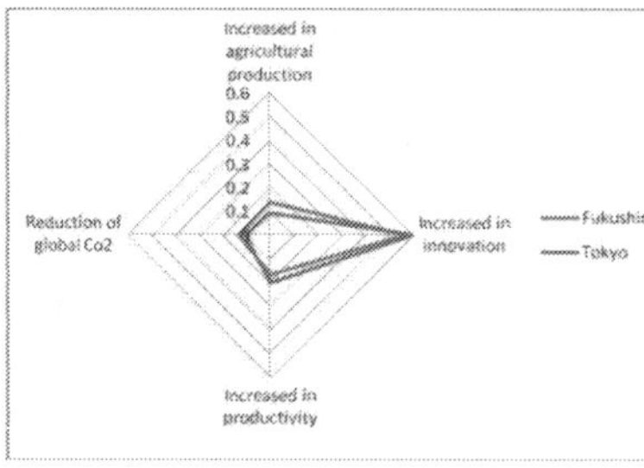

Figure 8b: 'integrated functional approach' comparison.

As it can be described, while Fukushima generally ranks lower in most areas, Tokyo seems to excel particularly in performance. Whilst Fukushima seems to be scoring higher on the national and regional planning along with rural and urban linkages; both cities score exceptional in increased innovation. Although, such outcome is likely due to Japan's reputation as a world technology leader; as a consequence of its geographical location, Fukushima needs to align its regional functionalities more broadly. Hence a higher national and regional planning along with rural and urban linkages score is observed.

For both cities, increase in innovation is extreme. This is considered a likely result of Japanese national and international efforts in research and development. Moreover, Tokyo mainly uses the most advanced and innovative technologies and thus is leading 'performance' not only with other Japanese cities but many cities around the globe. On the other hand, while for both cities and their regional areas, the reduction of global CO_2 is seemingly comparable. However, the increase in agricultural production is slightly higher in Fukushima. This is due to Fukushima's vast available area dedicated specifically for agricultural industry.

Overall, the Japanese federal government continuously places significant importance for both areas of the nation. This is evident by the adaptation and utilization of ground-breaking ITS for both cities, but specifically Tokyo. Such outcomes would benefit Tokyo's transportation authority's allowing them to further improve and optimize the functionality of their regional transportation services.

5. Lessons to be learned

The comparison of the integrated functional approach for both Fukushima and Tokyo, further highlights the essential aspect of such an approach, specifically for the regional development. The successful implementation of Fukushima and Tokyo integrated functional approach highlights the significance of increased innovation and productivity as the result of successful application of ITS for regional development. Additionally, the case studies highlight such enhanced regional development specifically requires a cautious and gradual alignment of transportation functionality and integrated functional approach.

From the four integrated functional approaches for regional development, the inclusion of ITS specifically for regional development, requires developing and utilizing specific technologies that meet not only the nation, but also region's specific shortfalls. To improve such regional development, increasing innovation is highly desirable. Such innovation would not only increase the novelty of specific processes, but also lead to improve regional production and productivity rate. Additionally, such an improvement of integrated functional approach will also lead to supporting national advancements.

6. Conclusion

This research found the inclusion of ITS is paramount for a successful transportation functionality and this in turn is necessary for successful regional development. In addition, such ITS adaptation also requires an absolute integrated functional approach. To demonstrate such proclamation, this paper utilized the regional areas of Fukushima and Tokyo as case studies. To contrast these two cities and their regional area developments, their actual transportation system functionalities were succinctly reviewed. While the Fukushima's vast available area is ideal for the agricultural production, it is Tokyo, which sets benchmarks in particular as the basis of both increased in innovation and productivity. Due to Tokyo's national and international significance, this city clearly outperformed Fukushima for both innovation and productivity.

Nevertheless, as this paper illustrated, the inclusion of integrated functional approach is a dominant factor for any regional development. Such integration needs to focus on more homogeneous communities along with their needs. Such integration needs to focus on more homogeneous communities along with their needs. Without such consideration, successful application of ITS for regional development is detrimentally affected. Finally, as this paper highlighted, one great challenge to achieve effective regional development is the cautious alignment of transportation functionality and integrated functional approach.

7. Data Availability Statement

Some or all data, models, or code generated or used during the study are available from the corresponding author by request.

References

1. S. Hernandez, A. Monzon, R. De Ona, "Urban Transport Interchanges: A Methodology for Evaluating Perceived Quality." Transportation Research, Part A **84**, pp.31-43 (2016)
2. J. Wasson, G. Boruff, A. Hainen, S. Remias, E. Hulme, G. Farnsworth and D. Bullock, "Evaluation of Spatial and Temporal Speed Limit Compliance in Highway Work Zones". Transportation Research Record: Journal of the Transportation Research Board **2258**, (2258), pp 1–15 (2011)
3. D. Bullock, M. Martchouk, F. Mannering, "Analysis of Freeway Travel Time Variability Using Bluetooth Detection". Journal of Transportation Engineering, **137** (10), pp 697–704 (2011)
4. K. Gharehbaghi and K. Robson, 'The Exploitation of ICT in Optimization of PTS: Productivity and Performance Enhancement Methods', in Proceedings of the 11th International Conference on Computer Modeling and Simulation, Association for Computing Machinery, Melbourne, 16 - 18 January, pp. 180-183. (2019)
5. K. Gharehbaghi, I. Clarkson, N. Hurst and F. Rahmani, "Transportation development for regional infrastructure: Implications for Australian rural areas", Transportation Research Procedia (2020) in press.
6. R. Kummitha and N. Crutzen, "How do we understand smart cities? An evolutionary perspective", Cities, **67**, pp 43-52 (2017)
7. K. Gharehbaghi, and K. McManus, "TIS condition monitoring using ANN integration: an overview", Journal of Engineering, Design and Technology, **17**, (1), pp. 204-217 (2019)
8. Y. Shu and W. Yao-Jan, "Travel Mode Identification Using Bluetooth Technology", Journal of Intelligent Transportation Systems, DOI: 10.1080/15472450.2017.1384698 (2013)
9. A. Bhaskar and E. Chung, "Fundamental understanding on the use of Bluetooth scanner as a complementary transport data", Transportation Research Part C: Emerging Technologies, **37**, pp.42-72 (2013)
10. T. Krasemann, "Computational decision-support for railway traffic management and associated configuration challenges: An experimental study", Journal of Rail Transport Planning and Management, **5**, (3) pp 95-109 (2015)
11. J. Mathey, S. Rößler, J. Banse and I. Lehmann, , "Brownfields As an Element of Green Infrastructure for Implementing Ecosystem Services into Urban Areas", Journal of urban planning and development, **141**, (3) (2015)
12. Giacomo Cao and Roberto Orrù, "Current Environmental Issues and Challenges", Springer, pp 39-46 (2014)
13. J. Black, A. Paez and P. Suthanaya, "Sustainable Urban Transportation: Performance Indicators and Some Analytical Approaches", Journal of urban planning and development, **128**, (4) (2002)
14. T. Kim, "Integrated Urban Systems Modeling: Theory and Applications", Springer, pp 191-199. (2012),
15. Gharehbaghi, K., McManus K., and Robson, K., (2019), "Minimizing the environmental impacts of mega infrastructure projects: Australian public transport perspective", Journal of Engineering, Design and Technology, in press.
16. E. Boschmann and M. Kwan, "Toward Socially Sustainable Urban Transportation: Progress and Potentials", international journal of sustainable transportation, **2**, (3) pp 138-157 (2008)
17. K. Gharehbaghi and M. Georgy, "Utilization of Infrastructure Gateway System (IGS) as a Transportation Infrastructure Optimization Tool", International Journal of Traffic and Transportation Engineering, **4** (1), pp. 8-15 (2015)
18. K. Gharehbaghi and M. Myers, "Intelligent System Intricacies: Safety, Security and Risk Management Apprehensions of ITS", Proceedings of the 8th International Conference on Industrial Technology

and Management, IEEE, Cambridge, United Kingdom, United Kingdom, pp 37-40. (2019)

19. K. Gharehbaghi and F. Rahmani "Deterioration of Transportation Infrastructures: Corrosion of Reinforcements in Concrete Structures", Materials and Technologies in Engineering, **940**, pp 160 - 166 (2018)
20. K. Gharehbaghi and Farnes "Buried Transportation Structures: Fuzzy-Markov Computation Case Study", Proceedings of the 10th International Structural Engineering and Construction Conference: Interdependence between Structural Engineering & Construction Management, Chicago, Illinois, United States, pp 1-6 (2019)
21. S. Ghosh and T. Lee "Intelligent Transportation systems: smart and green infrastructure design", 2nd edition, Boca Raton, FL : CRC Press (2012)
22. K. Gharehbaghi, K. Robson and Rahmani "Mega Transportation Infrastructure: a sustainable foresight", Transportation Research Procedia (2020) in press
23. H. Gudmundsson, R. Hall, G. Marsden and J. Zietsman "Sustainable transportation : indicators, frameworks, and performance management", Springer (2016)
24. P. Hardisty, "Environmental and economic sustainability", CRC Press/Taylor and Francis (2010)
25. S. Thekdi and J. Lambert, "Decision analysis and risk models for land development affecting Infrastructure systems", Journal of Risk Analysis, July, **32** (7), pp 1253-1269 (2012)
26. A. Sladkowski and W. Pamula "Intelligent Transportation Systems - Problems and Perspectives", Springer (2015)
27. C. Kahraman and S. Cevik Onar, "Intelligent techniques in engineering management theory and application", Springer (2015)
28. K. Gharehbaghi and K. Farnes "Process automation in Intelligent Transportation System (ITS)", International Journal of Machine Learning and Computing, International Association of Computer Science and Information Technology, **8** (3), pp. 294-297 (2018)
29. K. Gharehbaghi, "Artificial Neural Network for Transportation Infrastructure Systems", MATEC web of conferences, pp. 1-5 (2016)
30. L. Mattsson and E. Jenelius, "Vulnerability and resilience of transport systems - A discussion of recent research", Transportation Research Part A: Policy and Practice, **81**, pp 16-34 (2015)
31. P. Centobelli, R. Cerchione and E. Esposito, "Environmental sustainability in the service industry of transportation and logistics service providers: Systematic literature review and research direction", Transportation Research Part D: Transport and Environment, **53**, pp 454-470 (2017)
32. A. May, "Urban Transport and Sustainability: The Key Challenges", International Journal of Sustainable Transportation, **7**, (3), pp 170-185 (2012)
33. F. Corman, A. D'Ariano, A. Marra, D. Pacciarelli and M. Sama, "Integrating train scheduling and delay management in real-time railway traffic control", Transportation Research Part E: Logistics and Transportation Review, **105**, pp 213-239 (2017)
34. Ministry of LITT, Japan Ministry of Land, Infrastructure, Transport and Tourism, White Paper on Land, Infrastructure, Transport and Tourism in Japan, http://www.mlit.go.jp/en/statistics/white-paper-mlit-index.html [viewed on 17/10/2017] (2017)
35. Citylab, https://www.citylab.com/transportation/2012/05/secret-tokyos-rail-success/2044/ [viewed on 15/09/2017]

Real-time Aggressive Driving Detection System based on In-vehicle Information using LoRa Communication

Yongsu Jeon[1], Chanwoo Kim[1], Hyunwook Lee[1] and Yunju Baek[1,a]

[1]*Dept. of Electrical and Computer Engineering, Pusan National University, Pusan, Korea*

Abstract. Safe driving has attracted a significant amount of attention in recent years owing to the increase in the complexity of the driving environment. There are many research studies focusing on detection of aggressive driving that may cause traffic accidents. In this paper, we propose a system for acquiring vehicles' interior data and thereby detecting dangerous driving conditions. The system is designed to transmit the information acquired to a data server using Long Range (LoRa) communication networks. Through experimentation, we confirm that the proposed system can detect aggressive driving behaviors in real time and store them on the data server through LoRa communication. We evaluated techniques for acquiring in-vehicle information on 14 vehicles and confirmed that data can be extracted from most of the commonly available vehicles.

1 Introduction

As society develops, the number of vehicles on the road increases and the driving environment becomes increasingly complex. According to the statistics of traffic accidents provided by the Korean National Police Agency, more than 200,000 traffic accidents and about 5,000 deaths occur each year. Many studies related to safe driving have been performed. Some examples include systems such as intelligent transportation system (ITS) and advanced driver assistant system (ADAS) which assist in driving by performing additional safe driving functions. Technologies that are currently under active research and development include forward collision warning (FCW), lane departure warning (LDW), and pedestrian detection systems.

Research is also underway to analyze the driving behaviors of drivers. This research aims to prevent traffic accidents by detecting driving behavior that is likely to cause accidents and notifying the driver of their occurrence. By applying the technique detailed in this study, the driving habits of drivers and the frequency of dangerous driving incidents can be analyzed. Car insurance companies already use this information to sell user-specific insurance products. In order to analyze a driver's driving behavior, various pieces of information are required such as vehicle environment/surrounding conditions, vehicle status and operation information. The vehicle's environment information includes data regarding the surrounding environment of the target vehicle, such as the number of lanes on the road, number of surrounding vehicles, and the speed of traffic. The vehicle state information includes the movement data about the vehicle, including the direction, speed, and acceleration of the vehicle. Finally, the vehicle operation information is information detailing how the driver controls the vehicle, and monitors acceleration pedal, brake pedal, steering wheel and turn signal operations.

In existing driver behavior analysis systems, owing to the difficulty in acquiring vehicle operation information, only the vehicle state information is often used. These systems use acceleration sensors to acquire the speed and acceleration information of the vehicle and detect dangerous driving behaviors, such as sudden starts or stops. However, certain types of aggressive driving behaviors, such as threatening lane changes, are difficult to detect. Certain applications such as user-specific insurance require only vehicle operation information. In the existing driver behavior analysis systems, the driver directly influences the results of the driving behavior analysis. Therefore, existing systems are not compatible with other application systems. In other words, drivers who want to take advantage of a variety of safe driving application systems are forced to use multiple devices or use only driver behavior analysis systems.

The proposed system not only periodically transmits vehicle operation information to the server using Long Range (LoRa) communication, but also analyzes the information and displays the level of dangerous driving on the driver's smartphone in real time. When certain dangerous driving behaviors are detected, an alarm is triggered to alert the driver of the driving behavior.

The second chapter of this paper introduces the research related to the analysis of existing driver behaviors. Section 3 describes the proposed system, and Section 4 evaluates the performance of the proposed system. In Section 5, we state our conclusions regarding our system, methods, and required future work.

[a] Corresponding author: yunju@pusan.ac.kr

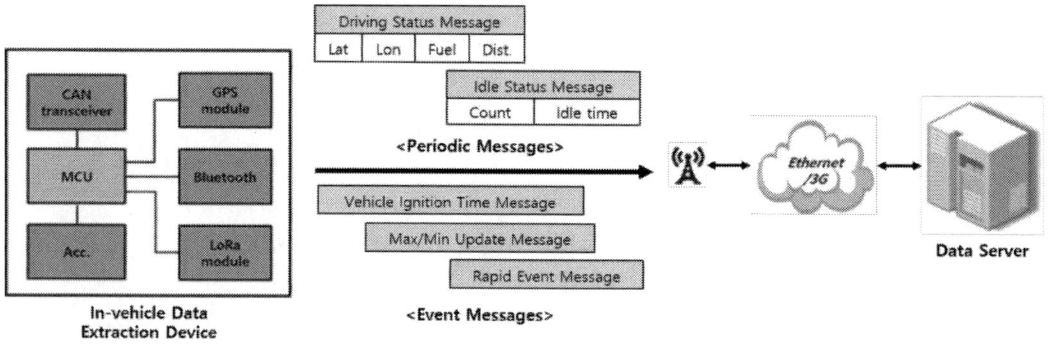

Figure 1. Proposed system architecture

2 Related work

Currently, there is a great amount of research being conducted that analyzes the driving behavior of drivers. Research studies focusing on driving behavior are actively being conducted using vehicle status information, machine learning and deep learning technologies. Lee [1] classified driving events using vehicle speed, acceleration, and yaw rate. To aid in classification, machine learning techniques such as auto encoder and self-organizing map (SOM) are used. This study clustered 32 driving styles and defined some of the specific clusters as dangerous driving behaviors. Chen [2] analyzed the driver's left turn and right turn characteristics by using an inertial measurement unit (IMU) sensor. They identified the driver by observing specific features during turning operations. Acceleration along the end-of-turn axis, deviation of acceleration along the end-of-turn axis, and deviation of the raw yaw rate were used to create features and utilize the Naïve Bayes classifier to classify drivers. The IMU sensor included in the smartphone was used to monitor and identify up to 12 drivers.

The technology that analyzes dangerous driving based on vehicle status information collects data through various sensors and analyzes driving behavior. However, this technology accurately analyzes driving behavior that moves mostly along the longitudinal direction of the road, but inaccurately for behavior that moves in the lateral direction.

On the other hand, vehicle operation information directly affected by the behavior of the driver is generally difficult to obtain from the vehicle. Han [3] extracted vehicle operation information from the on-board diagnostic system (OBD-II) in the vehicle and designed the system to send data to the driver's smartphone using bluetooth low energy (BLE) communication.

Martinelli [4] recognizes the driver's driving style by using vehicle status information and vehicle operation information. Vehicle status and operation information is obtained from the in-vehicle OBD-II system and includes 51 types of information, including fuel consumption, accelerator pedal position, steering angle, steering speed, and vehicle speed. Principal component analysis (PCA) was used to analyze the driver's disposition and to classify nine types of drivers.

As described above, in the case of the existing system using the vehicle status information and the operation information, not only is the driving behavior analysis not performed in real time, but it is also difficult to share data with other platforms due to the lack of complementary communication protocols.

3 Aggressive Driving Detection System

3.1 System architecture

The proposed system detects basic dangerous driving behaviors such as sudden starting, sudden stopping, and radical turning in real time based on vehicle status information and operation information. It also analyzes additional driving information such as idling time, mileage, travel time and top speed. This information is then transmitted to the control server using LoRa, a low power long distance communication technology. Figure 1 shows the proposed system's architecture.

The proposed system consists of an information collection device, data server, and the driver's smartphone. The collection device collects vehicle state information and vehicle operation information through the OBD-II system of vehicle.

Figure 2 shows a hardware block diagram of the proposed device. The information collecting device includes an MCU, a controller area network (CAN) communication unit for communicating with the OBD-II, a sensor unit for collecting status data of the vehicle, a Bluetooth communication unit for communicating with external components, and a LoRa communication unit.

The device transmits to the external component through the Bluetooth communication module and LoRa communication module according to the type of generated data. It transmits the information to the driver's smartphone through the Bluetooth communication unit and the information to the data server via the LoRa unit.

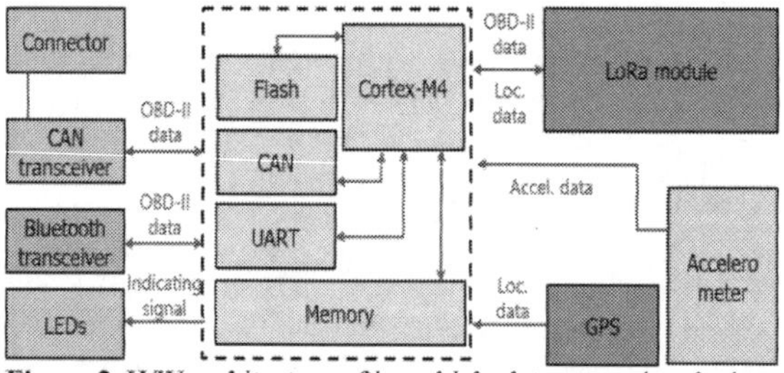

Figure 2. H/W architecture of in-vehicle data extraction device

It uses LoRaWAN protocol to send information to data server. The system requires a backend platform such as ThingPlug.

The data server receives aggressive and other driving information from the information collecting device. The server has a lightweight data exchange interface, such as JSON, which allows other applications to request and utilize the information received by the data server. The device can also transmit information directly to the target server being used by other applications.

The driver's smartphone receives driving information analyzed from the information collection device in real time using the Bluetooth interface of the smartphone. This allows the driver to be alerted to dangerous driving information in real time, and to expect immediate driving behavior correction through an alarm.

3.2 In-Vehicle data extraction algorithm

Two approaches can be used to obtain vehicle operation data from the in-vehicle CAN network. The first method is to directly communicate with internal electronic control units (ECUs) to acquire data, and the second is to monitor and analyze the communication between ECUs in the CAN bus contained in the OBD-II terminal of the vehicle. The first method is not suitable for practical applications as parts of the vehicle must be disassembled prior to connection. Many studies have already used the second approach successfully.

Representatively, Lim [5], acquires data through a multi-level filtering-based method of communicating data in the CAN bus. The first phase is the candidate collection phase, which collects all packets on the CAN bus for a specified period. The collected CAN packets are defined as the candidate data. During this period, packets are collected repeatedly for all states of the target vehicle. For example, in the gear position data, there are six states of parking gear, reverse gear, neutral gear, forward gear, second gear, and first gear based on a typical automatic transmission vehicle. That is, in this stage, every gear position is collected in all packets in the CAN bus for the collection period. The second phase is a data filtering phase, in which a packet indicating an operation state is selected from among the packets collected in the previous step. The data filtering phase consists of several filtering steps. In invariant filtering, duplicate packets among the collected packets in the same state are filtered. In distinct filtering, packets are compared for each state searching for packets having the same header information and different payload data. While a lot of operating information can be collected in the two steps above, some functions, such as turn indicators, need a further filter, such as a stripe filter. The turn indicator uses the flashing feature to turn on and search for packets with a 1:1 ratio of on and off.

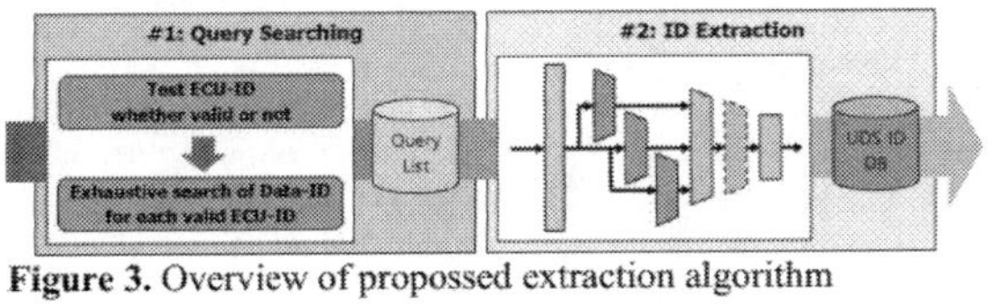

Figure 3. Overview of propossed extraction algorithm

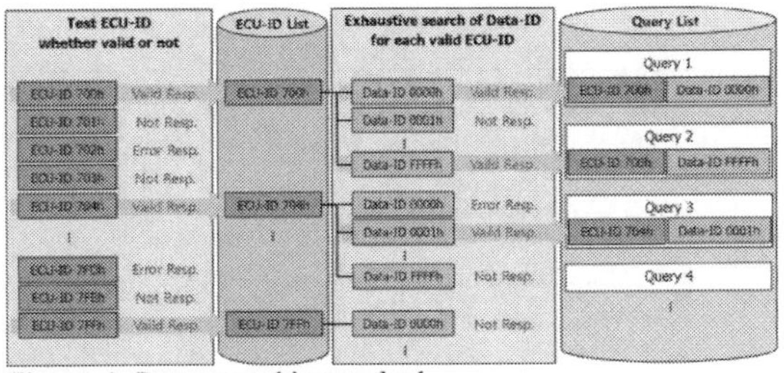

Figure 4. Query searching method

When applied to a real vehicle, this method has two limitations. First, not all vehicles have the same status for the same function. For example, in the case of a turn signal, some vehicles have one state of off and one of on, and some other vehicles have three states of one of off, one of on-light on, and one of on-light off. This problem can be overcome by employing several types of filtering orders. Secondly, in recent vehicles, unified diagnostic services (UDS) are installed so that communication packets between different ECUs cannot be intercepted via the CAN network connected to the OBD-II. In the case of vehicles with UDS, the existing research technology cannot be used natively. UDS operates in a query-response format, in which the ECU-ID and Data-ID in the query message for each function to be collected must be known in advance. In addition, since the response format is different for each vehicle, a separate filtering algorithm is required for each installation type.

In this study, we design an algorithm with the structure depicted in Figure 3 and propose a method to extract the operation information of the vehicle in which the UDS is in use. The proposed algorithm consists of two stages: query searching and ID extraction.

The query searching step consists of searching for a valid ECU-ID and then finding a valid Data-ID for each ECU-ID. First, the ECU-ID is searched within the range of 700h to 7FFh using the ECU test message provided by the UDS standard. Sending a query with a valid ECU-ID returns a valid response, otherwise there is either no response or an error response. The next step works similarly. As the valid ECU-IDs are known from the previous step, we put all possible Data-IDs for each case into the query message for each ECU-ID and observe the response. Data IDs have a total of 65,536 cases between 0000h to FFFFh. After testing every case, the ECU-IDs and corresponding Data-IDs can be used to create a valid query message data set. Figure 4 illustrates the process of query searching. The ID extraction step has some similarities to previous studies' methods as shown in Figure 5.

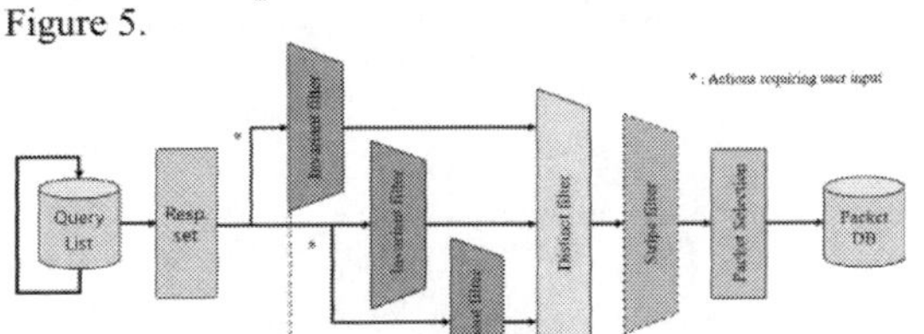

Figure 5. ID extraction method

Table 1. Driver behavior detection item list including aggressive driving

Name	Description
High RPM	Times the engine RPM rises above threshold
Max speed	Highest speed value while driving
Min/Max battery voltage	Min/Max value of battery voltage while driving
Min/Max coolant temp.	Min/Max value of coolant temperature while driving
On/Off start-up time	Start-up on/off time
Idle count/time	Idle driving count and time
Rapid acceleration/departure	Rapid acceleration & departure
Rapid deceleration/stop	Rapid deceleration & stop
Odometer	Odometer value
Fuel consumption	Average fuel consumption

The query message dataset collected in the query searching step is repeatedly transmitted. After that, using the algorithm derived in previous research, the response packet is collected and filtered after a status change of the object being observed.

3.3 Aggressive driving detection method

There are 16 data points of operational information analyzed by the system proposed in this paper and they are shown in Table 1. Risky driving behaviors such as rapid acceleration, rapid start, sudden deceleration, sudden stop and sudden turn are analyzed continuously. However, the information indicating the highest / lowest are only the highest / lowest peak values during the driving period between starting and stopping the vehicle.

Therefore, the proposed data collection device should be able to detect when the vehicle is started and when it is turned off. The device should also supply time stamps of the events being recorded. Conversely, the device can detect the start-up is period is over when no more packets are sent to the CAN bus. So, when the device recognizes that it has turned on, it initializes the high and low information measurements and updates the high and low values until it is turned off.

Figure 6. Prototype of proposed data collection device

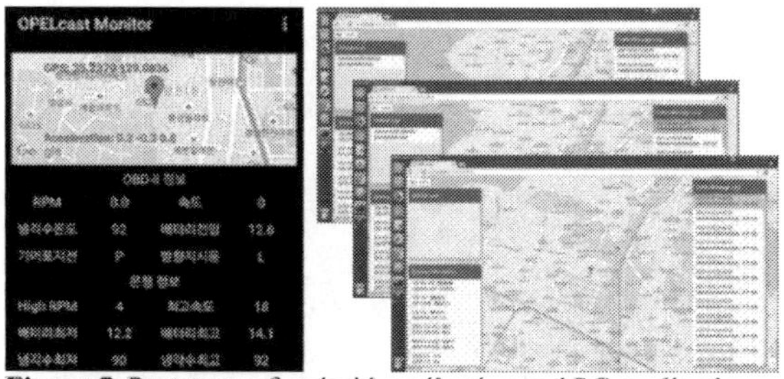

Figure 7. Prototype of android application and PC application

Dangerous driving behavior was designated by referring to the criteria of 11 dangerous driving behaviors provided by Korea Highway Traffic Authority. Rapid acceleration is considered to occur when the vehicle accelerates 8 km/h per second at the speed of 6.0 km/h or more. Rapid departure is defined as accelerating more than 10 km/h per second starting from the speed below 5.0 km/h. Rapid deceleration is when the speed is decelerated by more than 14 km/h and the speed is more than 6.0 km/h. A sudden stop is defined as the case when the speed decreases by more than 14 km/h per second and becomes less than 5.0 km/h. In cases of rapid rotation, the speed must be over 30 km/h and the cumulative rotation angle is between 60 ~ 120 degree within 3 s.

To detect these conditions, acceleration data is obtained via the acceleration sensor of the collecting device, speed data information is obtained from the OBD-II inside the vehicle, and cumulative rotation angle data is calculated using the steering wheel data of the OBD-II. In general, since the steering wheel ranges from -540 degrees to 540 degrees, the cumulative rotation angle information for 3 s can be calculated using this.

In the case of mileage information, it is simply calculated using time and speed data inside the vehicle, and fuel consumption is calculated by using the technique described by Park [6] based on the mass air flow (MAF) data of the vehicle.

4 Experimental results

In this paper, we verify the operation of the proposed system and evaluate the communication performance and the performance of the in-vehicle information extraction algorithm by experiment. To this end, we designed and implemented a device for collecting information. The MCU of the device was the STM32F429 based on Cortex-M4 and equipped with an MCP2551 transceiver for CAN communication. SoluM's TLS01CS1 was installed as a LoRa communication module and a FB153BC was used as a Bluetooth communication module. An LIS331 was used for the acceleration sensor. Figure 6 is a picture of the prototype device. In addition, an Android application for driver notification and monitoring services was developed, and a data server and control application was designed and based on Ubuntu 16.04 as shown in Figure 7.

The system allowed real-time acquisition of dangerous driving indicators such as rapid acceleration, rapid start, sudden deceleration, sudden stop and sudden turn, as well as driving information such as the number of times the

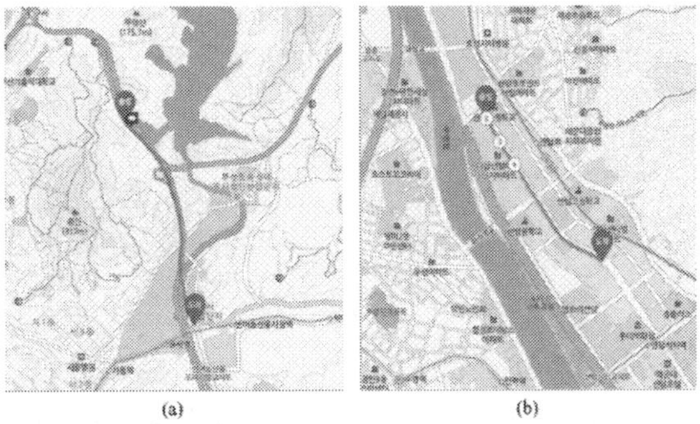

Figure 8. Driving course of experiment; (a) highway, (b) urban

vehicle idled, running time, mileage, driving during driving, and LoRa communication. The performance of the system was evaluated by transmitting the data to the data server.

The experiment was conducted on two real roads and the driving course is shown in Figure 8. Each driving course consisted of a highway environment and a city road environment and was completed in about 10 minutes. Four male and four female drivers in their twenties and thirties with a year or more of driving experience participated. Data was collected during 10 repetitions of driving each course by each participant.

Figure 9 is a confusion matrix for the success rate of detecting rapid acceleration, rapid start, rapid deceleration and sudden stop information. Events were detected at 1 s and 1.5 s intervals, respectively.

In the case of 1 s period, the recognition rate of the quick start event was 85%, then in the case of the 1.5 s detection period, it was confirmed that there were no misrecognition events. It was found that the detection success rate varies depending on the detection period and considering that the recorded events are affected by the acceleration change amount, it is necessary to adjust the recognition period according to the acceleration/deceleration performance of the vehicle. In the case of a rapid turn event, it was confirmed that there is no misrecognition because it uses directly measured steering information from the vehicle operation information.

As a result of the experiment, it is confirmed that there is an average error of 2.0%, which is small enough to be acceptable considering the error of the mileage value in the instrument panel is considered as a ground truth. The error rate was calculated by setting the measurements displayed on the dashboard of the vehicle as ground truth. It was also confirmed that no errors occurred in the detection and recording of the number of idling periods and time. The driving time was also calculated based on the start on / off time and confirmed that there was no significant error in its measurement either.

We evaluated the communication performance by examining the operating information which was sent to the control data server through LoRa communication. In the experiment, the information packet was transmitted every 10 s while driving on the course of Figure 8.

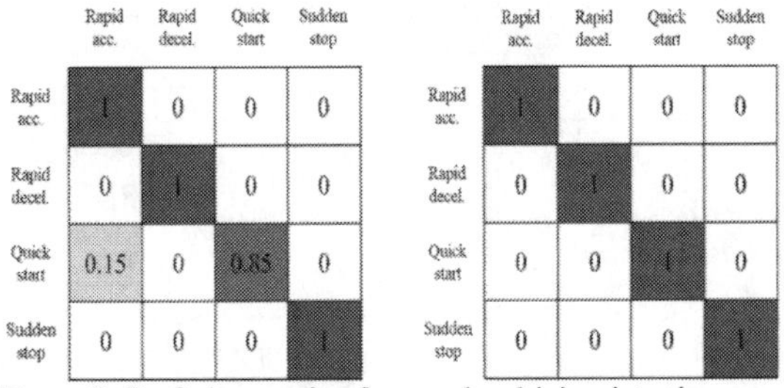

Figure 9. Confusion matrix of aggressive driving detection

Tx power was set to 14 dBm with a spreading factor of 12. Up to 65 transmission attempts were made, depending on travel time. Experiments were carried out with different antenna types, and the results are shown in Table 2. We confirmed that the Mono antenna performed best in the highway environment. However, we found that coil antennas have the lowest PER on city roads.

In order to evaluate the performance of the in-vehicle extraction algorithm proposed in this paper, a total of 14 vehicles were tested. Seven parameters of the internal information were examined including vehicle speed, engine RPM, steering angle, accelerator pedal depth, brake pedal depth, gear position and integration distance.

The experimental results are shown in Table 3. The section labelled UDS in Table 3 is data that could be obtained by the extraction algorithm proposed in this paper, and not by the existing extraction algorithms. For Hyundai, Kia and BMW, it is often found that information is needed through UDS on the vehicles manufactured after 2017. It was observed that the steering angle and integration distance information of some models were difficult to obtain. For this reason, the query message data set could not be saved for these measurements due to insufficient memory of the embedded device. Since the algorithm proposed in this paper conducts exhaustive search, a limitation exists in that all messages cannot be saved when the number of valid messages which must be retained are too large. It is also confirmed that the ECU-ID and Data-ID pair deduced through this extraction algorithm are not specified for the integration distance information of Renault SM6 vehicle. Finally, it is confirmed that an extraction time between 6 to 16 hours is required depending on the vehicle make/model and the information to be acquired when using the proposed algorithm.

Table 2. Evaluation results according to antenna type

		PER (%)	RSSI (dBm)	SNR (dB)	Data Rate (bps)
High way	Mono	21.4	-89.2	27.1	297.9
	Coil	25.0	-92.2	22.6	177.6
	Chip	28.6	-104.6	13.7	178.0
	Patch	23.1	-87.2	29.3	369.6
City	Mono	20.0	-97.2	10.0	198.9
	Coil	15.0	-101.2	7.5	236.3
	Chip	30.0	-96.1	17.1	160.9
	Patch	11.1	-96.7	0.6	167.4

Table 3. Evaluation results of data extraction algorithm

	Spd.	RPM	Accel.	Steer	Brake	Gear	Odo.
Avante	O	O	O	O	O	O	O
Granduer	O	O	O	O	O	O	O
Santefe	O	O	O	O	O	O	O
Starlex	O	O	O	UDS	O	O	UDS
EQ 900	O	O	O	UDS	UDS	UDS	UDS
Morning	O	O	O	O	O	O	O
K3	O	O	O	O	O	O	O
K5	O	O	O	O	O	O	O
Carnival	O	O	O	O	O	O	O
Spark	O	O	O	O	O	O	O
Cruse	O	O	O	X	O	O	O
Malibu	O	O	O	O	O	O	O
SM6	O	O	O	O	O	O	UDS
Tivoli	O	O	O	O	O	O	X
430i	O	O	O	O	X	O	X

5 Conclusion

In this paper, we designed and implemented a system that can extract driving information including dangerous driving behaviors in real time and transmit it to a data server through LoRa communication in real time. We also propose an internal information extraction algorithm that supports the UDS protocol to obtain the vehicle's interior information.

The performance of the proposed system was evaluated through experiments on real roads. It showed that the driving information can be detected in real time, and the information can be transmitted to the data server through LoRa communication. It was confirmed that the recognition performance of some operation information differs depending on the measurement cycle duration.

We verified the proposed algorithm for extracting the in-vehicle information through various methods and discovered difficulties in the extraction due to issues encountered such as insufficient memory of the prototype device for some of the tested vehicles. We also identified the limitations of the extended time required for extraction. In the future, we plan to reduce the extraction time and the size of the query message set.

Acknowledgment

This work was supported by the Center for Integrated Smart Sensors funded by the Ministry of Science, ICT & Future Planning as Global Frontier Project(CISS-2011-0031863)

References

1. J. Lee, K. Jang, A framework for evaluating aggressive driving behaviors based on in-vehicle driving records, *Transportation Research Prat F: Traffic Psychology and Behaviour*, **65**, 640-619, (2019)
2. D. Chen, K. Cho, K. G. Shin, Mobile IMUs reveal driver's identity from vehicle turns, arXiv preprint, arXiv :1710.04578, (2017)
3. D. Han, B. Kim, T. Kim, Y. Kim, Y. Baek, Design and implementation of BLE communication-based vehicle information recording and monitoring system, *The Proceedings of Korea Information and Comm unication Society*, **2018**(6), 36-37, (2018)
4. F. Martinelli, F. Mercaldo, A. Orlando, V. Nardone, A. Santone, K. A. Sangaiah, Human behavior characterization for driving style recognition in vehicle system, *Computers & Electrical Engineering*, (2018)
5. J. Lim, D. Han, Y. Jeon, Y. Baek, ID extraction OBD-II vehicle operation for driver behavior analysis, *The Proceedings of Korea Information and Communication Society*, **2018**(1), 100-101, (2018)
6. J. Park, K. Jung, K. Um, Estimation of fuel consumption vehicle diagnosis data, Journal of the Korea Institue of Information and Communication Engineering, **15**(12), 2582-2589, (2011)

An Overview of the Latest Progress and Core Challenge of Autonomous Vehicle Technologies

Zongwei Liu[1,2], Hao Jiang[1,2], Hong Tan[1,2] and Fuquan Zhao[1,2,a]

[1]*State Key Laboratory of Automotive Safety and Energy, Tsinghua University, Beijing 100084, China*
[2]*Tsinghua Automotive Strategy Research Institute, Tsinghua University, Beijing 100084, China*

Abstract. The mass production of autonomous vehicle is coming, thanks to the rapid progress of autonomous driving technology, especially the recent breakthroughs in LiDAR sensors, GPUs, and deep learning. Many automotive and IT companies represented by Waymo and GM are constantly promoting their advanced autonomous vehicles to hit public roads as early as possible. This paper systematically reviews the latest development and future trend of the autonomous vehicle technologies, discusses the extensive application of AI in ICV, and identifies the key problems and core challenges facing the commercialization of autonomous vehicle. Based on the review, it forecasts the prospects and conditions of autonomous vehicle's mass production and points out the arduous, long-term and systematic nature of its development.

1 Introduction

The fully interconnected and highly intelligent ICV industry ecology is emerging. As one of the most important manifestations of ICV, autonomous vehicle technology has made rapid progress in recent years. Automotive enterprises, parts and components enterprises, IT companies, and scientific institutions continue to increase investment in autonomous driving research, some of which have also proposed timetables for its commercialization. For example, Waymo was approved in January 2018 to operate as a Transportation Network Company (TNC) in Arizona and launched self-driving ride-hailing service in December. General Motors plans to roll out their public ride-hailing services with self-driving vehicles starting in 2019. And Uber also launched the autonomous taxi test service in San Francisco in March 2018 [1-2].

Meanwhile, technology is still the main bottleneck of autonomous vehicle ' s commercialization. Generally speaking, the technology of autonomous driving involves three aspects: perception, decision, and action, the connotation and relationship of which are shown in Figure 1. Based on the current status of autonomous vehicle technologies, this paper analyzes the latest development and future trend of them and then discusses the extensive application of AI in ICV area and the key problems and core challenges of the comprehensive commercialization. Following that, it points out the key points that the industry and academia must tackle to provide reference to the R&D of autonomous vehicle products and technology in the future.

2 Perception

2.1 Environment sensing

2.1.1 Automotive radar

Radar is an electronic device that detects targets by electromagnetic wave. It can be divided into narrow band and wide band by bandwidth and the larger offers higher range resolution. 24 GHz and 77 GHz are two dominant bands: the former is for short and medium range detection of 5-70m and the latter is mainly for medium and long range detection of 100-250m. With the improvement of accuracy requirements of automotive radar system, 77 GHz with ultra-high frequency band will become the mainstream. Radar can also be divided into short-range radar (SRR), medium-range radar (MRR) and long-range radar (LRR) by covered area. SRR has higher range resolution and wider field of view (FoV $\pm 65° \sim \pm 80°$), which is used for collision avoidance and object detection, and its typical applications include blind spot surveillance, parking assist and ACC support. MRR (typical FoV $\pm 40° \sim \pm 50°$) is used for LCA; LRR has high gain and narrow beam for long detection range and accurate direction measurement, with smaller FoV around $\pm 10°$ and more range resolution tolerance, which is mainly used for ACC, FCW, and AEB [3-4]. Key radar sensor manufacturers in the world include Bosch, Delphi, Continental, Denso, Valeo, ZF-TRW, etc. Freescale, Infineon, ST, Texas Instruments and ADI are the key chip suppliers.

[a] Corresponding author: zhaofuquan@tsinghua.edu.cn

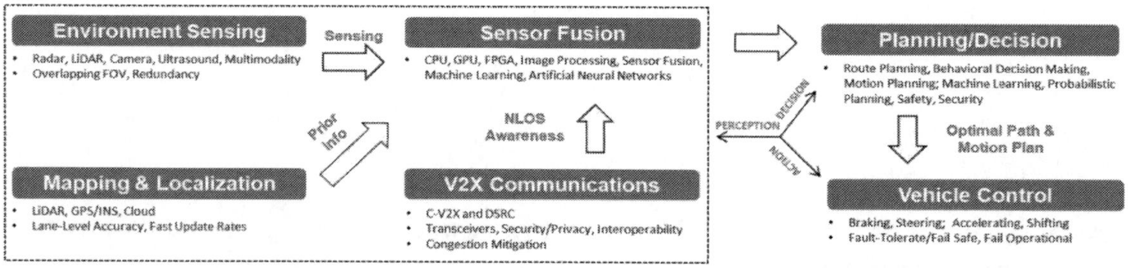

Figure 1. Classification, Connotation and Interrelation of Autonomous Driving Technologies.

Automotive radar has the following technology development trends in order to satisfy the autonomous driving requirements of high range resolution, high angular resolution, high distance /speed/angle accuracy, long detection range, wide opening angle, high ASIL, low power consumption, low cost and small size. (1) Moving to higher frequency (77 ~ 81 GHz): High-resolution millimeter-wave radar that operates in the 79 GHz band is expected to replace 24 GHz SRR completely in the future, with combination of high transmit power and high bandwidth allowing for simultaneous long range and high range resolution, smaller antenna size for a given bandwidth, better angular resolution, and smaller sensors reducing volume/weight related costs. (2) Chip technology roadmap: GaAs has high output power, high gain, very low noise, good linearity, but low integration level. SiGe has an adequate performance at mmWave bands, size and performance advantages, high temperature tolerance. CMOS is cheaper, allowing for higher levels of integration, lower power consumption. It is expected that chips will develop along the route from GaAs to SiGe and then to CMOS. (3) Bistatic over monostatic: Bistatic Radars have higher transmitter output power, better noise performance, and improved conversion gain of the receiver. (4) Multi-mode sensors: Multimode sensors usually support both MRR and LRR configurations in one sensor module, thus providing a significant cost reduction to automakers.

2.1.2 LiDAR

LiDAR is a kind of integrated optical detection and measurement system working in optical frequency band, current sensor performance of which is range 100~300 m, accuracy ±2.0~±5.0 cm, FoV 120~360° (horizontal), 25~120 ° (vertical), angular resolution 0.1~0.4 ° (horizontal), 0.4~2.0° (vertical). The advantages lie in its ability to create high-resolution 3D image, mid-range, accurate depth information and target discrimination, wide FoV, high angular resolution, independent of light conditions, and to detect lane markers and street signs, but it cannot transmit in rain, snow, fog or dust, and at present it cannot meet the performance, size and cost requirements of mass-produced autonomous vehicles at the industrial level simultaneously. Typical applications include mapping & navigation, obstacle detection and in-cabin monitoring.

LiDAR can be divided into mechanical scanning and solid-state LiDAR according to the presence or absence of mechanical rotating parts. Mechanical LiDAR has rotating parts to control the angle of laser emission to achieve wide HFoV 360° , while mass production is not feasible due to insufficient durability [5], large volume and relatively expensive price of mobile parts. Solid-state LiDAR relies on electronic components to control the angle of laser emission. Because of the narrow HFoV, multiple sensors are required for 360° view. However, its integrated design is more durable, compact and cost-effective. Therefore, solid-state is one of the most important trends in this field [6-7].

Solid-state LiDAR, including OPA, flash and MEMS, is the key technology to autonomous driving, in which area bigger vendors are working to make sensors cost-effective and startups continue to pop up around the world, some could emerge as industry leaders, given enough time and money. Such as Quanergy, which was invested by Aptiv in 2015, is in the development of OPA LiDAR; ASC, which was acquired by Continental in 2016, is committed to the development of 3D flash LiDAR and the products are expected to be available in two to three years; in addition, MEMS LiDAR will also be put into production, and Valeo is cooperating with LeddarTech in the development of related products.

2.1.3 Camera (Image sensor)

Camera converts an optical image into electronic signals, including CCD and CMOS. CCD transfers each pixel' s charge packet sequentially to a common output structure to convert charge to voltage, with less noise, generally higher quality images particularly in low light conditions, high dynamic range, better depth of color, higher resolution and light sensitivity; Charge-to-voltage conversion of CMOS takes place in each pixel. It has the advantages of lower power consumption, high integration, smaller size, easy manufacture and lower price, improved quality since its adoption.

Automotive camera can be divided into front view cameras, surround view cameras, rear view cameras and in-cabin cameras. Key tier 1 camera manufacturers include Bosch, Aptiv, Continental, Denso, Valeo, etc. Key image sensor suppliers are Infineon, Freescale, Texas Instruments, ST, Panasonic and so on. Key optics suppliers are Sony, Panasonic, Sharp, Konica Minolta, Omron and other enterprises.

The development trends of automotive camera include range improvement, modularization, scalability and interconnection, side or surround view cameras to expand rear view scope and replace side mirrors, in-cabin ToF camera (or 3D Flash LiDAR) for presence detection & gesture recognition and all-weather sensing (time-gated sensors). All-weather sensing needs night vision solution , in which regard IR camera is an important solution to increase driver seeing distance in darkness/poor weather and make up for the fatal weakness of the camera that can not work properly under low illumination [8]. It ' s currently only offered in luxury brands due to the high cost. Autoliv is the key system supplier. IR cameras are mainly divided into two categories: far infrared (FIR, passive, e.g. BMW Night Vision System) and near infrared (NIR, active, e.g. Mercedes Night Vision System).

2.1.4 Ultrasonic sensors

Ultrasonic sensors have a close detection range (0.15-6m), excellent very-near-range 3D-mapping and repeatable linear response, high precision & resolution due to slow travel of sound wave, and work equally well in snow, fog, rain, and dust due to short range. They are small and cheap as well, however not useful for gauging speed due to short range, and susceptible to temperature fluctuations, humidity, wind and EMI. So sensors need temperature and humidity compensations and proper shield. Its main applications include space detection, parallel parking, reverse parking, blind-spot detection, kick-to-open liftgates. In this field, Bosch Gen 6 Ultrasonic Sensors are the representatives, which are available in three mechanically compatible sensor variants.

2.2 Mapping & Localization

Based on the mapping by LiDAR and other systems and localization technologies such as GPS/INS and cloud, vehicles can obtain prior information beyond visual range and ensure redundancy to a certain extent.

The development of autonomous vehicles requires high-resolution map, which needs the vehicle coordinates with higher accuracy, more detailed information (lane markings, street signs, traffic signals, potholes, height of curb, etc.), and live update in path planning. The functions of HD map is to provide priori information for a level of redundancy, to reduce sensor loads so that sensors can focus on processing moving objects, extreme situation aid when road markings wear away or disappear under snow, precise localization, to identify other dynamic entities such as other cars and pedestrians, and incorporating human psychology into visual display [9].

The collection of HD maps mainly includes fleet survey and crowdsourcing. Some cartographic companies and technology giants adopt the former mode, that is, sending a fleet of probe vehicles to survey selected areas. Its advantages are high efficiency, comprehensive road information acquisition, but the task of data acquisition is arduous, the cost of mapping is high and it is hard to frequently update. Most enterprises adopt the low-cost crowdsourcing, that is, to fit sensing system on OEM cars/trucks to transmit their precise locations and images of the roads they are traveling, and to synthesize captured data into one giant digital map for near-real-time update [10]. At present, there is no formal industry standard for HD maps in the world, and the mapping methods and characteristics of all participants are different. For example, Waymo map paints 3D portraits of the world with the help of LiDAR, and extracts road features manually and using computer algorithms; Here map is based on satellite/aerial imagery and fleet probe data, while using computer vision algorithms and manual labor to extract road features; TomTom map is a detailed depth map, capturing unique pattern of roadway scenery instead of feature extraction, which is insensitive to environment changes; Civil Maps map is a crowdsourced 3D semantic map (machine vision data converted to map vectors), which combines a live video feed of road with map data using AR display.

Localization technologies, including absolute localization, dead reckoning (relative localization) and SLAM, work together to achieve accurate results. GPS is not accurate enough for AVs due to variable radio wave speed, blockage by canyons/tunnels, multi-path errors, radio interference, etc. Odometry and IMU are used for relative positioning. Using wheel encoders, odometry integrates wheel speeds over time to produce incremental vehicle motions and sums the vehicle movements to produce a position estimate. The errors mainly include systematic errors associated with imperfect kinematic vehicle model and wheel base and wheel diameter measurement errors, and random errors due to the slip between wheels and ground, loss of contact between wheels and ground caused by uneven or rocky terrain. IMU, consisting of 3 orthogonal accelerometers and 3 laser ring gyroscopes to provide 3D acceleration measurements and angular rate measurements about each axis, integrates accelerations twice and angular rates once to determine incremental vehicle movements. Compared with the odometry, it is not susceptible to wheel slip, but angle measurements drift with time and errors quickly accumulate. SLAM builds/updates a map of a vehicle's surroundings while keeping the vehicle located within the map. Bayesian filter is a framework to reduce the location uncertainty by using recursive state estimation, which calculates the probabilities of multiple beliefs to allow an AV to infer its position and orientation, and allows vehicles to continuously update their most likely position within a coordinate system, based on the most recently acquired sensor data [11]. LiDAR sensors are mainly divided into two categories: LiDAR and vision, among which the former is more mature.

2.3 V2X communications

Complementary to other sensors, V2X communications support 360° non-line-of-sight (NLOS) awareness and extend a vehicle's ability to see, hear and understand the environment down the road, at blind intersections, or in bad weather conditions, which cover a variety of application scenarios of V2V, V2I, V2P and V2N.

DSRC and C-V2X are two main technical routes of V2X [12]. DSRC is a two-way half-duplex short-to-medium-range (<1 km) wireless communication technology, with high data transmission (3 to 27 Mbps) critical in communications-based active safety applications, which can meet extremely low latency requirement for road safety messaging and control. DSRC has been basically finalized in 2014 after 10 years of R&D and testing and adopted by the US. There are relevant variant standards in Europe and Japan, and DSRC is likely to become mandated in the US for all light vehicles starting in as early as 2021. However, LOS technology imposes challenges in urban conditions where buildings situated at crossroads, and there is a lack of technology evolution path to meet more stringent autonomous vehicle requirements. In contrast, C-V2X is a vehicle wireless communication technology based on the evolution of cellular network communication technologies such as 4G/5G, which works both in-coverage and out-of-coverage (direct communication) and has improved V2I & V2N communication, extended V2V range and better non-line-of-sight awareness. Today's 4G LTE is only suited for non-safety-related use-cases, but a clear technology evolution roadmap exists. 5G will provide scalable connectivity to support an extreme variation of requirements, such as extreme throughput in millimeter wave bands, edgeless connectivity, high reliability (1ms end-to-end latency), high availability, see-through augmented reality and vulnerable road user discovery, thus becoming the key to enabling fully autonomous driving capabilities [13].

2.4 Sensor fusion

Multi-sensor effective fusion for objects detection/classification/tracking, location and V2X communication, i.e. association, aggregation and integration of data, can achieve complementary multimodality-sensing, extended NLOS perception, redundancy/robustness for 360° situational awareness in all conditions, thus ensuring the speediness and correctness of decision-making, which is the inevitable trend of autonomous driving [14]. Sensor fusion is divided into two categories (homogeneous and heterogeneous) and includes four processing layers (signal level fusion, object level fusion, feature level fusion and decision level fusion), which can be realized by Kalman filter, fuzzy logic, Bayesian networks and CNN, etc.

The architectures of sensor fusion include fully distributed processing, centralized processing and hybrid processing. Fully distributed processing requires lower bandwidth and central processing power, but it is larger, requires more expensive sensor modules with application and has a higher functional security requirement. It is hard to synchronize signals across nodes and may lose context information as well. Today, distributed data processing is almost exclusively used to do feature-level and decision-level fusion, which is not enough for L4/5 autonomy due to its limited sense of environment and loss of contextual information. So centralized processing is needed to complete environmental information available, with sensor modules of small size, low cost and lower functional safety requirement, but the central processing power/speed and bandwidth requirements are high and the center is non-scalable. Thanks to the fast development of supercomputers, a centralized architecture is now entirely possible. However, it is difficult to manage multi-supplier coordination on one chip, and there are no common standards for languages, formats, protocols, and networking. Hybrid processing takes both advantages into account, by local processing with intelligent sensors, reducing bandwidth requirement and centralized processing for path-finding, maneuvering, and motion trajectory, which is flexible and scalable. Therefore, the development from distributed processing and centralized processing to hybrid processing has become the main trend of information fusion system for autonomous vehicle, with corresponding requirements for communication and computing power.

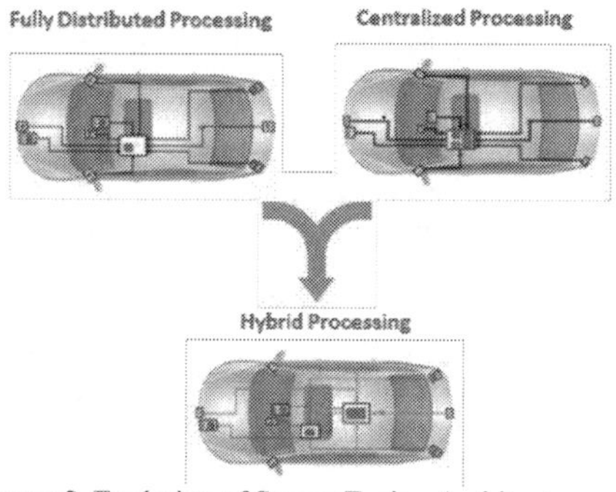

Figure 2. Evolution of Sensor Fusion Architectures.

3 Planning, decision making and vehicle control

The process of planning and decision-making includes route planning, behavioral decision-making and motion planning. Path planning is to select a route from the current position to the destination through the road network. By representing road network as a weighted directed graph corresponding to the cost of traversing a road segment, path planning can be converted into the problem of finding a minimum-cost path on the road network graph. Behavioral decision-making is to select the appropriate driving behavior at a specific time point based on the perceived behavior of other traffic participants (brake lights, turning lights, etc.), road conditions and infrastructure signals. In this process, machine learning based techniques (such as Gaussian mixture models, Gaussian process regression, etc.) and probabilistic planning (such as MDPs and their generalizations) are required to predict intention and estimate future trajectories of other traffic participants and then dictate driving actions. Motion planning is to translate selected driving behavior into a path/trajectory, to ensure dynamic feasibility and efficiency for vehicle, to ensure safety and comfort for passengers, and to avoid collisions with detected obstacles. Numerical approximation methods such as non-linear optimization in function space, graph-search approaches for the shortest path and incremental tree-based approaches to select the best branch are widely used.

The core function of vehicle control is path tracking/trajectory control, which executes the path/trajectory specified by the motion planning system, by selecting appropriate actuator inputs to carry out the planned motion and using a feedback controller to correct tracking errors, and ensures the reference path or trajectory stability in the presence of modeling errors and other uncertainties [15].

In the aspect of E/E architecture, distributed architecture is widely used, under which a large number of ECU units work together to provide various functions for drivers. With the demand for higher level automation (demanding an unprecedented increase in sensor technology, data bandwidth, processing performance, safety and security, etc.) and vehicle connectivity, from distributed processing to centralized processing gradually has become the main development trend of vehicle control, as shown in Table 1. Centralized E/E architecture will achieve higher overall system efficiency, faster system response time, and reduced performance requirements for lower level ECUs, with extremely high performance requirement for the high-level ECU and development/evaluation difficulties. The migration path of this process is firstly to add partial domain controllers to deal with sensor fusion and decision-making, so that more core functional modules are concentrated in the domain controllers, to achieve functional integration and domain fusion, and ultimately to realize a high-level vehicle controller managing overall autonomous strategy, passing decision information to lower level systems.

Table 1. E/E Architecture Development Trend [16].

Short-term: Distributed E/E Architecture	Medium-term: (Cross) Domain Centralized E/E Architecture	Long-term: Vehicle Centralized E/E Architecture
Modular: Each function has its ECU Integration: Functional Integration	Centralization: Central Domain ECUs Domain Fusion: Central Cross Domain ECUs	Vehicle Computer & Zone ECUs: Zone Oriented Architecture and Vehicle Computer Vehicle Cloud Computing: Vehicle functions in the cloud

4 Application and development of artificial intelligence

AI plays a decisive role in achieving autonomous driving and improving driving experience. AI can be divided into three levels according to the tasks it can accomplish: narrow or weak AI, general or strong AI, and super AI. The demand for data computation has been significantly increased from L3 to L5, with the increasing demand for AI. There is still a great controversy about whether or not super AI is achievable and what are its implications to mankind. Therefore, the current considerations mainly focus on narrow AI and general AI.

ANI refers to machine intelligence that implements functions in a limited and fully defined field. Today's AI systems are all ANI, based on which automobile is one of the most complex application cases to realize basic driving assistance functions. AGI refers to the realization of machine intelligence equal to human beings in different broad areas, which can perform comprehensive human perception tasks and is the main goal of AI research at present. As far as autonomous driving is concerned, L4 is facing 5-10 times more computational load than L3 and needs to achieve AGI basically to meet the needs; while L5 has a higher demand and needs to achieve AGI fully to cope with various complex application scenarios. Compared with traditional rule-based automatic driving decision-making, end-to-end deep learning is a possible method for realizing AGI, which gradually acquire the ability of autonomous driving decision-making by learning real driving data without manual processing, splitting and marking. The whole system is a black box of input perception and output decision-making. Its advantage is that the more scene data used in training, the more reliable the system will be. There is no need to mark and annotate scene data accurately, and the system is self-learning as a whole. But its disadvantage is that the decision-making process is not intuitively understood so companies are more comfortable to take traditional rule-based approach to obtain continuous data accumulation and capacity enhancement. In the future, the combination of deep learning with supercomputers will eventually bring self-driving cars to reality by training the system with a large amount of data.

AI has broad and important application prospects in vehicle safety and personalization, convenience/virtual assistance, besides autonomous driving functions. For example, fingerprint, iris, face and voice recognition can be accomplished by means of in-car optical scanner, NIR or visual camera, camera, audio transducer, to realize security & personalization like vehicle entry control, ignition switch control, mirror, seat, climate and music setting. Vital tracking and face/head/gaze TRK are carried out by cameras and on-board/wearable monitors for driver drowsiness/distraction/emotion monitoring and driver/passenger health condition monitoring. Voice/gesture activated HMI based on acoustic/language modeling, EEG analysis and other means, lip reading in noisy environments, and faster response with direct brain-to-vehicle interaction, will also be applied in ICV.

5 Key issues and core challenges

5.1 The utmost concern of Self-Driving vehicles – safety

With the great progress of sensor technology and computing power, safety is always the primary demand of ICV and the most concerned problem of autonomous vehicle.

It is necessary to incorporate safety into the design, to test and verify strictly, and to iterate between the two to achieve safe driving. To incorporate the safety into design, we should follow the systems engineering approach, adopt appropriate engineering tools such as FTA, FMEAs and HAZOP, adopt adaptive E/E architecture for service-oriented cloud services and OTA updates, and protect cybersecurity, including individual ECU protection, domain separation, internal network protection, intrusion detection, and gateways, firewalls, message authentication. Redundancy should be provided as well, such as redundant critical systems including computing, braking, steering, power source, communication path, collision detection and avoidance system and diagnostic monitoring, IMU and LiDAR for localization, multi-modality sensors of overlapping views. System self-detection, such as redundant system cross-detection. System self-check is also necessary, such as redundant system cross-check. The test/validation process includes not only conventional and SOTIF validations, but also the simulation/proving grounds/public road testing. There are many challenges, such as the driver is out of loop, the requirement is not always known, correct and complete or unambiguously specified, ISO26262 is not fully applicable, and behavior of probabilistic systems with non-deterministic algorithms is not repeatable even for identical initial conditions. It is also hard to verify inductive learning systems even with comprehensive training data, and complete testing is not feasible.

In addition, to ensure the safety of autonomous driving, technical support (MFG quality, cyber-security measures, rail-operational/fail-safe, HMI, etc.) and government laws are also needed, which are far more complex than traditional automobiles. Only by doing a good job in all aspects of the system engineering, can we finally win the public acceptance.

5.2 The real challenge -Edge cases handling

The edge case is a problem or situation that rarely occurs but can occur at an extreme operating parameter for autonomous vehicle. This is a special case usually involving input values that need special processing in program algorithms. In order to validate program behavior in this case, unit tests are needed to test the boundary conditions of algorithms, functions or methods. The edge case of autonomous driving involves handling unconventional events, including extreme weather conditions, other traffic participants' unpredictable behaviors, emergency vehicles, special road signs, etc. For example, how to identify pedestrians in wheelchairs on the sidewalk, how to judge whether children are chasing the ball behind when identifying football, and so on.

The extreme pursuit of safety makes edge cases especially challenging. Object classification and decision making rely on machine learning algorithms, the effectiveness of which is directly related to data training. However, these edge cases don't happen frequently in real life and are therefore difficult to gather data on, so specialized training is required and it's time-consuming. What is more serious is that the extreme nature of edge cases may lead to millions of combinations that algorithms have not learned, which will inevitably lead to potential safety hazards [17]. Compared with autonomous driving, human drivers can easily handle these edge cases by sending subtle signals to each other (such as common social interaction used in road negotiation), however, even AGI can hardly fully understand and deal with these situations correctly. Waymo is at the forefront of the industry in this respect, and its test mileage has exceeded 20 million kilometers, so its algorithm system has been better trained, which also explains why its vehicles' disengagement rate is significantly lower than competitors.

6 Conclusion

This study systematically analyzes the latest progress and trend of autonomous driving perception, decision and action technology, combined with the research on the application status and prospect of AI in ICV and the key problems and core challenges of autonomous driving, and makes a comprehensive judgment on the development of autonomous vehicle technologies. Autonomous vehicle technologies have made considerable progress that companies are comfortable to roll out L4 self-driving vehicles with careful design considerations including hardware and software redundancy, fail-operational/fail-safe measures, cybersecurity mechanisms, geo-fenced operation, remote support, etc.. Meanwhile, the nature of inductive learning systems makes autonomous vehicles hard to validate even with comprehensive sets of training data, not to mention in the endless edge cases where data is not easily or feasibly collected. Therefore, the first batch of L4 vehicles must be operated in their ODD, which defines the safe operating conditions including geographies, roadway types, speed range, environmental conditions, and state and local traffic laws. Looking forward to the future, the ODD will gradually be expanded as autonomous vehicle development progresses, meanwhile the leap from L4 to L5 is still a great challenge, because the core technology of super nonlinear enhancement, the huge amount of data required for AI training and the new transportation system matching with the autonomous vehicle are all waiting for a long time. In this sense, it may take years or even decades for us to see millions of self-driving cars running in unconstrained environments. The development of automobile technology is not only a problem of the automobile itself, but also a comprehensive consideration from the perspective of ITS and intelligent city in the future [18].

Acknowledgment

Resrach supported by National Natural Science Foundation of China (Grant No. U1764265).

References

1. Z. Liu, Fuquan Zhao's Insights on Automotive Industry (Vol. 1). China Machine Press: Beijing, China, 2017.
2. X. Kuang, F. Zhao, H. Hao, and Z. Liu, "Intelligent connected vehicles: the industrial practices and impacts on automotive valuechains in China," Asia Pacific Business Review, vol. 24, no. 1, pp. 1-21, 2018.
3. C. Ilas, " Electronic sensing technologies for autonomous ground vehicles: A review," 2013 8th International Symposium on Advanced Topics in Electrical Engineering (ATEE). IEEE, pp. 1-6, 2013.
4. M. E. Russell, A. Crain, A. Curran, R. A. Campbell, C. A. Drubin, and W. F. Miccioli, "Millimeter wave radar sensor for automotive intelligent cruise control (ICC)," 1997 IEEE MTT-S International Microwave Symposium Digest. IEEE. vol. 3, pp. 1257-1260
5. Velodyne LiDAR, " Puck real-time 3D LiDAR sensor, " Velodyne LiDAR, Inc., Morgan Hill, California, http://velodynelidar.com/docs/datasheet/63-9229_Rev-E_Puck%20_Spec%20Sheet_Web.pdf (2017).
6. R. Thakur, "Scanning LIDAR in Advanced Driver Assistance Systems and Beyond: Building a road map for next-generation LIDAR technology," IEEE Consumer Electronics Magazine, vol. 5, no. 3, pp. 48-54, 2016.
7. S. R. Davis, S. D. Rommel, D. Gann, B. Luey, J. D. Gamble, M. Ziemkiewicz, and M. Anderson, " A lightweight, rugged, solid state laser radar system enabled by non-mechanical electro-optic beam steerers, " In Laser Radar Technology and Applications XXI, International Society for Optics and Photonics, vol. 9832, pp. 98320K, 2016
8. J. E. Källhammer, "Night vision: Requirements and possible roadmap for FIR and NIR systems, " Photonics in the Automobile II, International Society for Optics and Photonics, vol. 6198, pp. 61980F, 2006
9. H. G. Seif and X. Hu, "Autonomous Driving in the iCity — HD Maps as a Key Challenge of the Automotive Industry," Engineering, vol. 2, no. 2, pp. 159-162, 2016.
10. K. Massow, B. Kwella, N. Pfeifer, F. Häusler, J. Pontow, I. Radusch, J. Hipp, F. Dölitzscher and M. Haueis, " Deriving HD maps for highly automated driving from vehicular probe data, " In 2016 IEEE 19th International Conference on Intelligent Transportation Systems (ITSC), IEEE, pp. 1745-1752, 2016.
11. H. Durrant-Whyte and T.Bailey, " Simultaneous localization and mapping: part I," IEEE robotics & automation magazine, vol. 13, no. 2, pp. 99-110, 2016
12. Z. Liu, X. Kuang and F. Zhao, " A survey on applications and development of V2X key technologies," Telecommunication Engineering, vol. 59, no. 1, pp. 117-124, 2019. [in Chinese]
13. S. Chen, J. Hu, Y. Shi, Y. Peng, J. Fang, R. Zhao and L. Zhao, "Vehicle-to-everything (v2x) services supported by LTE-based systems and 5G," IEEE Communications Standards Magazine, vol. 1, no. 2, pp. 70-76, 2017.
14. A. Ziebinski, R. Cupek, H. Erdogan, and S. Waechter, "A Survey of ADAS Technologies for the Future Perspective of Sensor Fusion," International Conference on Computational Collective Intelligence, Springer International Publishing, pp. 135-146, 2016.
15. B. Paden, M. Čáp, S. Z. Yong, D. Yershov and E. Frazzoli, "A survey of motion planning and control techniques for self-driving urban vehicles," IEEE Transactions on intelligent vehicles, vol. 1, no. 1, pp. 33-55, 2016.
16. Bosch, " E/E-ARCHITECTURE IN A CONNECTED WORLD. " [Online]. https://www.asam.net/index.php?eID=dumpFile&t=f&f=798&token=148b5052945a466cacfe8f31c44eb22509d5aad1
17. P. Lin, "Why ethics matters for autonomous cars," Autonomous Driving. Berlin: Springer, pp. 69-85, 2016.
18. F. Zhao and Z. Liu. " The Analysis of Strategic Values, Advantages and Disadvantages for China to Develop Intelligent Vehicles, " Modern Economic Research, no. 4, pp. 49–53, 2016. [in Chinese]

Hands on Wheel Classification Based on Depth Images and Neural Networks

Jan-Christoph Schmitz[a], Stephan Tilgner, Kathrin Kalischewski, Daniel Wagner and Anton Kummert

University of Wuppertal, School of Electrical, Information and Media Engineering, 42119 Wuppertal, Germany

Abstract. This paper describes a system to automatically observe if the driver has his hands on the wheel, which is important to know that he can intervene if necessary. To accomplish this an artificial neural network is used, which utilizes depth information captured by a camera in the roof module of the car. This means that the driver and the steering wheel are viewed from above. The created classification system is described. It is designed to require as little computational effort as possible, since the target application is on an embedded system in the car. A dataset is presented and the effect of a class imbalance that is incorporated in it is studied. Furthermore, it is examined which part, i.e. the depth or the intensity image, of the available data is important to achieve the best possible performance. Finally, by examining a learning curve, an experiment is made to find out whether the recording of further training data would be reasonable.

1 Introduction

As car automation progresses in the direction of autonomous driving, the driver is supported ever further. In an increasing number of situations cars can drive autonomously, which most probably will develop in a smooth transition to a point where the driver only has to intervene in certain rare situations, and ultimately no longer at all.

Especially in this transition period it is important that the vehicle's control unit has information about the driver's condition and activities, as it is particularly difficult for people to be alert over a longer period of time if they are not strongly challenged [1]. Knowing whether the driver has his hands on the steering wheel can be used in situations where the driver's attention is required. For example, autonomous driving should not be switched off if the driver cannot take control.

This work deals with the processing of image data from a depth camera, which is directed into the interior of the vehicle and is located in the roof module, using artificial neural networks. The aim is to find out whether it is possible to use this technique to classify if the driver has his hands on the steering wheel or not and to build a working system if possible. This includes finding out what part of the available data is needed for good performance. Due to the target application on an embedded system in the car, which limits the available computing power, rather small neural networks need to be used.

2 Related work

A lot of work is devoted to monitoring the driver's condition in order to further improve road safety or the user experience.

In one fundamental study, possibilities are considered to improve vehicle safety by means of image processing techniques [2]. For this purpose, both the environment as well as the condition of the vehicle and the driver himself are taken into account. Therefore, for example, the occupant position and posture, the intentions of the driver and the surrounding, e.g. the lanes and obstacles such as other vehicles, are analysed. Since the observation of the driver is an important aspect, [3] provides an overview of various studies dealing with the evaluation of the driver's condition, his behavior and the prediction of his intentions. In [4] this overview is extended by more recent research work, although the listed collection does not concentrate exclusively on the driver himself, but also contains studies which consider people outside the car.

Due to the low costs and the possibility to use depth image data, many studies rely on the *Microsoft Kinect* camera. A comparison of different machine learning methods concludes that artificial neural networks have advantages over other methods when it comes to identifying activities, such as using a mobile phone, on the basis of data obtained from the Kinect [5].

As a further research shows, the hands are a crucial feature to identify human activities [6]. In their study, only image sections around the hands are passed to a recurrent neural network which performs the evaluation.

In [7] the behavior of the driver is analysed on the basis of his hand activities, which includes the interaction with

[a] Corresponding author: jaschmitz@uni-wuppertal.

the steering wheel. Color and depth images are used, which are divided into different regions, such as the steering wheel or the gear lever. A Support Vector Machine distinguishes the different activities by recognizing in which of these regions the driver's hands are active. Occlusions of the hands are a serious problem, which is addressed in [8].

[9] uses the tool *XMOB*, which is presented in [10], to get the hand positions. If these are in a previously marked region, it is decided that the particular hand is on the steering wheel.

Of particular interest is [11]. Besides the question if the driver uses a mobile phone, the number of hands on the steering wheel is determined by artificial neural networks. Bounding boxes for the hands and the steering wheel are detected in a color image. If the intersection area between the bounding box of a hand and that of the steering wheel is greater than 5 % of the size of the hand bounding box, then the hand is classified as on the steering wheel. The detection method used to accomplish this is based on *Faster R-CNN* [12]. Its disadvantage, to not recognize very small objects, such as the hands in this case, is overcome by training the detection network with different scales of the Faster R-CNN's region proposals. Weaknesses of this method are, for example, the high computational effort required, and errors occurring under difficult lighting conditions.

Another approach uses crops of the hands to find out whether the shown hand grabs the steering wheel or a mobile phone [13]. To obtain the hand crops, a hand localization is performed using a CNN that extracts regions where a hand may be located. The next step is a pixel-wise skin classification to segment the hand so that a selection can be made from these regions. The *Histogram of Oriented Gradients* (HOG, [14]) method is used to extract important properties of the hand from these segmentations. Finally, an SVM uses these properties to classify whether the driver grabs one of the objects. All in all, this approach results in a very high computational effort, especially due to the large CNN, which is used for the hand localization.

Datasets developed especially for hand detection in cars are important for continuously improving algorithms and comparing them with each other. In [15] a dataset is presented which tries to represent driving scenarios that are as realistic as possible. It contains color images taken from different angles and with different illumination intensities. The annotations include bounding boxes for each hand as well as information on whether it is the left or right hand and whether it belongs to the driver or co-driver. In addition, there is information about the number of hands on the steering wheel.

Beside a neural network that detects hands or objects held by the driver, [16] introduces a simple procedure for semiautomatic annotation of hands. For this purpose, the driver wears green gloves and red wristbands. Chroma keying is used to detect the hand positions based on these colors. The data obtained in this way can then be used, for example, to train a hand detection method based on depth image data.

3 Hands on wheel classification system

3.1. System setup

The basis of the hands on wheel classification system is a *Time of Flight* (ToF) camera, which is integrated into the roof module of the car, so that the driver is seen from above, and provides intensity and depth information about the front part of the driver's cabin. This information is represented as images with a QVGA resolution. An example of these images is shown in Figure 1. Through an active lighting the images are independent of the ambient light. The intensity image Figure 1a results from measuring the reflected amount of this active lighting. Objects that poorly reflect this light are causing noise in the depth image Figure 1b.

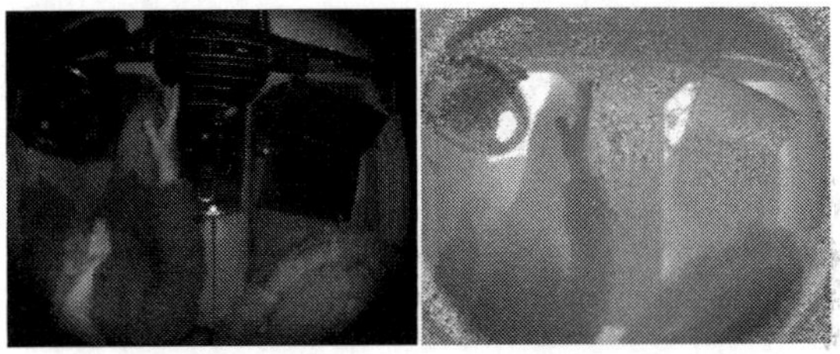

(a) Intensity image (b) Depth image

Figure 1. Example of the information captured by the ToF camera. The depth image contains high noise in some areas.

The images are recorded at a frame rate of 30 Hz, but only every third image is used to prevent the dataset from containing too many similar examples.

Only the left half of the images is used for the classification, since the other one contains no relevant information and the computational cost is lower the smaller the input image is.

The resulting dataset is saved as a multidimensional array. The first dimension represents the different examples, the second and third dimension contain the individual pixels and the fourth dimension distinguishes the intensity and depth information.

3.2 Network architecture

The network architecture of the hands on wheel classifier is depicted in Figure 2. The first three stages extract relevant features of the input images. Each of them contains a convolutional layer with size 3×3, which is followed by a ReLU activation function. Max pooling is applied during the first two stages to reduce the spatial size and thus the number of parameters in the fully connected layers. Dropout [17] is applied after the third stage to prevent overfitting. After that the extracted features are evaluated by two fully connected layers. The first one is followed by a ReLU activation function, whereas the second one is followed by a softmax activation function to predict a probability for each class.

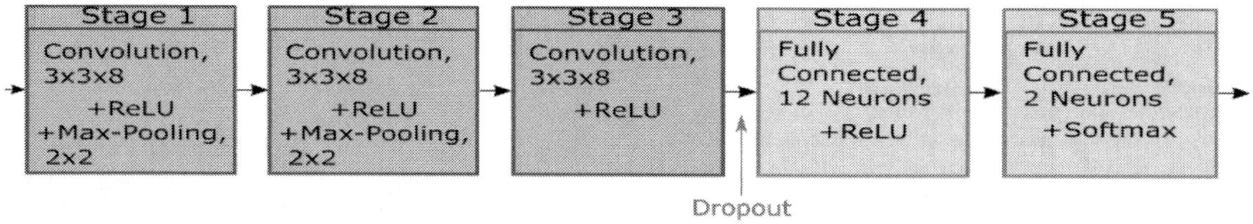

Figure 2. Architecture of the hands on wheel classifier. It consists of three convolutional layers, followed by two fully connected layers. Dropout [17] is applied between stage three and four to avoid overfitting. The last layer is followed by a softmax activation function to obtain probabilities for the occurrence of the classes.

3.3 Training procedure

The network is trained for 100 epochs, which consist of several mini-batches. Before the first epoch starts the weights are initialised to small random values based on a Gaussian distribution [18]. To calculate the difference between the predictions and labels, the cross-entropy loss function

$$H(y, \hat{y}) = -\sum_{k=1}^{K} y_k \log \hat{y}_k \qquad (1)$$

is used. $\hat{y}_i$ is the prediction of the classifier, y_i the value of the label and K is the number of different classes. For every mini-batch the average loss over all contained examples is calculated. *Weight decay* is added to this loss as a regularizer. The Adam optimizer [19] is used to update the weights dependent on the loss after each training step.

Several hyperparameters result from this architecture. These are the size of the mini-batches, the used amount of dropout and the amount of weight decay.

4 Experimental evaluation

4.1. Dataset

To evaluate the classifier, a dataset was created which contains a lot of variation to represent as many real situations as possible. For this purpose several video streams were recorded that contain different cars, several positions of the driver's seat and steering wheel, and varying drivers with diverse clothing. Figure 3 shows three different examples, in which the driver is wearing different clothes. Additionally two different versions of the ToF camera were used. In each stream only one class is executed, which simplifies the labelling procedure.

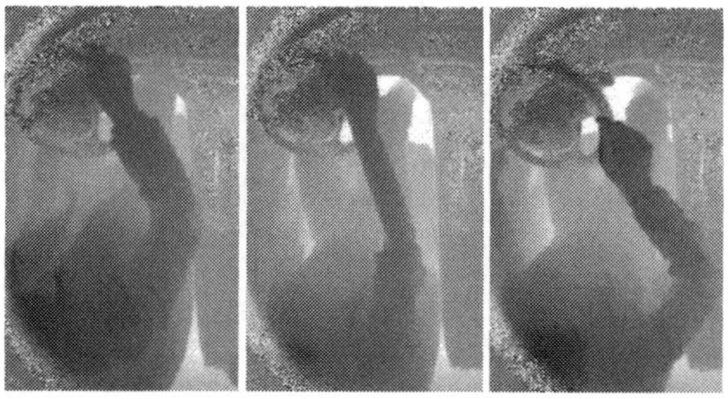

Figure 3. Examples for the variation in appearance of similar body postures caused by different clothes and body shapes.

Table I shows the composition of the dataset. The rows distinguish different experimental setups, for which the total number of images and the corresponding proportions are shown.

Table 1. Composition of the used dataset. Different experimental setups and the amount of corresponding images are shown in different rows. The dataset contains images from different versions of the ToF camera in car 1.

Setup	Proportion	Number of images
Car 1, Camera 1	65.12 %	58800
Car 1, Camera 2	3.99 %	3600
Car 2, Camera 1	30.89 %	27900
Total	100 %	90300

The distribution of the classes is depicted in Table II. It is recognizable that the dataset contains more examples in which the hands of the driver are not at the steering wheel.

Table 2. Distribution over the classes in the dataset. Only in about one third of the data the driver has his hands on the steering wheel.

Setup	Hands on wheel	Hands off wheel
Car 1, Camera 1	36.73 %	63.27 %
Car 1, Camera 2	50 %	50 %
Car 2, Camera 1	22.58 %	77.42
Total	32.89 %	67.11 %

To reasonably estimate the accuracy of the model, 5-fold cross-validation is carried out. In each run of the cross-validation, different 20 % of the examples are used in a separate validation set, which is not used during the training process. The presented results in this paper are the mean over the runs of the cross-validation. This procedure is intended to cover the generalization to all cases that occur in the dataset.

It is to mention that the dataset contains a lot of challenging cases, in which for example the driver is pretending to grasp the steering wheel. Therefore, the error on the dataset is expected to be higher than in a realistic scenario.

4.2. Analysis of the class imbalance

When a threshold is used to get a binary prediction from the classifier, the result is either right or wrong. Because of the class imbalance in the dataset (Table II) a simple threshold of 0.5 is not sufficient. Hence, the performance metric used in the evaluation is the area under

 https://doi.org/10.1051/matecconf/202030806003

the Precision-Recall Curve (PR AUC), which has a maximal value of 1.

To figure out how the class imbalance is effecting the model, the false negative rate, false positive rate and total error are plotted against the threshold in Figure 4. The total error is the proportion of misclassified examples. At a threshold of approximately 0.2 the false negative and false positive rate are the same. This means that the classifier is better in detecting the class in which the driver is not grasping the steering wheel. The minimum total error is approximately 5.4 %. Thus, at a threshold of 0.2, with a value of about 6 %, the total error is not far from the minimum, which means that the change of the threshold can counter the effect of the class imbalance in the dataset.

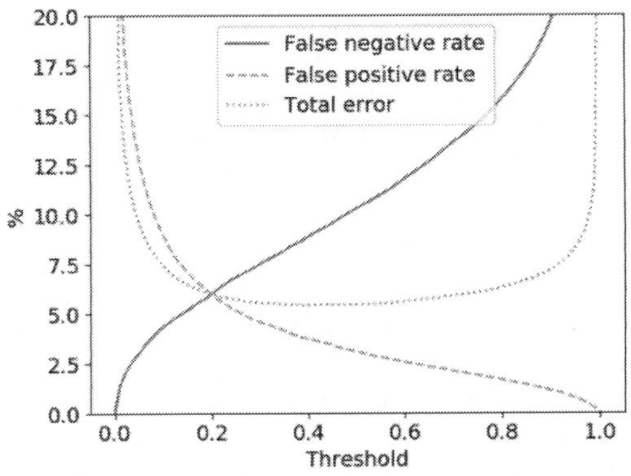

Figure 4. Classification results with different thresholds. The intersection point is at a threshold of approximately 0.2 and the total error at this point is not much higher than the minimum total error.

4.3. Input analysis

In order to obtain an efficient classification system, it is important to know which of the possible inputs contribute to the result. Therefore, it is investigated to what extent the intensity and depth images contain complementary information for the hands on wheel classification.

Figure 5 shows a comparison of the PR AUC for the classifier when different inputs are used. The different inputs are the depth image, the intensity image and both images, where the images are stacked in the fourth dimension. When the input is the intensity image alone, the results of the classifier are significantly worse than the results achieved by using the other two input methods. Since using both images as input does not induce an improvement over using the depth image alone, it can be concluded that no relevant additional information is obtained from the intensity image. Therefore, the depth image as input is sufficient and used for the further experiments.

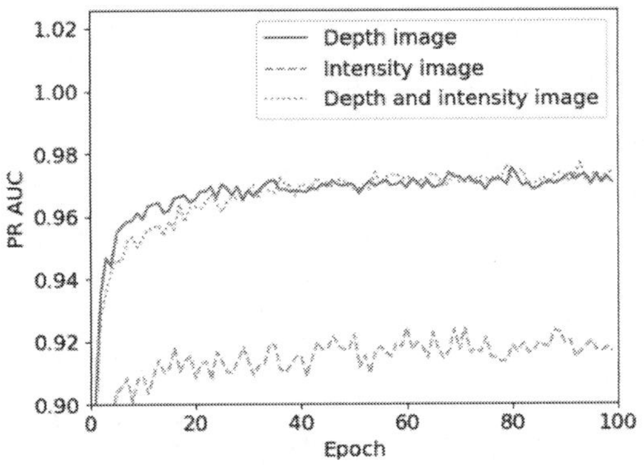

Figure 5. Comparison of the classifier using the depth image, the intensity image and both images as input. Using only the intensity image, a significantly worse result is achieved. There is no improvement over training on the depth image alone if the classifier is trained on both images.

4.4. Examining the learning curve

In order to find out whether an increase in the amount of training data would lead to a better result, the learning curve was examined. This is interesting, because the recording of new training data is a great effort and should therefore only be performed if an improvement in performance is likely. However, this method does not provide any information on how the result will behave if the new data covers certain cases that do not yet exist in the training data. This means that it can still be useful to generate new training data for cases where the system encounters many errors. In order to obtain the learning curve, the artificial neural net is trained multiple times. Initially, only a small proportion of the available training data is used. This percentage is increased in each training run so that finally all data is used. The result is a curve in which the error depends on the amount of training data. If this has already converged during the last training runs, it is only worth recording more data if it covers specific cases in which the performance is unsatisfactory. For each training run, the best value occurring during the training is used in the learning curve.

With this procedure the curves shown in Figure 6 are produced. During the generation of these curves, 20 % of the data set was used as a fixed test set. It is interesting to note that the result on the training data, after a certain point, remains constant as the size of the data set increases. This suggests that the network architecture of the classifier does not allow memorization. Otherwise, the result would become worse with a larger amount of training data, because more and more would have to be learned by heart.

The test data set shows a strong increase in performance at the beginning. After a slump at about 65 % of the training set, there is another positive trend towards the end. This means that a larger amount of training data could be helpful for the classifier to generalize to the used test data.

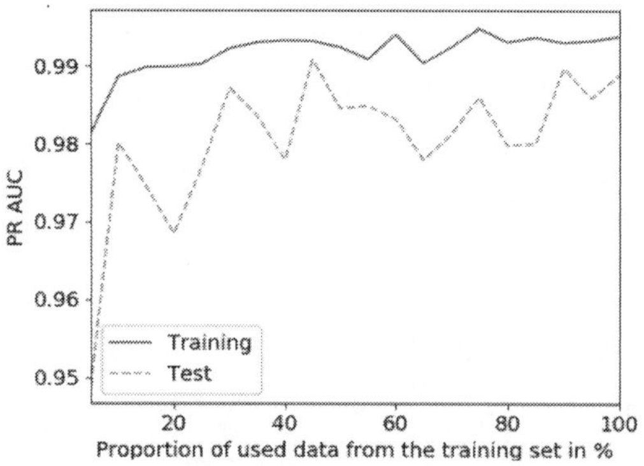

Figure 6. Learning curves of the hands on wheel classifier. A larger portion of the training data is used for each training run. With a larger number of training examples the result improves especially at the beginning. At the end a positive trend on the test data can be seen again.

5 Conclusion

In this paper it is investigated to which extent the information obtained from the images of a ToF camera as the input for an artificial neural network can be used to detect whether the driver's hands are on the steering wheel. This camera observes the interior of the vehicle and is mounted in the roof module. The resulting system performs well on a challenging dataset, which means that better results can be expected in practice. In addition the frame-wise predictions can be filtered to reduce the effect of outliers. However, the remaining error is still a little too high to completely rely on the classifier for the safety-relevant question of whether the driver has his hands on the steering wheel. The remaining errors occur in situations where the driver's hand is above his thigh or when the left arm is not in the image while the left hand touches the steering wheel in the outermost part of the image. To avoid the latter error, a larger field of view would be necessary. But these remaining errors are false negatives rather than false positives. This means that the system does not often mistakenly think the driver is ready to drive.

An analysis of the effect that arises from an imbalance in the dataset demonstrates that the effect of this imbalance can be reduced by choosing an appropriate threshold for the binary classification. Since it is more important to detect if the hands are off the wheel, this threshold could be used as well to get a better prediction on the corresponding class while accepting worse results on the other class. An alternative to adjusting the threshold would be balancing the dataset by generating more examples of the smaller class or weighting the loss function according to the class ratio.

Additionally, it is shown that the use of the depth image is sufficient to obtain a good classification result with the used system, as an additional use of the intensity image does not lead to any improvement.

Furthermore, an examination of the learning curve showed that an extension of the data set could lead to better performance.

For the future, the differentiation of more classes, which are combined afterwards, could have a positive effect. For example, recognizing a blocked view could lead to greater stability.

Acknowledgment

Special thanks goes to Aptiv for making this work possible by providing the experimental setup and constant support. In particular we thank David Schiebener and Alexander Barth for helpful discussions and feedback.

References

1. M. Kyriakidis, J. C. F. Winter, N. Stanton, T. Bellet, B. Arem, K. Brookhuis, M. H. Martens, K. Bengler, J. Andersson, N. Merat, N. Reed, M. Flament, M. Hagenzieker and R. Happee, TIES **53**, 1-27 (2017)
2. M. M. Trivedi, T. Gandhi and J. McCall, TITS **8**, 108-120 (2007)
3. C. Tran and M. M. Trivedi, *Visual Analysis of Humans*, 597-614 (2011)
4. E. Ohn-Bar and M. M. Trivedi, TITS **1**, 90-104 (2016)
5. Y. Xing, C. Lv, Z. Zhang, H. Wang, X. Na, D. Cao, E. Velenis and F.-Y. Wang, TCSS **5**, 95-108 (2018)
6. F. Baradel, C. Wolf and J. Mille, ICCVW, 604-613 (2017)
7. E. Ohn-Bar and M. Trivedi, IV, 1034-1039 (2013)
8. A. Rangesh, E. Ohn-Bar and M. M. Trivedi, CVPRW, 1224-1231 (2016)
9. C. Tran and M. M. Trivedi, ICVES, 97-101 (2009)
10. C. Tran and M. M. Trivedi, ISM, 446-447 (2009)
11. T. H. N. Le, Y. Zheng, C. Zhu, K. Luu and M. Savvides, CVPRW, 46-53 (2016)
12. S. Ren, K. He, R. Girshick and J. Sun, TPAMI **39**, 1137-1149 (2017)
13. Siddharth, A. Rangesh, E. Ohn-Bar and M. M. Trivedi, ITSC, 2545-2550 (2016)
14. N. Dalal and B. Triggs, CVPR, 886-893 (2005)
15. N. Das, E. Ohn-Bar and M. M. Trivedi, ITSC, 2953-2958 (2015)
16. A. Rangesh and M. M. Trivedi, CVPRW (2018)
17. N. Srivastava, G. Hinton, A. Krizhevsky, I. Sutskever and R. Salakhutdinov, JMLR **15**, 1929-1958 (2014)
18. X. Glorot and Y. Bengio, PMLR **9**, 249-256 (2010)
19. D. P. Kingma and J. Ba, ICLR (2015)

AUTHOR INDEX